(MILE END)

PRINCIPLES OF ENVIRONMENTAL PHYSICS

Second Edition

J.L. Monteith and M.H. Unsworth

Edward Arnold
A division of Hodder & Stoughton
LONDON NEW YORK MELBOURNE AUCKLAND

First published in Great Britain 1973
Second edition first published 1990

Distributed in the USA by Routledge, Chapman and Hall, Inc.
29 West 35th Street, New York, NY 10001

British Library Cataloguing in Publication Data

Monteith, J.L. (John Lennox), *1929–*
Principles of environmental physics.
2nd edn
1. Biophysics
I. Title II. Unsworth, M.H.
574.19′1

ISBN 0–7131–2931–X

Typeset in 10/11 pt Times by Wearside Tradespools, Fulwell, Sunderland
Printed and bound in Great Britain for Edward Arnold, the educational, academic and medical publishing division of Hodder and Stoughton Limited, 41 Bedford Square, London WC1B 3DQ by Biddles Limited, Guildford and King's Lynn

CONTENTS

PREFACE TO THE SECOND EDITION

Readers concerned with the evolution of environmental physics may draw numerous conclusions from differences between the second edition of this book and the first, published in 1973. Because the principles of the subject have not altered, fewer changes have been made in the first three chapters, which review the scope of environmental physics, than in any of the chapters which follow. In terms of practice, however, the subject has moved forward over a broad front and several bridgeheads have been established. The second edition takes account of this progress in various ways.

The original chapter on basic physics has been expanded and split in two (Chapters 2 and 3) to provide a less superficial introduction to the subject. Elsewhere, the sequence of material has been changed to present topics in a more logical, and, we hope, more comprehensible way. For example, net radiation is now treated as a climatological quantity in Chapter 4 and as a microclimatological quantity in Chapter 6. The subject of steady-state heat balances, previously split into 'dry' and 'wet' systems, has now been divided into separate chapters (11 and 12) for plants and animals but we hope that plant ecologists will find some interesting ideas in the animal chapter and *vice versa*.

In Chapters 4, 5 and 6, dealing with radiation, account has been taken of the significance of the so-called red/far-red ratio which controls the rate of some developmental processes in plants. A similar ratio is used to establish plant biomass or ground cover using multi-spectral scanners mounted in aircraft or in satellites. The principles of this technique are explained.

The physics of particle movement in the atmosphere has been introduced by enlarging Chapter 7 on momentum transfer and by introducing a new chapter (10) on the role of particles in mass transfer at the earth's surface.

Chapter 13 contains a new treatment of heat budgets in systems where temperature is changing systematically, using relations familiar to physicists and engineers, to describe processes which affect both plants and animals. The two final chapters contain material which has been expanded and rewritten to take account of major progress, both experimental and theoretical, in the study of transfer processes within and above plant canopies.

Because the subject of environmental physics has matured since 1973, the second edition inevitably contains more theoretical material than the first and the use of calculus is more prominent. However, we believe that very few sections are beyond the capability of students of biology or agriculture who have taken a good first-year course in mathematical methods.

To match the introduction of new material, references have been thoroughly revised. As in the first edition, they have been confined mainly to seminal papers and books from which readers can find their way into secondary literature if they wish. About half of the references were published after the first edition. Similarly, complete revision of the Bibliography was needed to take account of many new textbooks and conference proceedings dealing with aspects of environmental physics and physiology. Most of these books have come from workers in plant physiology, plant ecology, soil science and related branches of agricultural science who now use a common vocabulary and who have almost standardized the symbols which they use. A very small number of environmental physicists in Britain and in the USA have successfully applied the same concepts and terminology to livestock and wild animals. Even fewer workers have attempted to apply the same principles to man, as we have done here. Researchers in human physiology have evolved their own way of treating the subject of environmental physics, almost independent of parallel developments in plant and animal ecology.

A much more serious gulf lies between environmental physics as applied to biology and 'modern' physics as now taught in university departments. How many physics undergraduates are aware that the concepts of physics and sophisticated physical equipment are now widely used in agricultural science and ecology and that the theoretical basis for this work lies in unfashionable aspects of classical physics which are disappearing from the syllabus—properties of matter, for example? We know only one or two university physics departments in Britain whose undergraduate and research students are concerned with the problems discussed in this book. Some of these problems deal with the immediate environment of the student and therefore have considerable didactic value—human comfort, for example, in relation to weather, exercise and the efficiency of domestic heating systems. Other problems have global dimensions and will require intensive interdisciplinary research for many years ahead: the dispersion of atmospheric pollution and its agricultural consequences; the possible impact of climatic change on food production, habitability and natural ecosystems; the interpretation of satellite images in terms of water supplies and crop production.

We therefore hope that some university physics departments as well as biological departments, recognizing the challenge of environmental science, will start to use this book to plan new courses. Feed-back, however critical, will be helpful in preparing a third edition!

A number of authors and publishers kindly gave permission for the reproduction of diagrams. We are grateful to Dr. S.T. Henderson and Adam Hilger Publishers (Fig. 4.1); Dr. M.D. Steven and the Royal Meteorological Society (4.3); Mr. F.E. Lumb (4.6); Deutscher Wettendienst (4.11 and 13.5); Taylor and Francis Ltd (5.8) from *Ergonomics*; Dr. E.L. Deacon and Elsevier Publishing Co. (6.1 and 13.7); Dr. S.A. Bowers and Williams and Wilkins Co. (6.3) from *Soil Science*; Dr. K.J. McCree and Elsevier Publishing Co. (6.4) from *Agricultural Meteorology*; Dr. G. Stanhill and Pergamon Press (6.6); Professor L.E. Mount and Edward Arnold (6.9, 12.3 and 12.4); Dr. J.C.D. Hutchinson and Pergamon Press (6.10) from *Comparative Biochemis-*

try and Physiology; Dr. J. Grace and Oxford University Press (7.3) from *Journal of Experimental Botany*; the Royal Meteorological Society (7.4, 7.5, 11.6, 11.7, 11.8, 11.9, 13.3, 15.9, and 15.10); Dr. T.T. Mercer and Academic Press (7.6); Dr. D. Aylor and the American Society of Plant Physiologists (7.7) from *Plant Physiology*, and Pergamon Press (10.5) from *Atmospheric Environment*; Dr. R.H. Shaw and Elsevier Publishing Co. (7.10, 7.11) from *Agricultural Meteorology*; Dr. J.A. Clark and D. Reidel Publishing Co. (8.3 and Cover Illustration) from *Boundary Layer Meteorology*; Dr. S. Vogel and Clarendon Press (8.4); Dr. A.J. McArthur and the Royal Society of London (8.6, 8.12) from the *Proceedings of the Royal Society*; Dr. R.P. Clark and *The Lancet* (8.8, 8.10), and Cambridge University Press (8.7) from *Journal of Physiology*; Dr. P.F. Scholander and the Marine Biological Laboratory (8.11); Dr. T. Haseba and the Society of Agricultural Meteorology in Japan (9.3) from *Journal of Agricultural Meteorology*; Professor N.A. Fuchs and Pergamon Press (10.1); Dr. A.C. Chamberlain and Academic Press (10.3), and D. Reidel Publishing Co. (15.4) from *Boundary Layer Meteorology*; Dr. P. Little and Pergamon Press (10.6) from *Atmospheric Environment*; Dr. K. Raschke and Springer-Verlag (11.5) from *Planta*; Dr. A.M. Hemmingsen (12.2); Dr. D.M. Gates and Springer-Verlag (13.2); Dr. W.R. van Wijk and North Holland Publishing Co. (13.6); Academic Press (14.4 and 14.5); Blackwells Scientific (15.7, 15.8); Dr. M.R. Raupach and Annual Reviews Inc. (15.12) from *Annual Review of Fluid Mechanics*. Dr. A.I. Fraser and the Forestry Commission gave permission for the reproduction of Fig. 7.12. Dr. J.V. Lake and Dr. W. Porter provided chart records for Figs. 4.5 and 6.12 respectively, and Dr. I. Impens allowed us to use unpublished measurements in Fig. 9.1.

Finally, we gratefully acknowledge help in various forms: from Margaret Wilson and Caroline Reffin who typed the manuscript; Dr. A. C. Chamberlain, Dr. R. Leuning, Dr. A.J. McArthur and Dr. A.E. Wheldon who read sections of the text and suggested many improvements; Professor A. Willis who edited the final draft with the same meticulous care that he gave to the first edition; and the staff of Edward Arnold for their patience, courtesy and skill in producing this book.

J.L. Monteith
M.H. Unsworth
Edinburgh and Sutton Bonington
1989

SYMBOLS

The main symbols used in this book are arranged here in a table containing brief definitions of each quantity. A few of the symbols are universally accepted (e.g. R, g), some have been chosen because they appear very frequently in the literature of environmental physics (e.g. r_s, z_o, K_M), and some have been devised for the sake of consistency. In particular, the symbols **S** and **L** are used for flux densities of short- and long-wave radiation with subscripts to identify the geometrical character of the flux, e.g. $\mathbf{S}_d$ for the flux of diffuse short-wave radiation from the sky.

Flux densities of momentum, heat and mass are printed in bold case throughout the book (e.g. $\boldsymbol{\tau}$, **E**) and so is the latent heat of vaporization of water $\boldsymbol{\lambda}$, partly to distinguish it from wavelength λ and partly because it is often associated with **E**. Upper case subscripts are used to refer to momentum, heat, vapour, carbon dioxide, etc., e.g. r_V, K_M; most other subscripts are lower case, e.g. c_p for the specific heat of air at constant pressure.

The complete set of symbols represents the best compromise that could be found between consistency, clarity and familiarity.

ROMAN ALPHABET

A	area; azimuth angle with respect to south
A_b	area of solid object projected on a horizontal plane
A_p	area of solid object projected on plane perpendicular to solar beam
$A(z)$	amplitude of soil temperature wave at depth z
$\mathbf{B}$	total energy emitted by unit area of full radiator or black body
B	wet-bulb depression
$\mathbf{B}(\lambda)$	energy per unit wavelength in spectrum of full radiator or black body
c	volume fraction at CO_2 (e.g. v.p.m.); fraction of sky covered by cloud; velocity of light; mean velocity of gas molecules
c_d	drag coefficient for form drag and skin friction combined
c_f	drag coefficient for form drag
c_l	specific heat of a liquid
c_p	specific heat of air at constant pressure; efficiency of impaction of particles
c_s	specific heat of solid fraction of a soil
c_v	specific heat at constant volume

c'	bulk specific heat of soil
C	flux of heat per unit area by convection in air
$\mathscr{C}$	heat capacity of an organism per unit surface area
d	zero plane displacement
D	saturation vapour pressure deficit; diffusion coefficient for a gas in air (subscripts V for water vapour; C for CO_2); damping depth $(=(2\kappa'/\omega)^{1/2})$
e	partial pressure of water vapour in air
$e_s(T)$	saturation vapour pressure of water vapour at temperature T
δe	saturation deficit, i.e. $e_s(T) - e$
E_q	energy of a single quantum
E	flux of water vapour per unit area; evaporation rate
$\mathbf{E}_r$	respiratory evaporation rate of animal
$\mathbf{E}_s$	rate of evaporation from skin
$\mathbf{E}_t$	rate of evaporation from vegetation
F	generalized stability factor $(\phi_v\phi_m)^{-1}$
F	drag force on a particle; retention factor
F	mass flux of a gas per unit area; flux of radiant energy
g	acceleration by gravity (9.81 m s^{-2})
G	flux of heat by conduction, per unit area
h	Planck's constant (6.63×10^{-34} J s); relative humidity of air; height of cylinder, crop, etc.
H	total flux of sensible and latent heat, per unit area
i	intensity of turbulence, i.e. root mean square velocity/mean velocity
I	intensity of radiation (flux per unit solid angle)
J	rate of change of stored heat per unit area
k	von Karman's constant (0.41); thermal conductivity of air; attenuation coefficient; Boltzmann constant (1.38×10^{-23} J K^{-1})
k'	attenuation coefficient; thermal conductivity of a solid
K	diffusion coefficient for turbulent transfer in air (subscripts H for heat, M for momentum, V for water vapour, C for CO_2)
$\mathscr{K}$	canopy attenuation coefficient
$\mathscr{K}_s$	ratio of horizontally projected area of an object to its plane or total surface area
l	mixing length, stopping distance; length of plate in direction of airstream
L	leaf area index; Monin-Obukhov length
L	flux of long-wave radiation per unit area (subscript u for upwards; d for downwards; e from environment; b from body)
m	mass of a molecule or particle; air mass number
M	rate of heat production by metabolism per unit area of body surface
M	grammolecular mass (subscripts a for dry air, v for water vapour)
n	represents a number or dimensionless empirical constant in several equations
N	Avogadro constant (6.02×10^{23}); number of hours of daylight
N	radiance (radiant flux per unit area per unit sold angle)
P	latent heat equivalent of sweat rate per unit body area
p	total air pressure; interception probability in hair coats
q	specific humidity of air (mass of water vapour per unit mass of moist air)

$\mathbf{Q}$	rate of mass transfer
r	radius; resistance to transfer (subscripts M momentum, H heat, V water vapour, C for CO_2); usually applied to boundary layer transfer
r_a	resistance to transfer in the atmosphere (subscripts M, H, V, C as above)
r_b	boundary layer resistance of crop for mass transfer
r_c	canopy resistance
r_d	thermal resistance of human body
r_f	thermal resistance of hair, clothing; resistance for forced ventilation in open-top chambers
r_h	resistance of hole (one side) for mass transfer
r_i	incursion resistance for open-top chambers
r_p	resistance of pore for mass transfer
r_s	resistance of a set of stomata
r_t	total resistance of single stoma
r_H	resistance for heat transfer by convection, i.e. sensible heat
r_R	resistance for radiative heat transfer ($\rho c_p/4\sigma T^3$)
r_{HR}	resistance for simultaneous sensible and radiative heat exchange, i.e. r_H and r_R in parallel
r_V	resistance for water vapour transfer
R	Gas Constant (8.31 J mol^{-1} K^{-1})
$\mathbf{R}_n$	net radiation flux density
$\mathbf{R}_{ni}$	net radiation absorbed by a surface at the temperature of the ambient air
s	amount of entity per unit mass of air
S	gas concentration
$\mathbf{S}_d$	diffuse solar irradiance on horizontal surface
$\mathbf{S}_e$	solar radiation received by a body, per unit area, as a result of reflection from the environment
$\mathbf{S}_p$	direct solar irradiance on surface perpendicular to solar beam
$\mathbf{S}_b$	direct solar irradiance on horizontal surface
$\mathbf{S}_t$	total solar irradiance (usually) on horizontal surface
t	diffusion pathlength
T	temperature
T_a	air temperature
T_b	body temperature
T_c	cloud-base temperature
T_d	dew point temperature
T_e	equivalent temperature of air ($T+(e/\gamma)$)
T_{e*}	apparent equivalent temperature of air ($T+(e/\gamma^*)$)
T_f	effective temperature of ambient air
T_s, T_o	temperature of surface losing heat to environment
T_v	virtual temperature
T'	thermodynamic wet bulb temperature
T^*	standard temperature for vapour pressure specification
u	optical pathlength of water vapour in the atmosphere
$u(z)$	velocity of air at height z above earth's surface
u_*	friction velocity
v	molecular velocity

v_d deposition velocity
v_s sedimentation velocity
V_m molar volume at STP (22.4 L)
$\dot{V}$ volume
w vertical velocity of air; depth of precipitable water
W body weight of animal
x volume fraction (subscripts s for soil; l for liquid; g for gas); ratio of cylinder height to radius
z distance; height above earth's surface
z_o roughness length
Z height of equilibrium boundary layer

GREEK ALPHABET

α absorption coefficient (subscripts p for photosynthetically active; T for total radiation; r for red; i for infra-red)
$\alpha(\lambda)$ absorptivity at wavelength λ
β solar elevation; ratio of observed Nusselt number to that for a smooth plate
γ psychrometer constant ($=c_p p/\boldsymbol{\lambda}\varepsilon$)
γ^* apparent value of psychrometer constant ($=\gamma r_v/r_H$)
Γ dry adiabatic lapse rate, DALR (9.8×10^{-3} K m^{-1})
δ depth of a boundary layer
Δ rate of change of saturation vapour pressure with temperature, i.e. $\partial e_s(T)\partial T$
ε ratio of molecular weights of water vapour and air (0.622)
ε_a apparent emissivity of the atmosphere
$\varepsilon(\lambda)$ emissivity at wavelength λ
θ angle with respect to solar beam; potential temperature
κ thermal diffusivity of still air
κ' thermal diffusivity of a solid, e.g. soil
λ wavelength of electromagnetic radiation
$\boldsymbol{\lambda}$ latent heat of vaporization of water
μ coefficient of dynamic viscosity of air
ν coefficient of kinematic viscosity of air; frequency of electromagnetic radiation
ρ reflection coefficient (subscripts p for photosynthetically active; c for canopy; s for soil; r for red; i for infra-red; T for total radiation); density of a gas, e.g. air including water vapour component
ρ_a density of dry air
ρ_c density of CO_2
ρ_l density of a liquid
ρ_s density of a solid component of soil
ρ' bulk density of soil
ρ reflection coefficient
$\rho(\lambda)$ reflectivity of a surface at wavelength λ
σ Stefan-Boltzmann constant (5.67×10^{-8} W m^{-2} K^{-4})
Σ the sum of a series
$\boldsymbol{\tau}$ flux of momentum per unit area; shearing stress

τ	fraction of incident radiation transmitted, e.g. by a leaf; relaxation time, time constant, turbidity coefficient
ϕ	mass concentration of CO_2, e.g. g m^{-3}; angle between a plate and airstream
$\mathbf{\Phi}$	radiant flux density
χ	absolute humidity of air
$\chi_s(T)$	saturated absolute humidity at temperature T (°C)
ψ	angle of incidence
ω	angular frequency; solid angle

NON-DIMENSIONAL GROUPS

Le	Lewis number (κ/D)
Gr	Grashof number
Nu	Nusselt number
Pr	Prandtl number (ν/κ)
Re_*	Roughness Reynolds number ($u_* z_o/\nu$)
Re	Reynolds number
Ri	Richardson number
Sc	Schmidt number (ν/D)
Sh	Sherwood number
Stk	Stokes number

LOGARITHMS

ln	logarithm to the base e
log	logarithm to the base 10

1

THE SCOPE OF ENVIRONMENTAL PHYSICS

To grow and reproduce successfully, organisms must come to terms with their environment. Some microorganisms can grow at temperatures between −6 and 100°C and, when they are desiccated, can survive even down to −272°C. Higher forms of life on the other hand have adapted to a relatively narrow range of environments by evolving sensitive physiological responses to external physical stimuli. The physical environment of plants and animals has five main components which determine the survival of the species:

(i) the environment is a source of radiant energy which is trapped by the process of photosynthesis in green cells and stored in the form of carbohydrates, proteins and fats. These materials are the primary source of metabolic energy for all forms of life on land and in the oceans;

(ii) the environment is a source of the water, nitrogen, minerals and trace elements needed to form the components of living cells;

(iii) factors such as temperature and daylength determine the rates at which plants grow and develop, the demand of animals for food and the onset of reproductive cycles in both plants and animals;

(iv) the environment provides stimuli, notably in the form of light or gravity, which are perceived by plants and animals and provide frames of reference both in time and in space. They are essential for resetting biological clocks, providing a sense of balance, etc.;

(v) the environment determines the distribution and viability of pathogens and parasites which attack living organisms, and the susceptibility of organisms to attack.

To understand and explore relationships between organisms and their environment, the biologist should be familiar with the main concepts of the environmental sciences. He must search for links between physiology, biochemistry and molecular biology on the one hand and meteorology, soil science and oceanography on the other. One of these links is environmental physics—the measurement and analysis of interactions between organisms and their physical environment. The presence of an organism modifies the environment to which it is exposed, so that the physical stimulus received *from* the environment is partly determined by the physiological response *to* the environment.

When an organism interacts with its environment, the physical processes involved are rarely simple and the physiological mechanisms are often imperfectly understood. Fortunately, physicists are trained to use Occam's Razor when they interpret natural phenomena in terms of cause and effect: they observe the behaviour of a system and then seek the simplest way of describing it in terms of governing variables. Boyle's Law and Newton's Laws of Motion are classic examples of this attitude. More complex relations are avoided until the weight of experimental evidence shows they are essential. Many of the equations discussed in this book are approximations to reality which have been found useful to establish and explore ideas. The art of environmental physics lies in choosing robust approximations which maintain the principles of conservation for mass, momentum and energy.

It has become fashionable to describe approximations as *models*. These models may be either theoretical or experimental and both types are found in this book. To underline the significance of models which are deliberate simplifications of reality, temperature distributions over a real bean leaf, discussed on p. 129, were reproduced on the cover of the book. We have not considered models of plant or animal systems based on computer simulations. They can rarely be *tested* in the sense that physicists use the word because so many variables and assumptions are deployed in their derivation. Consequently, although they can be useful for identifying the sensitivity of systems to environmental variables, they seldom seem to us to contribute to an understanding of the *principles* of environmental physics.

Several volumes would be needed to review all the principles of environmental physics and the definite article was deliberately omitted from the title of this book because it makes no claim to be comprehensive. However, the topics which it covers are central to the subject: the exchange of radiation, heat, mass and momentum between organisms and their environment. Within these topics, similar analysis can be applied to a number of closely related problems in plant, animal and human ecology. The short bibliography at the end of the book can be consulted for more specialized treatments and for accounts of other branches of the subject such as the physics of soil water.

The lack of a common language is often a barrier to progress in interdisciplinary subjects and it is not easy for a physicist or meteorologist with no biological training to communicate with a physiologist or ecologist who is fearful of formulae. Throughout the book, simple electrical analogues are used to describe rates of transfer and exchange between organisms and their environment, and calculus has been kept to a minimum. The concept of *resistance* has been familiar to plant physiologists for many years, mainly as a way of expressing the physical factors that control rates of transpiration and photosynthesis, and animal physiologists have used the term to describe the insulation provided by clothing, coats, or by a layer of air. In micrometeorology, aerodynamic resistances derived from turbulent transfer coefficients can be used to calculate fluxes from a knowledge of the appropriate gradients, and resistances which govern the loss of water from vegetation are now incorporated in models of the atmosphere that include the behaviour of the earth's surface. Ohm's Law has therefore become an important unifying principle of environmental physics; the basis of a common language for biologists and physicists.

The choice of units was dictated by the structure of the Système International, modified by retaining the centimetre. For example, the dimensions of leaves are quoted in mm and cm. To adhere strictly to the metre or the millimetre as units of length often needs powers of 10 to avoid superfluous zeros and sometimes gives a false impression of precision. As most measurements in environmental physics have an accuracy between ± 1 and $\pm 10\%$ they should be quoted to 2 or at most 3 significant figures, preferably in a unit chosen to give quantities between 10^{-1} and 10^3. The area of a leaf would therefore be quoted as 23.5 cm^2 rather than 2.35×10^{-3} m^2 or 2350 mm^2.

Conversions from SI to c.g.s. are given in the Appendix, Table A.1.

2

GAS LAWS

The physical properties of gases are involved in many of the exchanges that take place between organisms and their environment. Therefore the relevant equations for air form an appropriate starting point for an environmental physics text. They also provide a basis for discussing the behaviour of water vapour, a gas whose significance in meteorology, hydrology and ecology is out of all proportion to its relatively small concentration in the atmosphere.

PRESSURE, VOLUME AND TEMPERATURE

The observable properties of a gas such as its temperature and pressure can be related to the mass and velocity of its constituent molecules by the Kinetic Theory of Gases which is based on Newton's Laws of Motion. Newton established the principle that when force is applied to a body, its momentum, the product of mass and velocity, changes at a rate proportional to the magnitude of the force. Appropriately, the unit of force in the Système International is a newton: and the unit of pressure (force per unit area) is a pascal — from the name of another famous natural philosopher.

The pressure p which a gas exerts on the surface of a liquid or solid is a measure of the rate at which momentum is transferred to the surface from molecules which strike it and rebound. On the assumption that the kinetic energy of all the molecules in an enclosed space is constant and by making further assumptions about the nature of a *perfect* gas, a simple relation can be established between pressure and kinetic energy per unit volume. When the density of the gas is ρ and the mean square molecular velocity is $\bar{v}^2$, the kinetic energy per unit volume is $\rho\bar{v}^2/2$ and

$$p = \rho\bar{v}^2/3 \tag{2.1}$$

implying that pressure is two-thirds of the kinetic energy per unit volume.

Although equation (2.1) is central to the Kinetic Theory of Gases, it has little practical value. A number of congruent but more useful relations can be derived from the observations of Boyle and Charles whose gas laws can be combined to give

$$pV \propto T \tag{2.2}$$

where V is the volume of a gas at an absolute temperature T (K). To establish a constant of proportionality, a standard amount of gas is defined by the volume V_m occupied by a mole at standard pressure and temperature (101.3 kPa and 273.2 K) which is 0.0224 m^3 (22.4 litres). Then

$$pV_m = RT \tag{2.3}$$

where $R = 8.314$ J mol^{-1} K^{-1}, called the molar **gas constant**, has the dimensions of a molecular specific heat.

Since the pressure exerted by a gas is a measure of its kinetic energy per unit volume, pV_m is proportional to the kinetic energy of a mole. A mole of any substance contains N molecules where $N = 6.02 \times 10^{23}$ is the **Avogadro constant**. It follows that the mean energy per molecule is proportional to

$$pV_m/N = (R/N)T = kT \tag{2.4}$$

where k is the **Boltzmann constant**.

Equation (2.3) is sometimes used in the form

$$p = \rho RT/M \tag{2.5}$$

obtained by writing the density of a gas as its molecular mass divided by its molecular volume, i.e.

$$\rho = M/V_m \tag{2.6}$$

For unit mass of any gas with volume V, $\rho = 1/V$ so equation (2.5) can also be written in the form

$$pV = RT/M \tag{2.7}$$

Equation (2.7) provides a general basis for exploring the relation between pressure, volume and temperature in unit mass of gas and is particularly useful in four cases:

(1) constant volume—p proportional to T
(2) constant pressure (isobaric)—V proportional to T
(3) constant temperature (isothermal)—V inversely proportional to p
(4) constant energy (adiabatic)—p, V and T may all change.

When the molecular weight of a gas is known, its density at STP can be calculated from equation (2.6) and its density at any other temperature and pressure from equation (2.5). Table 2.1 contains the molecular weights and densities at STP of the main constituents of dry air. Multiplying each density

Table 2.1 Composition of dry air

Gas	Molecular weight (g)	Density at STP (kg m^{-3})	Per cent by volume	Mass concentration (kg m^{-3})
Nitrogen	28.01	1.250	78.09	0.975
Oxygen	32.00	1.429	20.95	0.300
Argon	38.98	1.783	0.93	0.016
Carbon dioxide	44.01	1.977	0.03	0.001
Air	29.00	1.292	100.00	1.292

by the appropriate volume fraction gives the concentration of each component and the sum of these concentrations is the density of dry air. From a density of 1.292 kg m^{-3} and from equation (2.5) the effective molecular weight of air (in g) is 28.94 or 29.00 within 0.2%.

SPECIFIC HEATS

When unit mass of gas is heated but not allowed to expand, the increase in total heat content per unit increase of temperature is known as the specific heat at constant *volume*, usually given the symbol c_V. Conversely, if the gas is allowed to expand in such a way that its *pressure* stays constant, additional energy is needed for expansion so that the specific heat at constant pressure c_p is larger than c_V.

To evaluate the *difference* between c_p and c_V, the work done by expansion can be calculated by considering the special case of a cylinder with cross-section A fitted with a piston which applies a pressure p on gas in the cylinder, thereby exerting a force pA. If the gas is heated and expands to push the cylinder a distance x, the work done is the product of force pA and distance x or pAx which is also the product of the pressure and the change of volume Ax. The same relation is valid for any system in which gas expands at a constant pressure.

The work done for a small expansion dV of unit mass of gas at constant pressure p is found by differentiating equation (2.7) to give

$$pdV = (R/M)dT \qquad (2.8)$$

As the *difference* between the two specific heats is the work done in expansion per unit increase of temperature, it follows that

$$c_p - c_V = R/M \qquad (2.9)$$

The *ratio* of c_p to c_V for gases depends on the energy associated with the vibration and rotation of its molecules and therefore depends on the number of atoms forming the molecule. For diatomic molecules such as nitrogen and oxygen which are the major constituents of air, the theoretical value of c_p/c_V is 7/5, in excellent agreement with experiment. Because $c_p - c_V = R/M$, it follows that $c_p = (7/2)R/M$ and $c_V = (5/2)R/M$. In the natural environment, most processes involving the exchange of heat in air occur at a pressure (atmospheric) which is effectively constant and since the molecular weight of air is 28.94,

$$c_p = (7/2) \times (8.31/28.94) = 1.01 \text{ J g}^{-1} \text{ K}^{-1}$$

LAPSE RATE

Meteorologists apply thermodynamic principles to the atmosphere by imagining that discrete 'parcels' of air are posted either vertically or horizontally by the action of wind and turbulence. Relations between temperature, pressure and height can be deduced by assuming that processes within a parcel are adiabatic, i.e. that the parcel neither gains energy from its environment (e.g. by heating) nor loses energy.

Suppose a parcel containing unit mass of air makes a small ascent so that it expands as external pressure falls by dp. If there is no external supply of heat, the energy for expansion must come from cooling of the parcel by an amount dT. It is convenient to treat this as a two-stage process:

(1) parcel ascends and cools at constant pressure and volume providing energy $c_V dT$;
(2) parcel expands against external pressure p requiring energy pdV.

For an adiabatic process, the sum of these quantities must be zero, i.e.

$$c_V dT + pdV = 0 \tag{2.10}$$

Differentiating equation (2.7) and putting $R/M = c_p - c_V$ gives

$$VdP + pdV = (c_p - c_V)d\mathrm{T} \tag{2.11}$$

Eliminating pdV from the last two equations gives

$$c_p dT = VdP \tag{2.12}$$

Integration is now possible by putting $V = RT/Mp$ to give

$$\frac{dT}{T} = \left(\frac{R}{Mc_p}\right)\frac{dp}{p} \tag{2.13}$$

from which

$$T = p^{(R/Mc_p)} \tag{2.14}$$

where

$$R/(Mc_p) = (c_p - c_V)/c_p = 0.29 \text{ for air.}$$

When the change of temperature with *pressure* for the adiabatic ascent of a parcel has been established, the change of temperature with *height* can be found when the change of pressure with height is known. A layer of atmosphere with thickness dz and density ρ exerts a pressure dp equal to the weight per unit area or $g\rho dz$ where g is gravitational acceleration. Because p decreases as z increases

$$dp = -g\rho dz$$

and substituting for ρ from equation (2.5) gives

$$\frac{dp}{p} = -\left(\frac{gM}{RT}\right)dz \tag{2.15}$$

Substitution in equation (2.13) then gives

$$\frac{dT}{dz} = -g/c_p \tag{2.16}$$

This quantity is known as the **Dry Adiabatic Lapse Rate** or DALR. When both g and c_p are expressed in SI units, the DALR is

$$\Gamma = 9.8\ (\mathrm{m\ s^{-2}})/1.01 \times 10^3\ (\mathrm{J\ kg^{-1}\ K^{-1}}) \approx 1\ \mathrm{K\ per\ 100\ m}$$

'Dry' in this context implies that no condensation or evaporation occurs within the parcel. A Wet Adiabatic Lapse rate operates within cloud where the lapse rate is smaller than the DALR because of the release of latent heat by condensation.

The difference between the actual lapse rate of air and the DALR is a measure of the vertical stability of the atmosphere. During the day, the lapse rate up to a height of at least 1 km is usually larger than the DALR and is many times larger immediately over dry sunlit surfaces. Consequently,ascending parcels rapidly become warmer than their surroundings and experience buoyancy which accelerates their ascent and promotes turbulent mixing so that the atmosphere is said to be *unstable*. Conversely, temperature usually increases with height at night ('inversion' of the day-time lapse) so that rising parcels become cooler than their environment and further ascent is inhibited by buoyancy. Turbulence is suppressed and the atmosphere is said to be *stable*.

WATER AND WATER VAPOUR

The evaporation of water at the earth's surface to form water vapour in the atmosphere is a process of major physical and biological importance because the latent heat of vaporization is large in relation to the specific heat of air. The heat released by condensing 1 g of water vapour is enough to raise the temperature of 1 kg of air by 2.5 K. Water vapour has been called the working substance of the atmospheric heat engine because of its role in global heat transport. On a much smaller scale, it is the amount of latent heat removed by the evaporation of sweat that allows man and many other mammals to survive in the tropics. Sections which follow describe the physical significance of different ways of specifying the amount of vapour in a sample of air and relations between them.

Vapour pressure

When both air and liquid water are present in a closed container, molecules of water continually escape from the surface into the air to form water vapour but there is a counter-flow of molecules recaptured by the surface. If the air is dry initially, there is a net loss of molecules recognized as 'evaporation' but as the partial pressure (e) of the vapour increases, the evaporation rate decreases, reaching zero when the rate of loss is exactly balanced by the rate of return. The air is then said to be 'saturated' with vapour and the partial pressure is the saturation vapour pressure of water (SVP), often written $e_s(T)$ because it depends strongly on temperature. When a surface is maintained at a lower temperature than the air above it, it is possible for molecules to be captured faster than they are lost and this net gain is recognized as 'condensation'. Rigorous expressions for the dependence of $e_s(T)$ on T are obtained by integrating the Clausius-Clapeyron equation, but as the procedure is cumbersome, a simpler (and unorthodox) method will be used here with the advantage that it relates vapour pressure to the concepts of latent heat and free energy.

Suppose that the evaporation of unit mass of water can be represented by the isothermal expansion of vapour at a fictitious and large pressure e_o to a much larger volume of saturated water vapour at a smaller pressure $e_s(T)$. If the work done during this expansion is identified as the latent heat of vaporization, λ, and v is the volume of the gas at any point during expansion

$$\lambda = \int_{e_o}^{e_s(T)} e dV \tag{2.17}$$

Differentiating equation (2.7) for an isothermal system and putting $p = e$ gives

$$e dV + V de = 0$$

so that from equation (2.7)

$$e dV = -(R/M_w) T de/e$$

where M_w is the molecular weight of water.

Substitution in equation (2.17) then gives

$$\lambda = -\frac{RT}{M_w} \int_{e_o}^{e_s(T)} \frac{de}{e} = \frac{RT}{M_w} \ln[e_o/e_s(T)] \tag{2.18}$$

Given values of T and e_s (T) as in Table A.4, the initial pressure e_o can now be calculated as

$$e_o = e_s(T) \exp(\lambda M_w/RT) \tag{2.19}$$

When a constant value is taken of $\lambda = 2.48$ kJ g^{-1} at 10°C, e_o is found to change very little with temperature: it is 2.076×10^5 MPa at 0°C and 2.077×10^5 MPa at 20°C. (In fact, the value of λ decreases with temperature by about 2.4 J g^{-1} K^{-1} and e_o decreases with temperature if this is taken into account.

Equation (2.19) therefore provides a simple expression for the dependence of $e_s(T)$ on T with the form

$$e_s(T) = e_o \exp(-\lambda M_w/RT) \tag{2.20}$$

where e_o can be regarded as constant, at least over a restricted temperature range. However, it is more convenient to eliminate e_o by normalizing the equation to express $e_s(T)$ as a proportion of the saturated vapour pressure at some standard temperature T^* so that

$$e_s(T^*) = e_o \exp(-\lambda M_w/RT^*) \tag{2.21}$$

Dividing equation (2.20) by (2.21) then gives

$$e_s(T) = e_s(T^*) \exp\{A(T - T^*)/T\} \tag{2.22}$$

where $A = \lambda M_w/RT^*$.

For $T^* = 273$ K (0°C), $\lambda = 2470$ J g^{-1} K^{-1} so that $A = 19.59$. But over the range 273 to 293 K, more exact values of $e_s(T)$ (to within 1 Pa) are obtained by giving A an arbitrary value of 19.65 (see Table A.4). Similarly, with $T^* = 293$, the calculated value of A is 18.3 but $A = 18.00$ gives values of $e_s(T)$ from 293 to 313 K. (The need to adjust the value of A arises from the slight dependence of e_o on temperature coupled with the sensitivity of $e_s(T)$ to the value of A).

An empirical equation introduced by Tetens (1930) has almost the same form as equation (2.22) and is more exact over a much wider temperature range. As given by Murray (1967), it is

$$e_s(T) = e_s(T^*) \exp\{A(T - T^*)/(T - T')\} \tag{2.23}$$

where $A = 17.27$
$T^* = 273$ K ($e_s(T^*) = 0.611$ kPa)
$T' = 36$ K

Values of saturation vapour pressure from the Tetens formula are within 1 Pa of the exact values given in Appendix A4 up to 35°C.

The rate of increase of $e_s(T)$ with temperature is an important quantity in micrometeorology (see Chapter 11) and is usually given the symbol Δ or s. Between 0 and 30°C, $e_s(T)$ increases by about 6.5% per °C whereas the pressure of unsaturated vapour of any ideal gas increases by only 0.4% (1/273) per °C. By differentiating equation (2.20) with respect to T it can be shown that

$$\Delta = \lambda M_w e_s(T)/(RT^2) \tag{2.24}$$

and this expression is exact enough for all practical purposes up to 40°C.

Dew point

The dew point (T_d) of a sample of air with vapour pressure e is the temperature to which it must be cooled to become saturated, i.e. it is defined by the equation $e = e_s(T_d)$. When the vapour pressure is known, the dew point can be found approximately from tables of SVP or more exactly by inverting a formula such as equation (2.22) to obtain temperature as a function of vapour pressure, i.e.

$$T_d = \frac{T^*}{1 - \{\ln e/e_s(T^*)\}/A} \tag{2.25}$$

The specification of a dew point is most useful in problems of dew formation which occurs when the temperature of a surface is below the dew point of the ambient air.

Saturation vapour pressure deficit

The saturation vapour pressure deficit of an air sample (sometimes 'vapour pressure deficit' or just 'saturation deficit' for short) is the difference between the saturation vapour pressure and the actual vapour pressure i.e. $e_s(T) - e$. In ecological problems, it is often regarded as a measure of the 'drying power' of air and it plays an important part in determining the relative rates of growth and transpiration in plants. In micrometeorology the gradient of saturation deficit is a measure of the lack of equilibrium between a wet surface and the air passing over it (Chapter 11).

Specific and absolute humidity

Specific humidity (q) is the mass of water vapour per unit *mass* of moist air and is an appropriate quantity to use in problems of vapour transport in the lower atmosphere because it is independent of temperature (but see p. 241). It is closely related to *absolute* humidity (χ) or vapour density which is the mass of water vapour per unit *volume* of moist air. If the density of the moist air is ρ, $\chi = \rho q$. Figure 2.1 shows the relation between saturation vapour pressure, absolute humidity and temperature.

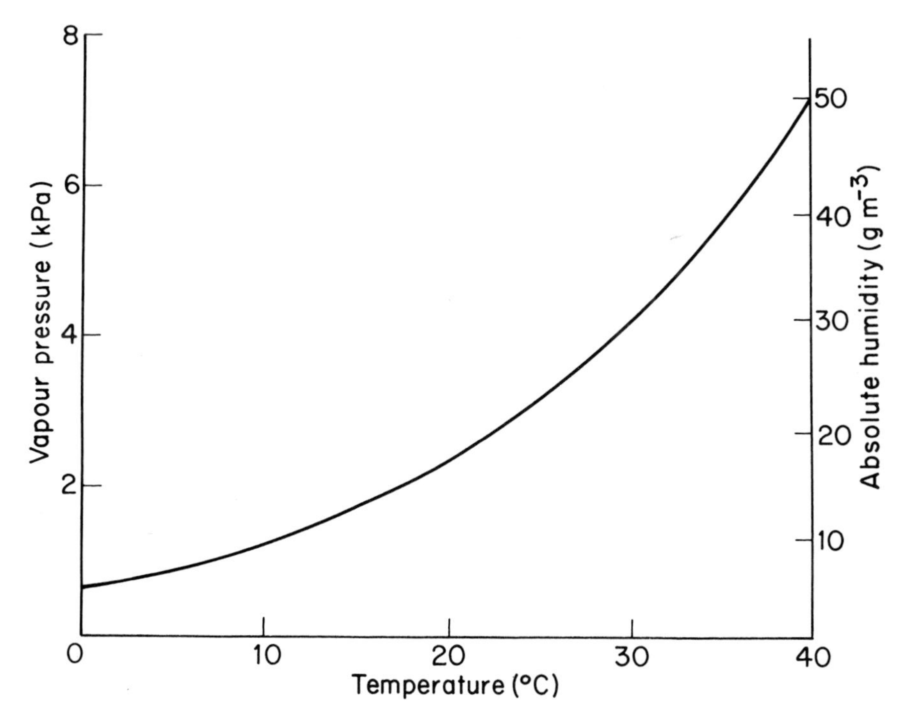

Fig. 2.1 The relation between saturation vapour pressure, absolute humidity, and temperature.

Both absolute and specific humidity can be expressed as functions of total air pressure p and vapour pressure e. For the vapour component only, equation (2.5) can be written in the form

$$e = \chi(R/M_w)T$$

so that with $M_w = 18 \text{ g mol}^{-1}$,

$$\chi(\text{g m}^{-3}) = \frac{M_w e}{RT} = \frac{2165e\ (\text{kPa})}{T\ (\text{K})} \qquad (2.26)$$

Similarly, the dry air component has a pressure $p-e$ and therefore has a density of

$$\rho_A = M_A(p-e)/RT$$

where the molecular weight of air is $M_A = 29$ g mol^{-1}.

The density of the moist air is the sum of the density of its components, i.e.

$$\rho = \{M_w e + M_A(p-e)\}/RT \tag{2.27}$$

and the specific humidity is therefore

$$q = \chi/\rho = \frac{\varepsilon e}{(p-e)+\varepsilon e} \approx \frac{\varepsilon e}{p} \tag{2.28}$$

where $\varepsilon = M_w/M_A = 0.622$. The approximation is usually valid in microclimatic problems because e is two orders of magnitude smaller than p.

The density of *dry* air with the same temperature and pressure as *moist* air is given by

$$\rho' = pM_A/RT \tag{2.29}$$

and substitution in equation (2.27) gives the density of moist air as

$$\rho = \rho'\{1 - e(1-\varepsilon)/p\} \tag{2.30}$$

Moist air is therefore less dense than dry air at the same temperature and is relatively buoyant. In problems involving the transfer of heat as a consequence of buoyancy (see Chapter 8), it is convenient to express the difference in density produced by water vapour in terms of a virtual temperature at which dry air would have the same density as a sample of moist air at an actual temperature T. Combination of equations (2.29) and (2.27) gives the virtual temperature as

$$T_v = T/\{1-(1-\varepsilon)e/p\} \approx T(1+(1-\varepsilon)e/p) \tag{2.31}$$

where the approximation is valid when e is much smaller than p.

Relative humidity

The *relative* humidity of moist air (h) is defined as the ratio of its actual vapour pressure to the saturation vapour pressure at the same temperature, i.e. $h = e/e_s(T)$. Although this quantity is frequently quoted in climatic statistics, and (like saturation deficit), is widely regarded as a measure of the drying power of air, its fundamental significance lies in the specification of thermodynamic equilibrium between liquid water and water vapour. For a surface of pure water, equilibrium is established when the air is saturated so that the relative humidity is unity. But when water contains dissolved salts (as in plant cells) or is held by capillary forces in a porous medium (such as soil), the equilibrium relative humidity is less than unity.

To take a homely example, samples of cake or biscuits contain water in equilibrium with a unique relative humidity. If a sample is exposed to air with a higher value of h it will absorb water, but it will lose water to air with a lower value of h. Because biscuits have much smaller pores than cakes, their equilibrium relative humidity is lower. The relative humidity of air in most kitchens is intermediate between the equilibrium value for biscuits and cakes, so biscuits left on the table usually absorb water, becoming soft, whereas cakes lose water, becoming crisp.

The physics of drying and wetting depends on the 'free energy' of water in the sample, defined as the amount of work needed to move unit mass of water to a pool of pure water at atmospheric pressure and at the same temperature. The relation between free energy and relative humidity is more useful than this abstract definition. To fix ideas, consider a salt solution in a closed, isothermal container above which is air with a vapour pressure $e < e_s(T)$. The energy needed to change unit mass of water from its liquid state to saturated vapour is simply the latent heat of vaporization, already evaluated in terms of expansion from an initial pressure e to $e_s(T)$. Because the system is closed (an essential point), further work is needed to expand the vapour from its volume at $e_s(T)$ to its larger volume e. By use of equation (2.18) again, this energy is

$$E = (R/M_w)T \ln \{e_s(T)/e\} \tag{2.32}$$

This expression gives the work in addition to latent heat needed to change liquid water from a state of equilibrium with air at $e_s(T)$ to vapour at a pressure $e = he_s(T)$. In other words, the water behaves as if it had a *negative* free energy of

$$-E = (R/M_w)T \ln h \tag{2.33}$$

Table 2.2 illustrates the range of free energies and equivalent relative humidities applicable to problems of environmental physics in which it is conventional to express energy per unit volume as a pressure (1 Pa $\equiv$ 1 J m^{-3}).

Table 2.2 Equilibrium relative humidity (at 20°C) and equivalent free energy of some common solutions and substances

	Relative humidity (%)	Free energy (Eq. 2.32) (MPa)
Plant cellular fluid	100 to 98	0 to −3
Ocean water	98	−3
Fresh salami	80	−30
Saturated NaCl	75	−39
Crisp cornflakes	20	−218
Saturated LiCl.H_2O	11	−299

The argument used to establish E as a function of h is not valid in an open system where pure water evaporates into unsaturated air because all parts of the system are exposed to the same pressure — atmospheric. The energy needed to evaporate unit mass of water is simply the latent heat of vaporization and is not augmented by an expansion term as physiologists believed for many years (Monteith, 1972). For similar reasons, the rate at which water evaporates depends on the difference between the saturation vapour pressure at the appropriate temperature and the vapour pressure of air moving over the water. The rate is not a direct function of the corresponding difference in free energy (Monteith and Campbell, 1980).

Wet-bulb temperature

An unsaturated parcel of air containing a small quantity of liquid water can be saturated adiabatically if all the energy used for evaporation is obtained by cooling the air. The minimum temperature achieved in an ideal process of this kind is called the thermodynamic wet-bulb temperature because it is closely related to the observed temperature of a thermometer covered with a wet sleeve. The derivation of the wet-bulb temperature and discussion of its environmental significance is deferred to Chapter 11.

Summary

To review several methods of specifying the amount of water vapour in air, Fig. 2.2 contains a point X representing a sample of air with temperature T and vapour pressure e. The change of saturation vapour pressure with temperature is given by the exponential curve SS′. Dew point temperature T_d is given by the point Z at which the curve intersects a line of constant vapour

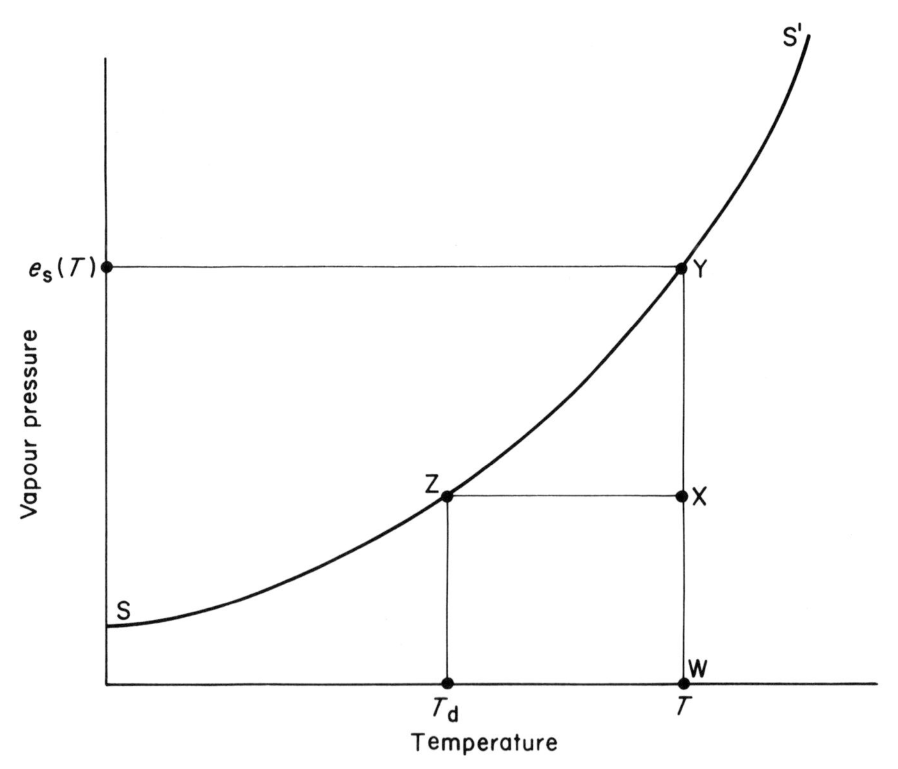

Fig. 2.2 Relation between dew point temperature, saturation vapour pressure deficit and relative humidity of air sample (see text and also Fig. 11.2).

pressure through X and saturation vapour pressure $e_s(T)$ is given by the intersection of the curve with a vertical line through X meeting the temperature axis at W. Saturation vapour pressure deficit $e_s(T) - e$ is $YW - XW = YX$ and relative humidity is XW/YW.

OTHER GASES

In Table 2.1, the percentage by volume of CO_2 in the atmosphere was shown as approximately 0.03. In 1985, the global average percentage was estimated to be 344 volumes (parts) of CO_2 per million volumes (parts) of air (vpm or ppm). The mass concentration is found by multiplying this figure by the density of 1.98 kg m^{-3} (at 0°C) to give 0.68 g m^{-3} and dividing by the density of air gives the specific mass as 0.53 mg CO_2 per g air.

Mainly as a consequence of burning fossil fuel, the global concentration of CO_2 is now rising at about 1.4 vpm per year and is expected to reach twice its pre-industrial value of about 270 vpm by the middle of the 21st century. Implications for global thermodynamics and therefore for all forms of life are the subject of vigorous controversy (Bolin, 1981).

Other gases associated with pollution are sulphur dioxide, principally from coal burning, and ozone, formed by the action of sunlight on nitrogen oxides and hydrocarbons from motor vehicle exhausts. Both gases also have natural sources. Volume concentrations of pollutant gases are often expressed in parts per 10^9 (i.e. North American billion — ppb). Concentrations of sulphur dioxide range from less than 1 ppb in very clean air to 100 ppb close to centres of industry. The corresponding range for ozone in the lower atmosphere is from about 20 ppb in clean air to about 200 ppb in conditions of photochemical smog.

3

TRANSPORT LAWS

The last chapter was concerned primarily with ways of specifying the state of the atmosphere in terms of properties such as pressure and temperature. To complete this introduction to some of the major concepts and principles which environmental physics depends on, we now consider how the transport of entities such as mass, momentum and energy is determined by the state of the atmosphere and the corresponding state of the surface involved in the exchange, whether soil, vegetation or the coat of an animal.

Two main types of transport system are considered: the first depends on the motion of molecules or particles and the second on electromagnetic waves.

GENERAL TRANSFER EQUATION

A simple general equation can be derived for transport within a gas by 'carriers', which may be molecules or particles or eddies, capable of transporting units of a property P such as momentum or heat or mass. Even when the carriers are moving randomly, net transport may occur in any direction provided P decreases with distance in that direction. The carrier can then unload its excess of P at a point where the local value is less than at the starting point.

To evaluate the net flow of P in one dimension, consider a volume of gas with unit horizontal cross-section and a vertical height l assumed to be the mean distance for unloading a property of the carrier (Fig. 3.1). Over the plane defining the base of the volume, P has a uniform value $P(\mathrm{O})$ and if the vertical gradient (change with height) of P is dP/dz, the value at height l will be $P(\mathrm{O})+l dP/dz$. Carriers which originate from a height l will therefore have a load corresponding to $P(\mathrm{O})+l dP/dz$ and if they move a distance l vertically downwards to the plane where the standard load corresponds to $P(\mathrm{O})$, they will be able to unload an excess of $l dP/dz$.

To find the rate of transport equivalent to this excess, assume that n carriers per unit volume move with a mean *random* velocity c so that the number moving towards one face of a rectangular volume at any instant is $nc/6$ per unit area. The downward flux of P (quantity per unit area and per unit time) is therefore given by $(ncl/6)\, dP/dz$. However, there is a corresponding upward flux of carriers reaching the same plane from below after setting off with a load given by $P - l dP/dz$. Mathematically, the upward flow of a deficit is

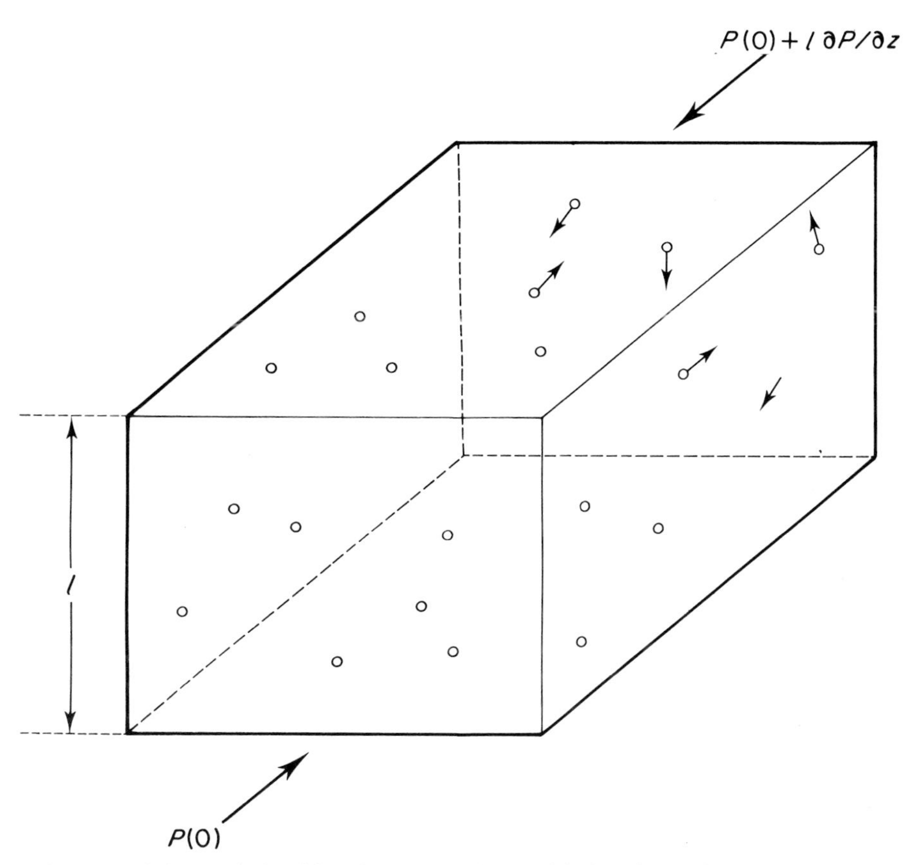

Fig. 3.1 Volume of air with unit cross-section and height l containing n carriers per unit volume moving with random velocity c (see text).

equivalent to the downward flow of an excess so the net *downward* flux of P is

$$\mathbf{F} = (ncl/3)\, dP/dz \tag{3.1}$$

For the application of equation (3.1) to the transport of entities by individual molecules, the mean velocity w is often related to the root mean square molecular velocity c by assuming that, at any instant, one-third of the molecules in the system are moving in the z direction so that $w = c/3$.

Micrometeorologists, on the other hand, are concerned with a form of equation (3.1) in which P is replaced by the amount of an entity per unit mass of air (s) rather than per unit volume (since the latter depends on temperature) and the two quantities are related by $nP = \rho s$, where ρ is air density. The flux equation therefore becomes

$$\mathbf{F} = -\overline{\rho w l}\,(d\bar{s}/dz) \tag{3.2}$$

The minus sign is needed to indicate that the flux is downwards if s increases upwards; averaging bars are a reminder that both w and s fluctuate over a wide range of time-scales as a consequence of turbulence. In this context, the quantity l is known as the 'mixing length' for turbulent transport.

It is also possible to write the instantaneous values of s and w as the sum of mean values $\bar{s}$ or $\bar{w}$ and corresponding deviations from the mean s' and w'. The net flux then becomes

$$\overline{\rho ws} \equiv \overline{(\rho\bar{w}+\rho w')(\bar{s}+s')} \equiv \overline{\rho w's'} \qquad (3.3)$$

where $\bar{s}'$ and $\bar{w}'$ are zero by definition and $\bar{w}$ is assumed to be zero near the ground when averaging is performed over a period long compared with the lifetime of the largest eddy (say 10 minutes). This relation provides the 'eddy correlation' method of measuring vertical fluxes discussed in Chapter 14.

MOLECULAR TRANSFER PROCESSES

According to equation (2.1), the mean square velocity of molecular motion in an ideal gas is $\overline{v^2} = 3p/\rho$. Substituting $p = 10^5$ N m^{-2} and $\rho = 1.29$ kg m^{-3} for air at 0°C gives the root mean square velocity as $(\overline{v^2})^{1/2} = 480$ m s^{-1}. Molecular motion in air is therefore extremely rapid over the whole range of temperatures found in nature and this motion is responsible for a number of processes fundamental to micrometeorology: the transfer of momentum in moving air responsible for the phenomenon of viscosity; the transfer of heat by the process of conduction; and the transfer of mass by the diffusion of water vapour, carbon dioxide and other gases. Because all three forms of transfer are a direct consequence of molecular motion, they are described by similar relationships which will be considered for the simplest possible case of diffusion in one dimension only.

Momentum and viscosity

When a stream of air flows over a solid surface, its velocity increases with distance from the surface. For a simple discussion of viscosity, the velocity gradient $\partial u/\partial z$ will be assumed linear as shown in Fig. 3.2. A more realistic velocity profile will be considered in Chapter 7. Provided the air is isothermal, the velocity of molecular agitation will be the same at all distances from the surface but the horizontal component of velocity in the x direction increases with vertical distance z. As a direct consequence of molecular agitation, there is a constant interchange of molecules between adjacent horizontal layers with a corresponding vertical exchange of horizontal momentum.

The horizontal momentum of a molecule attributable to motion of the gas as distinct from random motion is mu so from equation (3.1) the rate of transfer of momentum, otherwise known as the **shearing stress**, can be written

$$\tau = (ncl/3)\, d(mu)/dz = (cl/3)\, d(\rho u)/dz \qquad (3.4)$$

as the density of the gas is $\rho = mn$. This is formally identical to the empirical equation defining the *kinematic viscosity* of a gas ν, viz.

$$\tau = \nu d(\rho u)/dz \qquad (3.5)$$

showing that ν is a function of molecular velocity and mean free path. Where

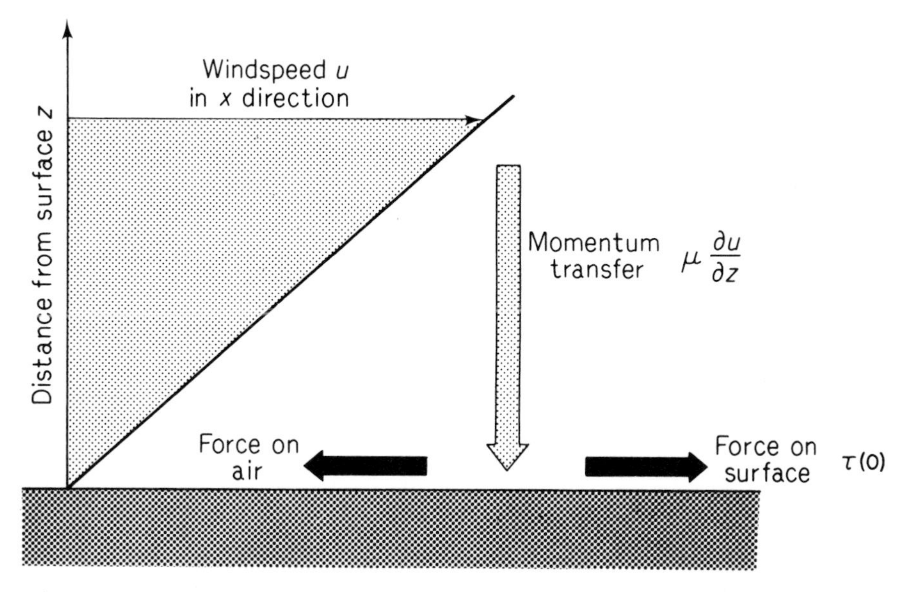

Fig. 3.2 Transfer of momentum from moving air to a stationary surface, showing related forces.

the change of ρ with distance is small, it is more convenient to write

$$\tau = \mu du/dz \tag{3.6}$$

where $\mu = \bar{\rho}\nu$ is the coefficient of *dynamic viscosity* and $\bar{\rho}$ is a mean density. By convention, the flux of momentum is taken as positive when it is directed *towards* a surface and it therefore has the same sign as the velocity gradient (see Fig. 3.2).

The momentum transferred layer by layer through the gas is finally absorbed by the surface which therefore experiences a frictional force acting in the direction of the flow. The reaction to this force required by Newton's Third Law is the frictional drag exerted on the gas by the surface in a direction opposite to the flow.

Heat and thermal conductivity

The conduction of heat in still air is analogous to the transfer of momentum. In Fig. 3.3, a layer of warm air makes contact with a cooler surface. The velocity of molecules therefore increases with distance from the surface and the exchange of molecules between adjacent layers of air is responsible for a net transfer of molecular energy and hence of heat. The rate of transfer of heat is proportional to the gradient of heat content per unit volume of the air and may therefore be written

$$\mathbf{C} = -\kappa d(\rho c_p T)/dz \tag{3.7}$$

where κ, the **thermal diffusivity** of air, has the same dimensions ($L^2\ T^{-1}$) as

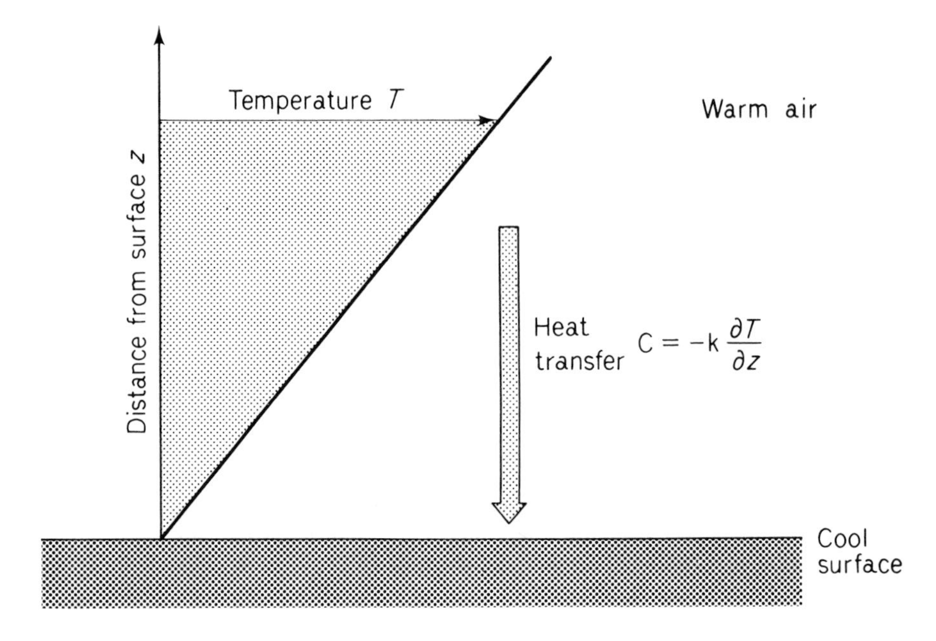

Fig. 3.3 Transfer of heat from still, warm air to a cool surface.

the kinematic viscosity and ρc_p is the heat content per unit volume of air relative to the value at $T = 0$ (°C or K). As in the treatment of momentum, it is convenient to assume that ρ has a constant value of $\bar{\rho}$ over the distance considered and to define a **thermal conductivity** as $k = \bar{\rho} c_p \kappa$ so that

$$\mathbf{C} = -k dT/dz \tag{3.8}$$

identical to the equation for the conduction of heat in solids.

In contrast to the convention for momentum, **C** is taken as positive when the flux of heat is *away* from the surface in which case dT/dz is negative. The equation therefore contains a minus sign.

Mass transfer and diffusivity

In the presence of a gradient of gas concentration, molecular agitation is responsible for a transfer of mass, generally referred to as 'diffusion' although this word can also be applied to momentum and heat. In Fig. 3.4, a layer of still air containing water vapour makes contact with a hygroscopic surface where water is absorbed. The number of molecules of vapour per unit volume increases with distance from the surface and the exchange of molecules between adjacent layers produces a net movement towards the surface. The transfer of molecules expressed as a mass flux per unit area **E** is proportional to the gradient of concentration and the transport equation analogous to equations (3.6) and (3.8) is

$$\mathbf{E} = -D d\chi/dz = -D d(\rho q)/dz = -\bar{\rho} D dq/dz \tag{3.9}$$

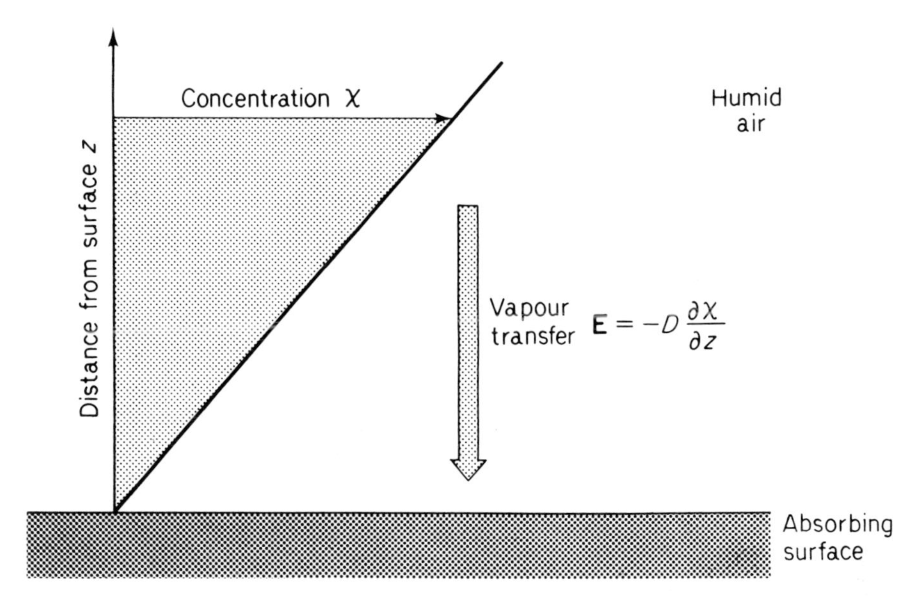

Fig. 3.4 Transfer of vapour from humid air to an absorbing surface.

where $\bar{\rho}$ is an appropriate mean density, D (dimensions $L^2 T^{-1}$) is the molecular diffusion coefficient for water vapour and other symbols are defined on p. 11. The sign convention in this equation is the same as for heat.

DIFFUSION COEFFICIENTS

Because the same process of molecular agitation is responsible for all three types of transfer, the diffusion coefficients for momentum, heat, water vapour and other gases are similar in size and in their dependence on temperature. Values of the coefficients at different temperatures calculated from the Chapman-Eskog kinetic theory of gases agree well with measurements and are given in Table A.3 (p. 267). The temperature dependence of the diffusion coefficients is usually expressed by a power law, e.g. $D(T) = D(0)\{T/T(0)\}^n$ where $D(0)$ is the coefficient at a base temperature $T(0)$ (K) and n is an index between 1.5 and 2.0. Within the limited range of temperature relevant to environmental physics, say -10 to 50°C, a simple temperature coefficient of 0.007 is accurate enough for practical purposes, i.e.

$$D(T)/D(0) = \kappa(T)/\kappa(0) = \nu(T)/\nu(0) = (1 + 0.007T)$$

where T is the temperature in °C and the coefficients in units of $m^2\ s^{-1}$ are

$$\begin{aligned}
\nu(0) &= 13.3 \times 10^{-6} \text{ (momentum)} \\
\kappa(0) &= 18.9 \times 10^{-6} \text{ (heat)} \\
D(0) &= 21.2 \times 10^{-6} \text{ (water vapour)} \\
&= 12.9 \times 10^{-6} \text{ (carbon dioxide)} \\
&= 11.2 \times 10^{-6} \text{ (sulphur dioxide)}
\end{aligned}$$

Resistances

Equations (3.4), (3.6) and (3.7) have the same form

$$\text{flux} = \text{diffusion coefficient} \times \text{gradient}$$

which is a general way of stating Fick's Law of Diffusion. This law can be applied to problems in which diffusion is a one-, two- or three-dimensional process but only one-dimensional cases will be considered in this book. Because the gradient of a quantity at a point is often difficult to estimate accurately, Fick's law is generally applied in an integrated form. The integration is very straightforward in cases where the (one-dimensional) flux can be treated as constant in the direction specified by the coordinate z, e.g. at right angles to a surface. Then the integration of (3.9) for example gives

$$\mathbf{E} = \frac{\rho q(z_1) - \rho q(z_2)}{\int_{z_1}^{z_2} dz/D} \tag{3.10}$$

where $\rho q(z_1)$ and $\rho q(z_2)$ are concentrations of water vapour at distances z_1 and z_2 from a surface absorbing or releasing water vapour at a rate $\mathbf{E}$. Usually $\rho q(z_1)$ is taken as the concentration at the surface so that $z_1 = 0$.

Equation (3.10) and similar equations derived by integrating equations (3.6) and (3.8) are analogous to Ohm's Law:

$$\text{current through resistance} = \frac{\text{potential difference across resistance}}{\text{resistance}}$$

Equivalent expressions for diffusion can be written as follows:

$$\text{rate of transfer of entity} = \text{potential difference} \div \text{resistance}$$

$$\text{momentum } \boldsymbol{\tau} = \rho u \div \int dz/\nu$$
$$\text{heat } \mathbf{C} = \rho c_p T \div \int dz/\kappa$$
$$\text{mass } \mathbf{E} = \rho q \div \int dz/D$$

Diffusion coefficients have dimensions of $(\text{length})^2 \times (\text{time})$ so the corresponding resistances have dimensions of (time)/(length) or 1/(velocity). In a system where rates of diffusion are governed purely by molecular processes, the coefficients can usually be assumed independent of z so that $\int_{z_1}^{z_2} dz/D$, for example, becomes simply $(z_2 - z_1)/D$ or (diffusion pathlength) ÷ (diffusion coefficient). When the molecular diffusion coefficient for water vapour in air is $0.25\ \text{cm}^2\ \text{s}^{-1}$, the resistance for a pathlength of 1 cm is $4\ \text{s cm}^{-1}$. It is often convenient to treat the process of diffusion in laminar boundary layers in terms of resistances and in Chapters 6, 8 and 9, the following symbols are used:

- r_M resistance for momentum transfer at the surface of a body
- r_H resistance for convective heat transfer
- r_V resistance for water vapour transfer
- r_C resistance for CO_2 transfer

The concept of resistance is not limited to molecular diffusion but is applicable to any system in which fluxes are uniquely related to gradients. In the atmosphere where turbulence is the dominant mechanism of diffusion, diffusion coefficients are several orders of magnitude larger than the corresponding molecular value and increase with height above the ground (Chapter 7). Diffusion resistances for momentum, heat, water vapour and carbon dioxide in the atmosphere will be distinguished by the symbols r_{aM}, r_{aH}, r_{aV} and r_{aC}; the measurement of these resistances is discussed in Chapters 7 to 9.

In studies of the deposition of radioactive material from the atmosphere to the surface, the rate of transfer is sometimes expressed as a deposition velocity which is the reciprocal of a diffusion resistance. In this case the surface concentration is often assumed to be zero and the deposition velocity is found by dividing the dosage of radioactive material by its concentration at an arbitrary height.

Brownian motion

The zig-zag motion of particles suspended in a fluid or gas was first described by the English botanist Brown in 1827, but it was nearly 80 years later that Einstein used the kinetic theory of gases to show that the motion was the result of multiple collisions with the surrounding molecules. He found that the mean square displacement $\overline{x^2}$ of a particle in time t is given by

$$\overline{x^2} = 2Dt \tag{3.11}$$

where D is a diffusion coefficient (dimensions $L^2\ T^{-1}$) for the particle, analogous to the coefficient for gas molecules. The quantity D depends on the intensity of molecular bombardment (a function of absolute temperature), and on the viscosity of the fluid, as follows.

Suppose that particles, each with mass m (kg), are dispersed in a container where they neither stick to the walls nor coagulate. The Boltzmann statistical description derived from kinetic theory requires that, as a consequence of the earth's gravitational field, the particle concentration n should decrease exponentially with height z(m) according to the relation

$$n = n(0) \exp(-mgz/kT) \tag{3.12}$$

where
$n = n(0)$ at $z = 0$
g = gravitational acceleration (m s^{-2})
T = absolute temperature (K)
k = Boltzmann's constant (J K^{-1})

If a horizontal area at height z within the container is considered, the flux of particles upwards by diffusion (cf. diffusion of gases) is:

$$\mathbf{F}_1 = -D\frac{dn}{dz} = n\frac{Dmg}{kT} \tag{3.13}$$

from equation (3.12). But because all particles tend to move downwards in response to gravity, there must be a downward flux through the area of

$$\mathbf{F}_2 = nV_s$$

where V_s is the 'sedimentation velocity' (see Chapter 10).

Since the system is in equilibrium, $\mathbf{F}_1 = \mathbf{F}_2$ and so

$$D = kTV_s/mg \tag{3.14}$$

For spherical particles, radius r, obeying Stokes' Law, the downward force mg due to gravity is balanced by a drag force $6\pi\mu rV_s$, and so

$$D = kT/6\pi\mu r \tag{3.15}$$

Thus D depends inversely on particle radius (Table A.6); its dependence on temperature is dominated by the rapid decrease in dynamic viscosity with increasing temperature.

Equations (3.11) and (3.15) show that the root mean square displacement of a particle by Brownian motion is proportional to $T^{0.5}$ and to $r^{-0.5}$. Surprisingly, $\overline{x^2}$ does not depend on the mass of the particle, a result confirmed by experiment.

RADIATION LAWS

The origin and nature of radiation

Electromagnetic radiation is a form of energy derived from oscillating magnetic and electrostatic fields and is capable of transmission through empty space where its velocity is $c = 3 \times 10^8$ m s^{-1}. The frequency of oscillation ν is related to the wavelength λ by the standard wave equation $c = \lambda\nu$ and the wave number $1/\lambda = \nu/c$ is sometimes used as an index of frequency.

The ability to emit and absorb radiation is an intrinsic property of solids, liquids and gases and is always associated with changes in the energy state of atoms and molecules. Changes in the energy state of atomic electrons are associated with line spectra confined to a specific frequency or set of frequencies. In molecules, the energy of radiation is derived from the vibration and rotation of individual atoms within the molecular structure and numerous possible energy states allow radiation to be emitted or absorbed over a wide range of frequencies to form band spectra. The principle of energy conservation is fundamental to the material origin of radiation. The amount of radiant energy emitted by an individual atom or molecule is equal to the decrease in the potential energy of its constituents.

Full or black body radiation

Relationships between radiation absorbed and emitted by matter were examined by Kirchhoff. He defined the absorptivity of a surface $\alpha(\lambda)$ as the fraction of incident radiation absorbed at a specific wavelength λ. The emissivity $\varepsilon(\lambda)$ was defined as the ratio of the actual radiation emitted at the wavelength λ to a hypothetical amount of radiant flux $\mathbf{B}(\lambda)$. By considering the thermal equilibrium of an object inside an enclosure at a uniform temperature, he showed that $\alpha(\lambda)$ is always equal to $\varepsilon(\lambda)$. For an object completely absorbing radiation at wavelength λ, $\alpha(\lambda) = 1$, $\varepsilon(\lambda) = 1$ and the emitted radiation is $\mathbf{B}(\lambda)$. In the special case of an object with $\varepsilon = 1$ at *all* wavelengths, the spectrum of emitted radiation is known as the 'full' or **black body spectrum**. Within the range of temperatures prevailing at the earth's surface, nearly all the radiation emitted by full radiators is confined to the waveband 3 to 100 μm, and most natural objects—soil, vegetation, water—behave radiatively almost like full radiators in this restricted region of the spectrum (but not in the visible spectrum). Even fresh snow, one of the whitest surfaces in nature, emits radiation like a black body between 3 and 100 μm. The statement 'snow behaves like a black body' refers therefore to the radiation *emitted* by a snow surface and not to solar radiation *reflected* by snow. The semantic confusion inherent in the term 'black body' can be avoided by referring to 'full radiation' and to a 'full radiator'.

After Kirchhoff's work was published in 1859, the emission of radiation by matter was investigated by a number of experimental and theoretical physicists. By combining a spectrometer with a sensitive thermopile, they established that the spectral distribution of radiation from a full radiator resembles the curve in Fig. 3.5 in which the chosen temperatures of 6000 and 300°K correspond approximately to the black body temperatures of the sun and the earth's surface.

Wien's Law

Wien deduced from thermodynamic principles that the energy per unit wavelength $\mathbf{E}(\lambda)$ should be a function of absolute temperature T and of λ such that

$$\mathbf{E}(\lambda) = f(\lambda T)/\lambda^5 \tag{3.16}$$

Deductions from this relation are:

(1) that when spectra from full radiators at different temperatures are compared, the wavelength λ_m at which $\mathbf{E}(\lambda)$ reaches a maximum should be inversely proportional to T so that $\lambda_m T$ is the same for all values of T. The value of this constant is 2897 μm K so, in Fig. 3.5, $\lambda_m = 0.48$ μm for a solar spectrum ($T = 6000$ K) and 9.7 μm for a terrestrial spectrum ($T = 300$ K);

(2) the value of $\mathbf{E}(\lambda)$ at λ_m is proportional to $\lambda_m{}^{-5}$ and therefore to T^5 because $\lambda_m \propto 1/T$. The ratio of $\mathbf{E}(\lambda_m)$ for solar and terrestrial radiation is therefore $(6000/300)^5 = 3.2 \times 10^6$.

Stefan's Law

Stefan and Boltzmann showed that the energy emitted by a full radiator was proportional to the fourth power of its absolute temperature: in symbols,

$$\mathbf{B} = \sigma T^4 \tag{3.17}$$

where $\mathbf{B}$ (W m^{-2}) is the flux emitted by unit area of a plane surface into an imaginary hemisphere surrounding it and the Stefan-Boltzmann constant is

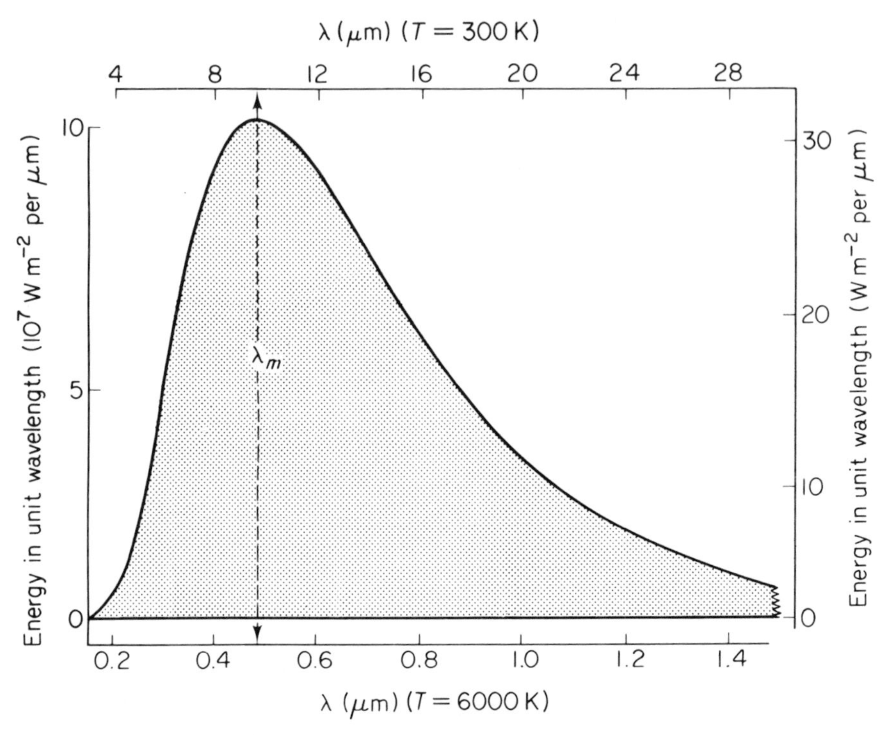

Fig. 3.5 Spectral distribution of radiant energy from a full radiator at a temperature of (a) 6000 K, left-hand vertical and lower horizontal axis and (b) 300 K, right-hand vertical and upper horizontal axis. About 10% of the energy is emitted at wavelengths longer than those shown in the diagram. If this tail were included, the total area under the curve would be proportional to σT^4 (W m^{-2}). λ_m is the wavelength at which the energy per unit wavelength is maximal.

5.67×10^{-8} W m^{-2} K^{-1}. The spatial distribution of this flux is considered on pp. 28–30.

As a generalization from equation (3.17), the radiation emitted from unit area of any plane surface with an emissivity of $\varepsilon(<1)$ can be written in the form

$$\Phi = \varepsilon \sigma T^n$$

where n is a numerical index. For a 'grey' surface whose emissivity is almost independent of wavelength, $n = 4$. When radiation is emitted predominantly at wavelengths less than λ_m, n exceeds 4 and conversely when the bulk of emitted radiation appears in a waveband above λ_m, n is less than 4.

Planck's Law

Prolonged attempts to establish the shape of the black-body spectrum culminated in a theory developed by Max Planck who laid the foundation for much of modern physics by introducing the quantum hypothesis. Planck found that the spectrum could not be predicted from classical mechanics because its principles imposed no restriction on the amount of radiant energy which a molecule could emit. He postulated that energy was emitted in discrete packets which he called 'quanta' and that the energy of a single quantum E_q was proportional to the frequency of the radiation, i.e.

$$E_q = h\nu \tag{3.18}$$

where $h = 6.63 \times 10^{-34}$ J s is **Planck's constant**.

The formula derived by Planck on the basis of this hypothesis was

$$E(\lambda) = \frac{8\pi hc\lambda^{-5}}{\exp\{hc/(k\lambda T) - 1\}} \tag{3.19}$$

where k is the Boltzmann constant (see p. 5). Differentiation of this equation with respect to λ gives the constant of Wien's Law and integration reveals that the Stefan-Boltzmann constant is a multiple of $k^4c^{-2}h^{-3}$.

Quantum unit

Because the amount of energy in a single quantum is inconveniently small, photochemists express the number of quanta in a reaction as a multiple of the Avogadro constant $N = 6.02 \times 10^{23}$. This number of quanta was originally called an 'Einstein' in recognition of Einstein's contribution to the foundations of photochemical theory. It is now usually called a mole. (Although the mole is formally defined as 'an amount of substance', its application in this context suggests that it should now be redefined simply as 'a number, namely, the number of molecules in 0.012 kg of carbon 12'.) A photochemical reaction requiring one quantum per molecule of a compound therefore requires one mole of quanta per mole of compound.

Small temperature differences

Although the total radiation emitted by a full radiator is proportional to T^4, the exchange of energy between radiators at temperatures T_1 and T_2 becomes nearly proportional to $(T_2 - T_1)$ when this difference is small, a common case in natural environments.

If we write $(T_2 - T_1) = \delta T$, the difference between full radiation at the two temperatures is

$$\begin{aligned} \mathbf{R} &= \sigma\{(T_1 + \delta T)^4 - T_1^4\} \\ &= \sigma\{4T_1^3\delta T + 6T_1^2\delta T^2 + \ldots\} \\ &\approx 4\sigma T_1^3\delta T\{1 + 6\delta T/4T_1\} \end{aligned} \tag{3.20}$$

where negligible terms in δT^3 and δT^4 are omitted. The value of $\mathbf{R}$ can therefore be taken as $4\sigma T_1^3\delta T$ with an error given by $-1.5\ \delta T/T$. For $T = 298$ K (25°C), the error is only 0.005 or 0.5% per degree temperature difference.

By analogy with the equation already derived for heat transfer in gases, a notional resistance to radiative transfer r_R may be derived by writing

$$\mathbf{R} = \rho c_p \delta T / r_R \tag{3.21}$$

where the introduction of a volumetric specific heat allows r_R to have the same dimensions as resistances for momentum, heat and mass transfer. From equations (3.20) and (3.21)

$$r_R \approx \rho c_p / (4\sigma T^3) \tag{3.22}$$

which has a value of almost exactly 300 s m^{-1} at 298 K.

Spatial relations

An amount of radiant energy emitted, transmitted, or received per unit time is known as a **radiant flux** and in most problems of environmental physics, the watt is a convenient unit of flux. The term **radiant flux density** means flux per unit area, usually quoted in watts per m^2. **Irradiance** is the radiant flux density incident on a surface and **emittance** (or radiant excitance) is the radiant flux density emitted by a surface.

For a beam of parallel radiation, flux density is defined in terms of a plane at right angles to the beam, but several additional terms are needed to describe the spatial relationships of radiation dispersing in all directions from a point source or from a radiating surface. Figure 3.6(a) represents the flux $d\mathbf{F}$ emitted from a point source into a solid angle $d\omega$ where $d\mathbf{F}$ and $d\omega$ are both very small quantities. The intensity of radiation $\mathbf{I}$ is defined as the flux per unit solid angle or $\mathbf{I} = d\mathbf{F}/d\omega$. This quantity may be expressed in watts per steradian.

Figure 3.6(b) illustrates the definition of a closely related quantity—radiance. An element of surface with an area dS emits a flux $d\mathbf{F}$ in a direction specified by an angle ψ with respect to the normal. When the element is projected at right angles to the direction of the flux it shrinks to an area $dS \cos \psi$ which is the apparent or effective area of the surface viewed from an angle ψ. The **radiance** of the element in this direction is the flux emitted in the direction per unit solid angle, or $d\mathbf{F}/\omega$, divided by the projected area $dS \cos \psi$. In other words, radiance is equivalent to the intensity of radiant flux observed in a particular direction divided by the apparent area of the source in the same direction. This quantity may be expressed in watts per m^2 per steradian.

The term 'intensity' is often used colloquially as a synonym for flux density and is also confused with radiance.

Cosine law for emission and absorption

The concept of radiance is linked to an important law describing the spatial distribution of radiation emitted by a full radiator which has a uniform surface temperature T. This temperature determines the total flux of energy emitted by the surface (σT^4) and can be estimated by measuring the radiance of the surface with an appropriate instrument.

As the surface of a full radiator will appear to have the same temperature whatever angle it is viewed from, the intensity of radiation emitted from a point on the surface and the radiance of an element of surface must both be

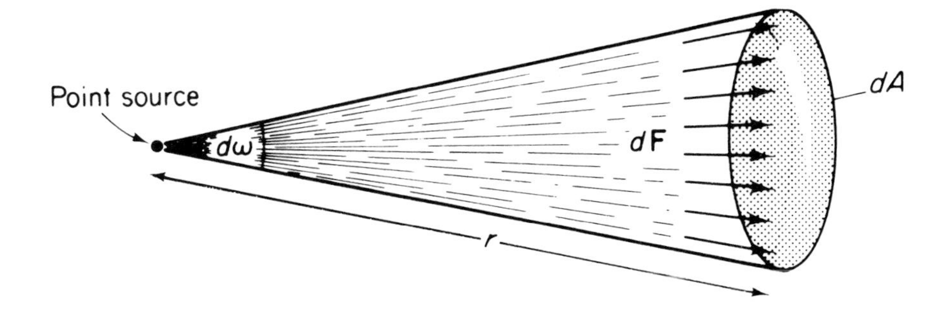

Solid angle $d\omega = dA/r^2$
Intensity $\mathbf{I} = d\mathbf{F}/d\omega$

(a)

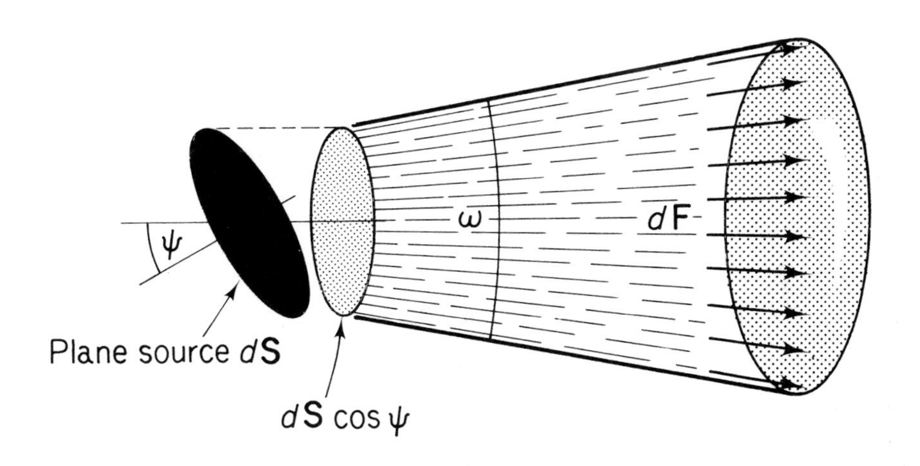

Intensity $d\mathbf{I} = d\mathbf{F}/\omega$
Radiance $= (d\mathbf{F}/\omega) \div d\mathbf{S}\cos\psi$
$= d\mathbf{I}/(d\mathbf{S}\cos\psi)$

(b)

Fig. 3.6 (a) Geometry of radiation emitted by a point source. (b) Geometry of radiation emitted by a surface element. In both diagrams a portion of a spherical surface receives radiation at normal incidence, but when the distance between the source and the receiving surface is large, it can be treated as a plane.

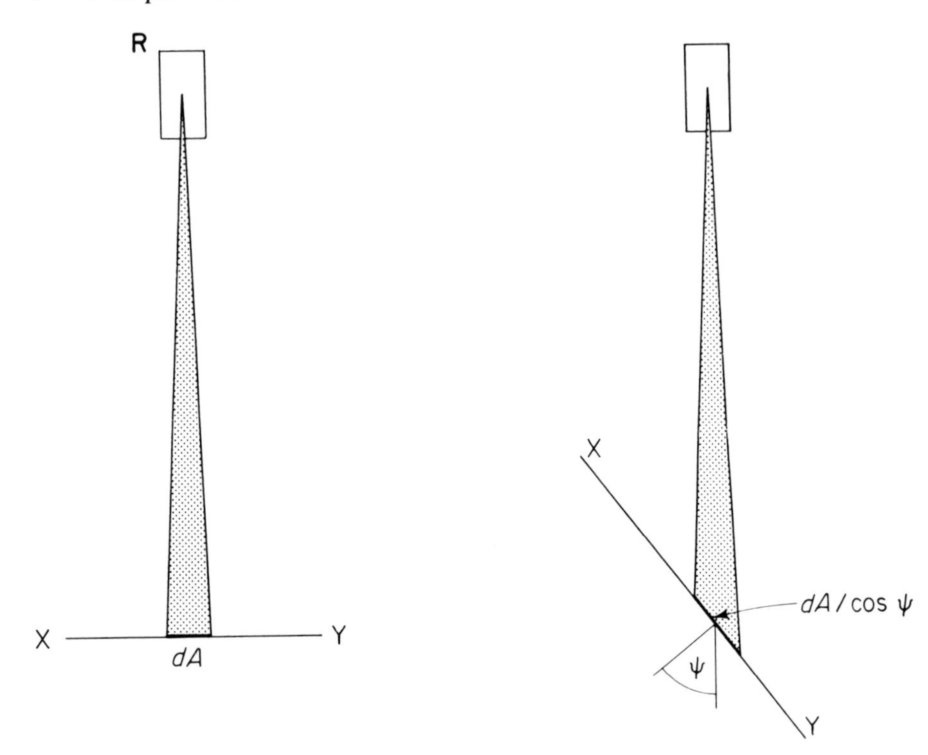

Fig. 3.7 The amount of radiation received by the radiometer from the surface XY is independent of the angle of emission, but the flux emitted per unit area is proportional to cos ψ.

independent of ψ. On the other hand, the flux per unit solid angle divided by the *true* area of the surface must be proportional to cos ψ.

Figure 3.7 makes this point diagrammatically. A radiometer R mounted vertically above an extended horizontal surface XY 'sees' an area dA and measures a flux which is proportional to dA. When the surface is tilted through an angle ψ, the radiometer now sees a larger surface $dA/\cos\psi$, but provided the temperature of the surface stays the same, its radiance will be constant and the flux recorded by the radiometer will also be constant. It follows that the flux emitted per unit area (the emittance of the surface) at an angle ψ must be proportional to cos ψ so that the product of emittance ($\propto \cos\psi$) and the area emitting to the instrument ($\propto 1/\cos\psi$) stays the same for all values of ψ.

This argument is the basis of **Lambert's Cosine Law** which states that when radiation is emitted by a full radiator at an angle ψ to the normal, the flux per unit solid angle emitted by unit surface is proportional to cos ψ. As a corollary to Lambert's Law, it can be shown by simple geometry that when a full radiator is exposed to a beam of radiant energy at an angle ψ to the normal, the flux density of the absorbed radiation is proportional to cos ψ. In remote sensing it is often necessary to specify the directions of both incident and reflected radiation and reflectivity is then described as 'bi-directional'.

Reflection

The reflectivity of a surface $\rho(\lambda)$ is defined as the ratio of the incident flux to the reflected flux at the same wavelength. Two extreme types of behaviour can be distinguished. For surfaces exhibiting specular or mirror-like reflection, a beam of radiation incident at an angle ψ to the normal is reflected at the same angle $(-\psi)$. On the other hand, the radiation scattered by a perfectly diffuse reflector is distributed in all directions according to the Cosine Law, i.e. the intensity of the scattered radiation is independent of the angle of reflection but the flux reflected from a specific area is proportional to $\cos \psi$.

The nature of reflection from the surface of an object depends in a complex way on its electrical properties and on the structure of the surface. In general, specular reflection assumes increasing importance as the angle of incidence increases and surfaces acting as specular reflectors absorb less radiation than diffuse reflectors made of the same material.

Most natural surfaces act as diffuse reflectors when ψ is less than 60 or 70°, but as ψ approaches 90°, a condition known as grazing incidence, the reflection from open water, waxy leaves and other smooth surfaces becomes dominantly specular and there is a corresponding increase in reflectivity. The effect is often visible at sunrise and sunset over an extensive water surface, or a lawn, or a field of barley in ear.

Radiance and irradiance

When a plane surface is surrounded by a uniform source of radiant energy, a simple relation exists between the irradiance of the surface (the flux incident per unit area) and the radiance of the source. Figure 3.8 displays a surface of unit area surrounded by a radiating hemispherical shell so large that the surface can be treated as a point at the centre of the hemisphere. The area dS (black) is a small element of radiating surface and the radiation reaching the centre of the hemisphere from dS makes an angle β with the normal to the plane. As the projection of unit area in the direction of the radiation is $1 \times \cos \beta$, the solid angle which the area subtends at dS is $\omega = \cos \beta / r^2$. If the element dS has a radiance $\mathbf{N}$, the flux emitted by dS in the direction of the disc must be $\mathbf{N} \times dS \times \omega = \mathbf{N} dS \cos \beta / r^2$. To find the total irradiance of the disc, this quantity must be integrated over the whole hemisphere, but if the radiance is uniform, conventional calculus can be avoided by noting that $dS \cos \beta$ is the area dS projected on the equatorial plane. It follows that $\int \cos \beta \, dS$ is the area of the whole plane or πr^2 so that the total irradiance at the centre of the plane becomes

$$(\mathbf{N}/r^2) \int \cos \beta \, dS = \pi \mathbf{N} \quad (3.23)$$

The irradiance expressed in watts m^{-2} is therefore found by multiplying the

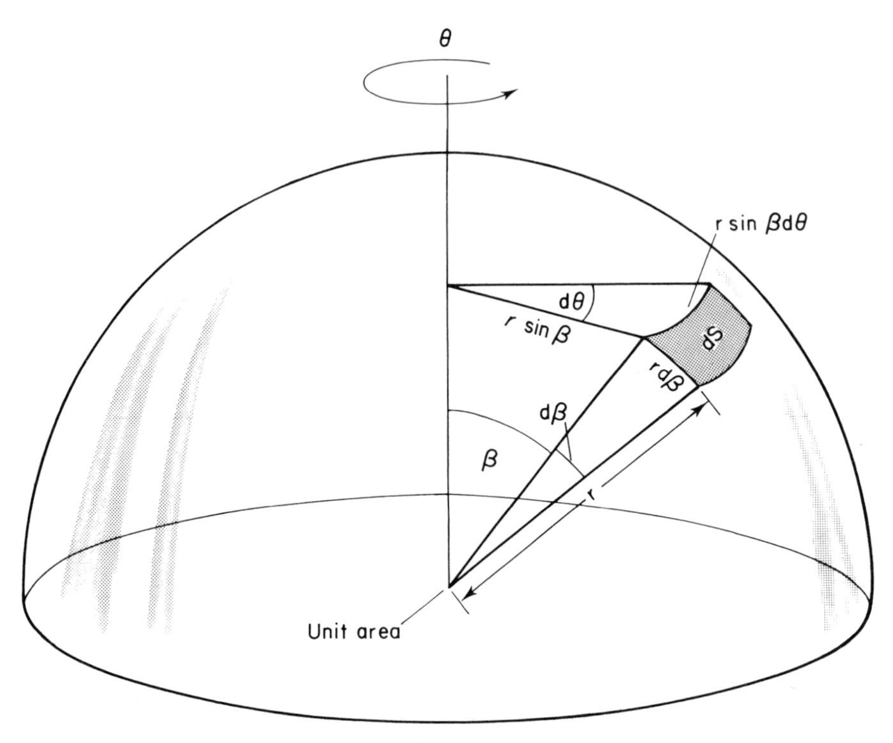

Fig. 3.8 Method for calculating irradiance at the centre of an equatorial plane from a surface element dS at angle β to vertical axis (see text).

radiance in watts m^{-2} steradian^{-1} by the solid angle π.

A more rigorous treatment is needed if the radiance depends on the position of dS with respect to the surface receiving radiation. It is necessary to treat dS as a rectangle whose sides are $rd\beta$ and $r \sin \beta d\theta$ where θ is an azimuth angle with respect to the axis of the hemisphere, radius r. If we write $dS = r^2 \sin \beta d\beta d\theta$ the integral becomes

$$\int_{\theta=0}^{2\pi} \int_{\beta=0}^{\pi/2} \mathbf{N}(\beta, \theta) \left(\frac{\cos \beta}{r^2} \right) r^2 \sin \beta d\beta d\theta$$

$$= 2\pi \int_{\beta=0}^{\pi/2} \mathbf{N} \sin \beta \cos \beta d\beta$$

$$= \pi \int_{\beta=0}^{\pi/2} \mathbf{N} \sin 2\beta d\beta \tag{3.24}$$

Attenuation of a parallel beam

When a beam of radiation consisting of parallel rays of radiation passes through a gas or liquid, quanta encounter molecules of the medium or particles in suspension. After interception, a quantum may suffer one of two fates: it may be absorbed, thereby increasing the energy of the absorbing molecule or particle; or it may be scattered, i.e. diverted from its previous course either forwards (within 90° of the beam) or backwards in a process akin to reflection from a solid. After transmission through the medium, the beam is said to be 'attenuated' by losses caused by absorption and scattering.

Beer's Law, frequently invoked in environmental physics, describes attenuation in a very simple system where radiation of a single wavelength is absorbed but not scattered when it passes through a homogeneous medium. Suppose that at some distance x into the medium the flux density of radiation is $\mathbf{\Phi}(x)$ (Fig. 3.9). Absorption in a thin layer dx, assumed proportional to dx and to $\mathbf{\Phi}(x)$, may be written

$$d\mathbf{\Phi} = -k\mathbf{\Phi}(x)dx \tag{3.25}$$

where the constant of proportionality k, described as an 'attenuation coefficient', is the probability of a ray being intercepted within the small distance dx. Integration gives

$$\mathbf{\Phi}(x) = \mathbf{\Phi}(0)\exp(-kx) \tag{3.26}$$

where $\mathbf{\Phi}(0)$ is the flux density at $x = 0$.

Beer's Law can also be applied to radiation in a waveband over which k is constant; or to a system in which the concentration of scattering centres is so small that a quantum is unlikely to be intercepted more than once ('single scattering').

The treatment of multiple scattering is much more complex for two main

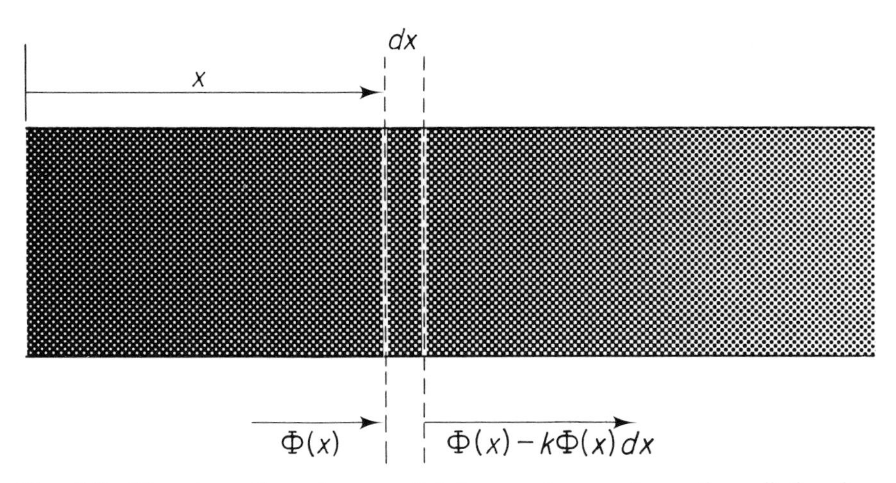

Fig. 3.9 The absorption of a parallel beam of monochromatic radiation in a homogeneous medium with an absorption coefficient k. $\mathbf{\Phi}(0)$ is the incident flux, $\mathbf{\Phi}(x)$ is the flux at depth x and the flux absorbed in a thin layer dx is $k\mathbf{\Phi}(x)dx$.

reasons: (a) radiation in a beam scattered backwards must be considered as well as the forward beam; (b) if k depends on beam direction, the angular distribution of scattering must be taken into account. For the simplest case where k is independent of beam direction, equations of a type developed by Kubelka and Munk are valid. They allow the attenuation coefficient to be expressed as a function of a reflection coefficient ρ which is the probability of an intercepted quantum being reflected backwards and τ which is the probability of forward scattering, implying that the probability of absorption is $\alpha = 1-\rho-\tau$.

In a system with multiple scattering, it is necessary to distinguish two streams of radiation. One moves *into* the medium, and at a distance x from the boundary has a flux density $\mathbf{\Phi}_+(x)$. The other, generated by scattering, moves out of the medium with a flux density $\mathbf{\Phi}_-(x)$. The inward flux is depleted by absorption and reflection but is augmented by reflection of a fraction of the outward stream. The *net* loss of inward flux at a depth x and in a distance dx is therefore

$$d\mathbf{\Phi}_+(x) = (-(\alpha+\rho)\mathbf{\Phi}_+(x)+\rho\mathbf{\Phi}_-(x))dx \tag{3.27}$$

where $(\alpha+\rho)$ is a probability of interception.

The outward stream is also weakened by absorption and reflection but is augmented by reflection of the inward stream to give a net outward flux of

$$d\mathbf{\Phi}_-(x) = -((\alpha+\rho)\mathbf{\Phi}_-(x)-\rho\mathbf{\Phi}_+(x))dx \tag{3.28}$$

where the minus sign in front of the brackets is a reminder that the outward flux is moving in a negative direction with respect to the x axis.

For a system in which the strength of the forward beam is reduced virtually to zero, the bulk reflection coefficient is given by

$$\rho' = (1-\tau-k')/\rho \tag{3.29}$$

where

$$k' = \{(1-\tau^2)-\rho^2\}^{0.5} \tag{3.30}$$

For the special case of uniform scattering in all directions (isotropic scattering), $\rho = \tau = \alpha/2$

$$\rho' = (1-\alpha^{0.5})/(1+\alpha^{0.5}) \tag{3.31}$$

and

$$k' = \alpha^{0.5} \tag{3.32}$$

(Note that these equations are not relevant to the limiting case $\alpha = 1$ when Beer's Law applies.)

When the forward beam strikes a boundary before it is completely attenuated, a fraction ρ_b may be reflected. The fluxes of radiation in the medium, both forwards and backwards, can then be expressed as functions of ρ_b, ρ, τ and of the concentration of the medium.

Complex numerical methods must be deployed to obtain fluxes when condition (b) is not satisfied, i.e. when k is a function of the direction of scattering (Chandrasekhar, 1960).

Examples are given later of the application of Beer's Law to the atmosphere (where the assumption of single scattering is usually valid) and to crop canopies where it is a useful working approximation. In Chapter 6, application of the Kubelka-Munk equations is discussed with reference to canopies and animal coats.

4

RADIATION ENVIRONMENT

All the energy for physical and biological processes at the earth's surface comes from the sun and much of environmental physics is concerned with ways in which this energy is dispersed or stored in thermal, mechanical or chemical form. This chapter considers the quantity and quality of solar (short-wave) radiation received at the gound and the exchange of terrestrial (long-wave) radiation between the ground and the atmosphere.

SOLAR RADIATION

Solar constant

At the mean distance of the earth from the sun, which is 1.50×10^8 m, the irradiance of a surface held perpendicular to the solar beam is known as the **Solar Constant**. The name is somewhat misleading because this quantity is known to change by small amounts over periods of weeks in response to changes within the sun and by amounts large enough to induce climatic change over much longer periods when the solar system passes through clouds of galactic dust.

Increasingly accurate determinations of the Solar Constant have been made from mountain tops, balloons, rocket aircraft flying above the stratosphere and satellites. Observations from a pyrheliometer on the Nimbus 7 satellite range between 1369 and 1375 W m^{-2} with a mean of 1373 W m^{-2} and the indication of a downward trend of about 0.02% per year (Hickey *et al.*, 1982).

The total energy emitted by the sun can be calculated by multiplying the Solar Constant by the area of a sphere at the earth's mean distance d, i.e. it is

$$E = 4\pi d^2 \times 1373 = 3.88 \times 10^{26} \text{ W}$$

The radius of the sun is $r = 6.69 \times 10^8$ m so, assuming the surface behaves like a full radiator (a close approximation), its effective temperature is given by

$$\sigma T^4 = 3.88 \times 10^{26}/(4\pi r^2)$$

from which the (rounded off) value of T is 5770 K.

Sun-earth geometry

Major features of radiation at the surface of the earth are determined by the earth's rotation about its own axis and by its elliptical orbit around the sun. The polar axis about which the earth rotates is fixed in space (pointing at the Pole Star) at a mean angle of 66.5° to the plane of its orbit but with a small top-like wobble. The angle between the orbital plane and the earth's equatorial plane therefore oscillates between a maximum of $90 - 66.5 = 23.5°$ in midsummer and a minimum of $-23.5°$ in midwinter with small deviations attributable to the wobble. This angle is known as the **solar declination** (δ) and its value for any date and year can be found from astronomical tables.

At any point on the earth's surface, the angle between the direction of the sun and a vertical axis depends on the latitude of the site, and on time t(h), most conveniently referred to the time when the sun reaches its zenith. The hour angle of the sun is the fraction of 2π which the earth has turned after local solar noon, i.e. $\theta = 2\pi t/24$. Three-dimensional geometry is needed to show that the zenith angle of the sun ψ at latitude φ is given by

$$\cos \psi = \sin \varphi \sin \delta + \cos \varphi \cos \delta \cos \theta \tag{4.1}$$

and that the azimuth angle A with respect to south is given by

$$\sin A = -\cos \delta \sin \theta / \sin \psi \tag{4.2}$$

The angle of the sun at its zenith is found by putting $\theta = 0$ in equation (4.1) to give

$$\cos \psi = \sin \varphi \sin \delta + \cos \varphi \cos \delta$$
$$= \cos (\varphi - \delta)$$

so

$$\psi = \varphi - \delta$$

The time from solar noon to sunset is found by putting $\psi = \pi/2$ in equation (4.1) to give

$$\cos \theta = -\tan \varphi \tan \delta$$

from which the daylength, defined as the period for which the sun is above the horizon, is

$$2t = 24\theta/\pi = (24/\pi) \cos^{-1} (\tan \varphi \tan \delta) \tag{4.3}$$

a quantity usually obtained from tables (e.g. List, 1966).

Some developmental processes in plants and animals respond to the very weak levels of radiation received during twilight before sunrise and after sunset. The biological length of a day may therefore exceed the daylength given by equation (4.3) to an extent which can be estimated from tables of *civil* twilight (sun dropping to 6° below horizon with a correction for refraction) or *astronomical* twilight (down to 18° below horizon). At the equator, the interval between sunset and civil twilight is close to 22 min throughout the year but at latitude 50° it ranges from 45 min at midsummer to 32 min at the equinoxes.

Quality

For biological work, the spectrum of solar radiation can be divided into three major wavebands shown in Table 4.1 with the corresponding fraction of the Solar Constant. Most measurements of solar energy at the ground are confined to the visible and near-infra-red wavebands which contain energy in almost equal proportions.

Table 4.1 Distribution of energy in the spectrum of radiation emitted by the sun

Waveband (nm)	Energy (%)
0– 300	1.2
300– 400 (ultra-violet)	7.8
400– 700 (visible/PAR)	39.8
700–1500 (near infra-red)	38.8
1500–∞	12.4
	100.0

The ultra-violet spectrum can be further subdivided into UVA (400–320 nm) which produces tanning in skin; UVB (320 to 290 nm) responsible for skin cancer and Vitamin D synthesis; and UVC (290 to 200 nm), potentially harmful but absorbed almost completely by ozone in the stratosphere.

The waveband to which the human eye is sensitive ranges from blue (400 nm) through green (550 nm) to red (700 nm). Photosynthesis is stimulated by radiation in the same waveband which is therefore referred to as Photosynthetically Active Radiation (PAR or PhAR), a misnomer because it is green cells which are active, not radiation. Initially, PAR was applied to radiation measured in units of energy flux density (W m^{-2}) but it is increasingly used for the corresponding quantum flux density (mol m^{-2} s^{-1}) because, when photosynthesis rates are compared for light of different quality (e.g. from the sun and from lamps), they are more closely related to the quantum content than to the energy content of the radiation. The fraction of PAR to total energy in the extraterrestrial solar spectrum is 0.45 (Moon, 1940).

Many developmental processes in green plants have been found to depend on the state of the pigment phytochrome which absorbs radiation in wavebands centred at 660 and 730 nm. Physiologists and ecologists now regard the ratio of spectral irradiance at these wavelengths, known as the red: far-red ratio, as an environmental signal of major significance for germination, survival and reproduction (Smith and Morgan, 1981).

ATTENUATION IN THE ATMOSPHERE

As the solar beam passes through the earth's atmosphere, it is modified in quantity, quality and direction by processes of scattering and absorption (Fig. 4.1).

Scattering has two main forms. First, individual quanta striking molecules of any gas in the atmosphere are diverted more or less uniformly in all

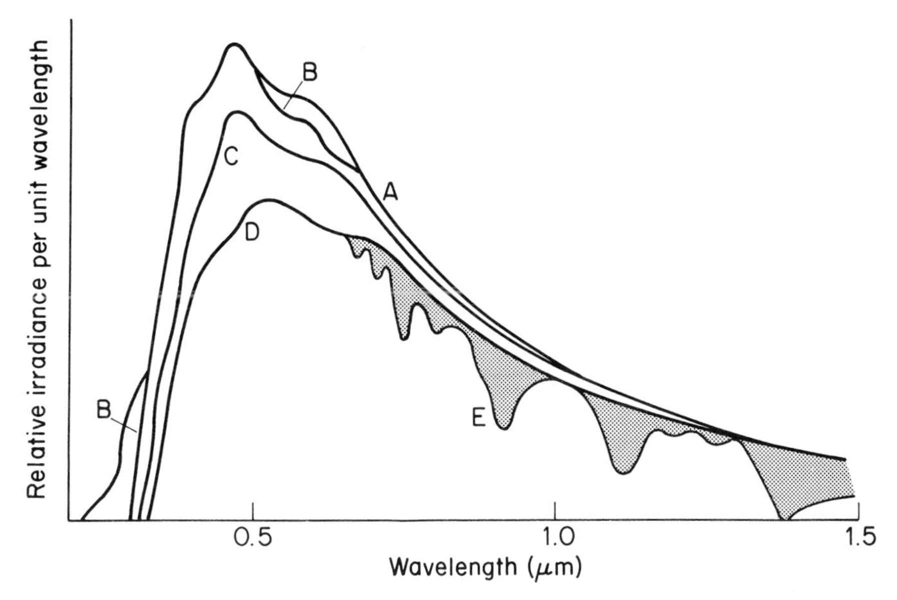

Fig. 4.1 Successive processes attenuating the solar beam as it penetrates the atmosphere. A—extraterrestrial radiation, B—after ozone absorption, C—after molecular scattering, D—after aerosol scattering, E—after water vapour and oxygen absorption (from Henderson, 1977).

directions, a process known as **Rayleigh scattering** after the physicist who showed theoretically that the effectiveness of molecular scattering was proportional to the inverse fourth power of the wavelength. The scattering of blue light ($\lambda = 400$ nm) therefore exceeds the scattering of red (700 nm) by a factor of $(7/4)^4$ or about 9. This is the physical basis of the blue colour of the sky as seen from the ground and the blue haze surrounding the Earth photographed by astronauts travelling to the Moon. The apparent redness of the sun's disc is further evidence that blue light has been removed from the beam preferentially.

Rayleigh scattering is confined to systems in which the diameter of the scatterer (d) is much smaller than the wavelength of the radiation. This condition is not met for particles of dust, smoke, pollen etc. in the atmosphere, referred to as 'aerosol', for which d and λ often have the same order of magnitude. Theory developed by Mie predicts that the wavelength dependence of scattering should be a function of d/λ in this case and that, for some values of the ratio, longer wavelengths should be scattered more than short—the reverse of Rayleigh scattering. This happens rarely, but in 1951, smoke with the appropriate size of particles drifted across the Atlantic from forest fires in Canada and caused much anxiety in Europe when sun and moon turned pale blue!

Usually, aerosol contains such a wide range of particle sizes that scattering is not strongly dependent on λ and a set of measurements in the English Midlands gave a wavelength dependence between $\lambda^{1.3}$ and λ^2 (McCartney and

Unsworth, 1978). In these measurements, the scattering was predominantly 'forwards', i.e. in the direction of the incident radiation, a characteristic of all scattering by aerosol.

The second process of attenuation is *absorption* by ozone (particularly in the ultra-violet spectrum), by water vapour (with strong bands in the infra-red), and by carbon dioxide, and oxygen. In contrast to scattering, which simply changes the direction of radiation, absorption removes energy from the beam so that the atmosphere is heated.

In the visible region of the spectrum, absorption by atmospheric gases is much less important than scattering in determining the spectral distribution of solar energy. In the infra-red spectrum, however, absorption is much more important than scattering and several atmospheric constituents absorb strongly, notably water vapour with absorption bands between 0.9 and 3 μm. The presence of water vapour in the atmosphere increases the amount of visible radiation relative to infra-red radiation.

The extent of absorption and scattering in the atmosphere depends partly on the pathlength of the solar beam and partly on the amount of the attenuating constituent present in the path. The pathlength is usually specified in terms of an air mass number which is the length of the path relative to the depth of the atmosphere. For values of zenith angle ψ less than 80°, the air mass number is simply $m = \sec \psi$, but for values between 80° and 90°, m is smaller than $\sec \psi$, because of the earth's curvature and values, corrected for refraction, can be obtained from tables (e.g. List, 1966).

The amount of water vapour in the atmosphere can be specified by a depth of 'precipitable water' defined as the amount of rain that would be formed if all the water vapour were condensed (between 5 and 50 mm at most stations). If the precipitable water is u, the pathlength for water vapour is um. The amount of ozone is specified by an equivalent depth of the pure gas at a pressure of one atmosphere (101.3 kPa).

Clouds, consisting of water droplets or ice crystals, scatter radiation both forwards and backwards but, when the depth of cloud is substantial, back-scattering predominates and thick stratus can reflect up to 70% of incident radiation, appearing snow-white from an aircraft flying above it. About 20% of the radiation may be absorbed leaving only 10% for transmission so that the base of such a cloud seems grey. At the edge of a cumulus, however, where the concentration of droplets is small, forward scattering is strong—the silver lining effect—and under a thin sheet of cirrus, the reduction of irradiance can be less than 30%, about half a stop to a photographer.

SOLAR RADIATION AT THE GROUND

As a consequence of attenuation, radiation has two distinct directional properties when it reaches the ground. *Direct* radiation arrives from the direction of the solar disc and includes a small component scattered directly forward. The term *diffuse* describes all other scattered radiation received from the blue sky (including the very bright aureole surrounding the sun) and from clouds, either by reflection or by transmission. The sum of the energy flux densities for direct and diffuse radiation is known as *total* or *global* radiation and for climatological purposes is measured on a horizontal surface.

Direct radiation

Direct radiation at the ground, measured at right angles to the beam, rarely exceeds 75% of the Solar Constant, i.e. about 1030 W m^{-2}. The minimum loss of 25% is attributable to molecular scattering and to absorption in almost equal proportions. Generally, aerosol is responsible for significant attenuation which increases the ratio of diffuse to total radiation and also changes spectral composition. Several forms of turbidity coefficient are used to quantify attenuation by aerosol. One of the most straightforward follows from applying Beer's Law to the atmosphere. If the irradiance of the direct beam below an aerosol-free atmosphere is $\mathbf{S}_p(0)$, radiation in the presence of aerosol can be written

$$\mathbf{S}_p(\tau) = \mathbf{S}_p(0) \exp(-\tau m) \tag{4.4}$$

where τ is a turbidity coefficient and m is air mass. The value of τ can be determined by measuring $\mathbf{S}_p(\tau)$ as a function of $m = \sec \psi$ and by calculating $\mathbf{S}_p(0)$ (also a function of m) from the properties of a clean atmosphere containing specified amounts of absorbing and scattering gases.

A series of measurements in Britain gave values of τ ranging from 0.07 in very clean air at the top of Ben Nevis to 0.6 beneath very polluted air in the English Midlands (Unsworth and Monteith, 1972). Corresponding values of $\exp(-\tau m)$ for $\psi = 30°$ are 0.92 and 0.50 implying losses of up to 50% of radiant energy.

The spectrum of direct radiation depends strongly on the pathlength of the beam and therefore on solar zenith angle. Figure 4.2 shows that the spectral

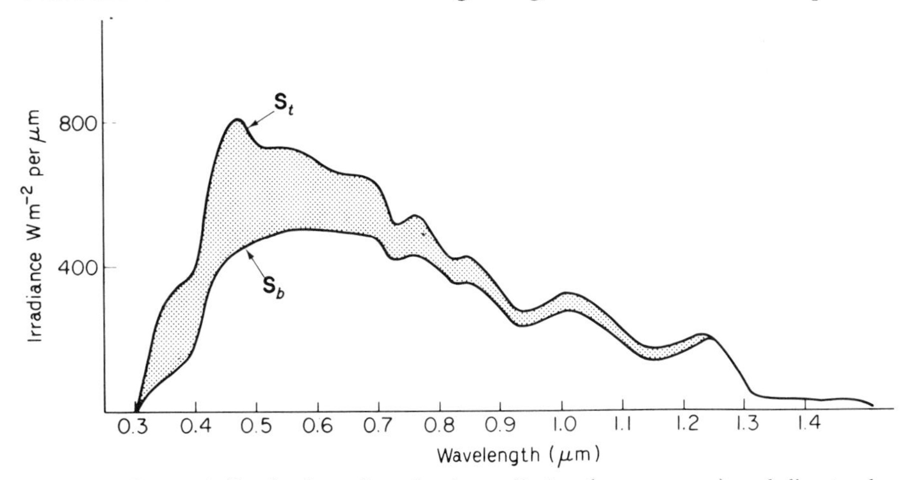

Fig. 4.2 Spectral distribution of total solar radiation (upper curve) and direct solar radiation (lower curve) calculated for a model atmosphere by Tooming and Gulyaev (1967). Solar elevation is 30° ($m = 2$) and precipitable water is 21 mm. The shaded area represents the diffuse flux which has maximum energy per unit wavelength at about 0.46 μm.

irradiance calculated for the solar beam is almost constant between 500 and 700 nm whereas, in the corresponding black-body spectrum, irradiance decreases markedly as wavelength increases beyond $\lambda_m \approx 500$ nm (Fig. 3.5). The difference is mainly a consequence of Rayleigh scattering which changes with wavelength in the opposite direction from the black-body spectrum; and ozone absorption is implicated too. As zenith angle increases, attenuation by scattering becomes very pronounced and the wavelength for maximum irradiance moves into the infra-red waveband when the sun is less than 20° above the horizon.

The measurements referred to above also showed that for zenith angles between 40° and 60°, the ratio of visible to total radiation in the direct beam decreased from a maximum of about 0.5 in clean air to about 0.4 in very turbid air. The maximum exceeds the figure of 0.45 for extraterrestrial radiation because losses of visible radiation by scattering are more than offset by losses of infra-red radiation absorbed by water vapour and oxygen (Fig. 4.1). The quantum content of *direct* radiation increased with turbidity from a minimum of about 2.7 μmol J^{-1} total radiation in clean air to about 2.8 μmol J^{-1} in turbid air. Few comparable figures are available for other sites but Weiss and Norman (1985) in Nebraska reported that the visible : total energy ratio was close to 0.46 for all sky conditions and zenith angles up to 80°. (Confusion has arisen in the literature where these values for *direct* radiation are compared with values referring to total radiation as given later.)

Diffuse radiation

Beneath a clean, cloudless atmosphere, the absolute amount of diffuse radiation increases to a maximum somewhat less than 200 W m^{-2} when ψ is less than 50° and the ratio of diffuse to total radiation falls between 0.1 and 0.15. As turbidity increases, so does $\mathbf{S}_d/\mathbf{S}_t$ and for $\psi < 60°$, observations in the English Midlands fit the relation

$$\mathbf{S}_d/\mathbf{S}_t = 0.68\tau + 0.10 \tag{4.5}$$

For $\psi > 60°$, $\mathbf{S}_d/\mathbf{S}_t$ is a function of ψ and is larger than equation (4.5) predicts.

With increasing cloud amount also, $\mathbf{S}_d/\mathbf{S}_t$ increases and reaches unity when the sun is obscured by dense cloud: but the absolute level of $\mathbf{S}_d$ is maximal when cloud cover is about 50%. The spectral composition of diffuse radiation is also strongly influenced by cloudiness. Beneath a cloudless sky, energy is predominantly within the visible spectrum (Fig. 4.2) but as cloud increases the ratio decreases towards the value of 0.5 characteristic of total radiation.

Angular distribution of skylight

Under an overcast sky, the flux of solar radiation received at the ground is almost completely diffuse. If it were perfectly diffuse, the radiance of the cloud base observed from the ground would be uniform and would therefore be equal to $\mathbf{S}_d/\pi$ from equation (3.23). The source providing this distribution is known as a **Uniform Overcast Sky** (UOS).

In practice, the average radiance of a heavily overcast sky is between two and three times greater at the zenith than the horizon (because multiple-scattered radiation is attenuated by an air mass depending on its perceived direction and so regions near the horizon appear relatively depleted). To allow for this variation, ambitious architects and pedantic professors describe the radiance distribution of overcast skies as a function of zenith angle given by

$$\mathbf{N}(\psi) = \mathbf{N}(0)(1 + b \cos \psi)/(1 + b)$$

This distribution defines a **Standard Overcast Sky**. Measurements indicate that the number $(1 + b)$ which is the ratio of radiance at the zenith to that at the horizon is typically in the range 2.1–2.4 (Steven and Unsworth, 1979), values supported by theoretical analysis (Goudriaan, 1977) which also shows the dependence on surface reflection coefficients. The value $(1 + b) = 3$, in common use, is based on photometric studies and significantly overestimates the diffuse irradiance of surfaces.

Under a cloudless sky, the angular distribution of skylight depends on the position of the sun and cannot be described by any simple relation. In

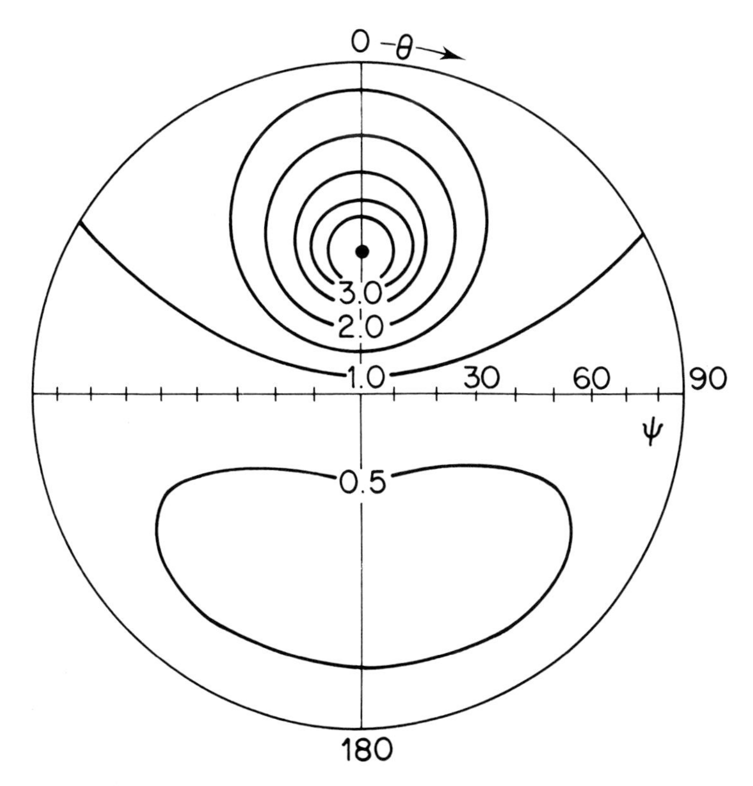

Fig. 4.3 Standard distribution for solar zenith angle 35° of normalized sky radiance $\pi\mathbf{N}/\mathbf{S}_d$ where $\mathbf{N}$ is the value of sky radiance at a point and $\pi\mathbf{N}$ is the diffuse flux which the surface would receive if the whole sky were uniformly bright (see p. 42) (from Steven, 1977).

general, the sky round the sun is much brighter than elsewhere because there is a preponderance of scattering in a forward direction but there is a sector of sky about 90° from the sun where the intensity of skylight is below the average for the hemisphere (Fig. 4.3). On average, the diffuse radiation from a blue sky tends to be stronger nearer the horizon than at the zenith. As the atmosphere becomes more dusty, the general effect is to reduce the radiance of the circumsolar region and to increase the relative radiance of the upper part of the sky at the expense of regions near the horizon. Consequently the angular distribution of radiance becomes more uniform as turbidity increases.

Total radiation

The total radiation on a horizontal surface is given formally by

$$\mathbf{S}_t = \mathbf{S}_p \cos\psi + \mathbf{S}_d$$
$$= \mathbf{S}_b + \mathbf{S}_d \qquad (4.6)$$

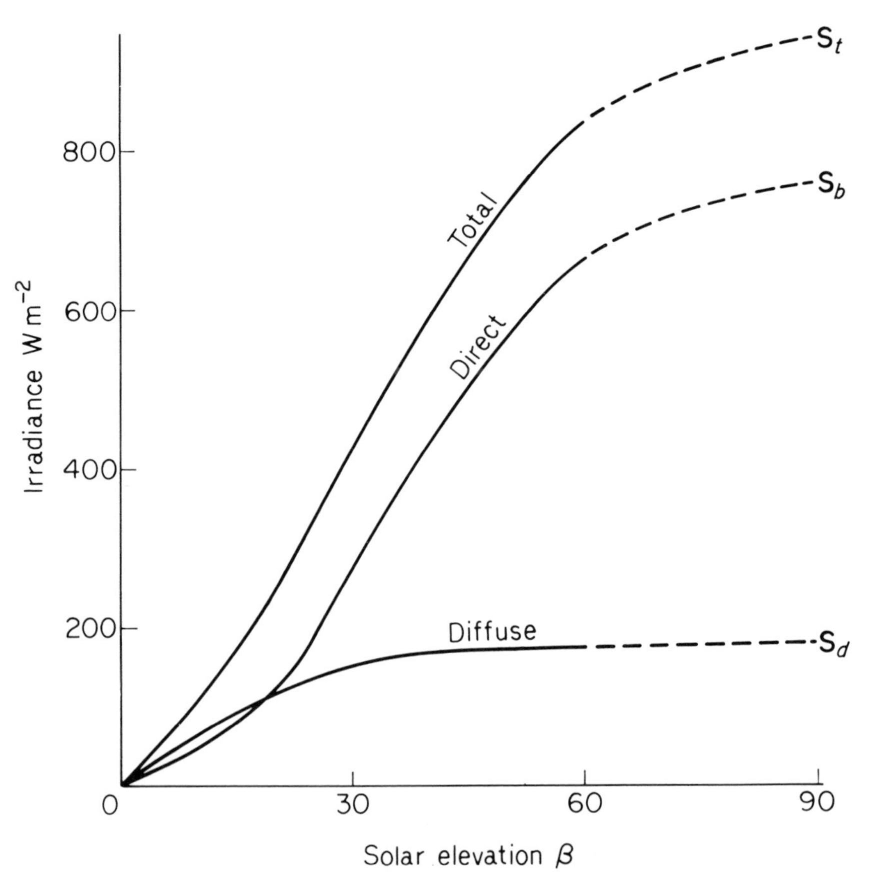

Fig. 4.4 Solar irradiance on a cloudless sky (16 July 1969) at Sutton Bonington (53°N, 1°W): $\mathbf{S}_t$ total flux; $\mathbf{S}_b$ direct flux on horizontal surface; $\mathbf{S}_d$ diffuse flux. Full line from measurements; dashed line extrapolated.

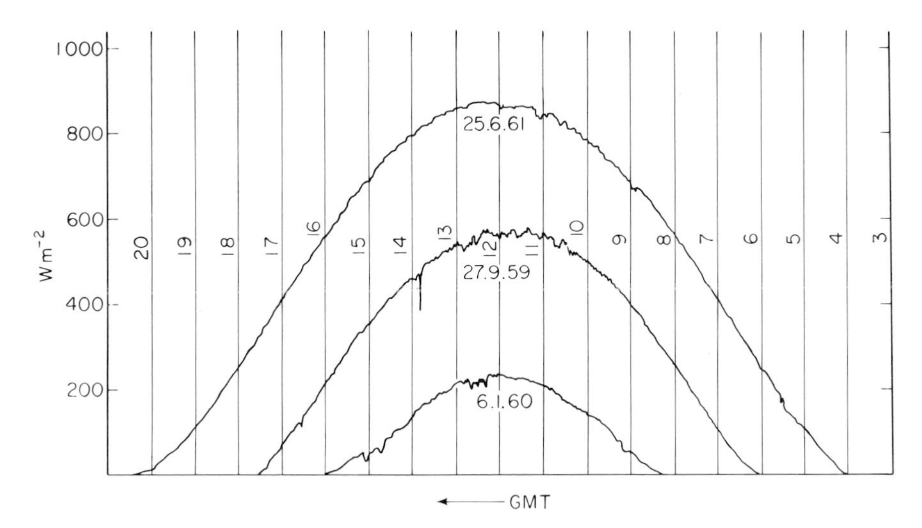

Fig. 4.5 Solar radiation on three cloudless days at Rothamsted (52°N, 0°W). During the middle of the day, the record tends to fluctuate more than in the morning and evening, suggesting a diurnal change in the amount of dust in the lower atmosphere, at least in summer and autumn. Three recorder charts were superimposed to facilitate this comparison.

where $\mathbf{S}_b = \mathbf{S}_p \cos\psi$ is the contribution from the direct beam. Fig. 4.4 contains characteristic values of $\mathbf{S}_t$ as a function of solar zenith angle and Fig. 4.2 gives the spectral distribution for $m = 2$.

On cloudless days, illustrated by Fig. 4.5, the change of $\mathbf{S}_t$ with time is approximately sinusoidal. This form is distorted by cloud but in most climates the degree of cloud cover, averaged over a period of a month, is almost constant throughout the day so the average change of irradiance over a day is again sinusoidal. In both cases, the irradiance at t hours after sunrise can be expressed as

$$\mathbf{S}_t = \mathbf{S}_{tm} \sin(\pi t/n) \tag{4.7}$$

where $\mathbf{S}_{tm}$ is the maximum irradiance at solar noon and n is the daylength in hours. This equation can be integrated to give an approximate relation between maximum irradiance and the daily integral of irradiation by writing

$$\int_0^n \mathbf{S}_t dt \approx 2\mathbf{S}_{tm} \int_0^{n/2} \sin(\pi t/n) dt = (2n/\pi)\mathbf{S}_{tm} \tag{4.8}$$

For example, over southern England in summer, $\mathbf{S}_{tm}$ may reach 900 W m^{-2} on a clear cloudless day and with $n = 16$ h $= 58 \times 10^3$ s, the insolation calculated from equation (4.8) is 33 MJ m^{-2} compared with a measured maximum of about 30 MJ m^{-2}. In Israel, $\mathbf{S}_{tm}$ reaches 1050 W m^{-2} in summer; for a daylength of 14 h the equation gives an insolation of 34 MJ m^{-2} compared with 32 MJ m^{-2} by measurement.

At higher latitudes in summer when dawn and dusk are prolonged, a full sine wave may be more appropriate than equation (4.8). If $\mathbf{S}_t$ is given by

$$\mathbf{S}_t \approx \mathbf{S}_{tm}(1 - \cos 2\pi t/n) = \mathbf{S}_{tm} \sin^2(\pi t/n) \tag{4.9}$$

integration yields

$$\int_0^n \mathbf{S}_t dt = \mathbf{S}_{tm} n/2 \tag{4.10}$$

Gloyne (1972) showed that the radiation regime at Aberdeen (57°N) was described best by the average of values given by equations (4.8) and (4.10).

In most climates, the daily receipt of total solar radiation is greatly reduced by cloud for at least part of the year. Figure 4.6 shows the extent to which the total irradiance beneath continuous cloud depends on cloud type and solar altitude ($\pi/2 - \psi$). The fraction of extraterrestrial radiation can be read from the full lines and the corresponding irradiance by interpolation between the dashed lines.

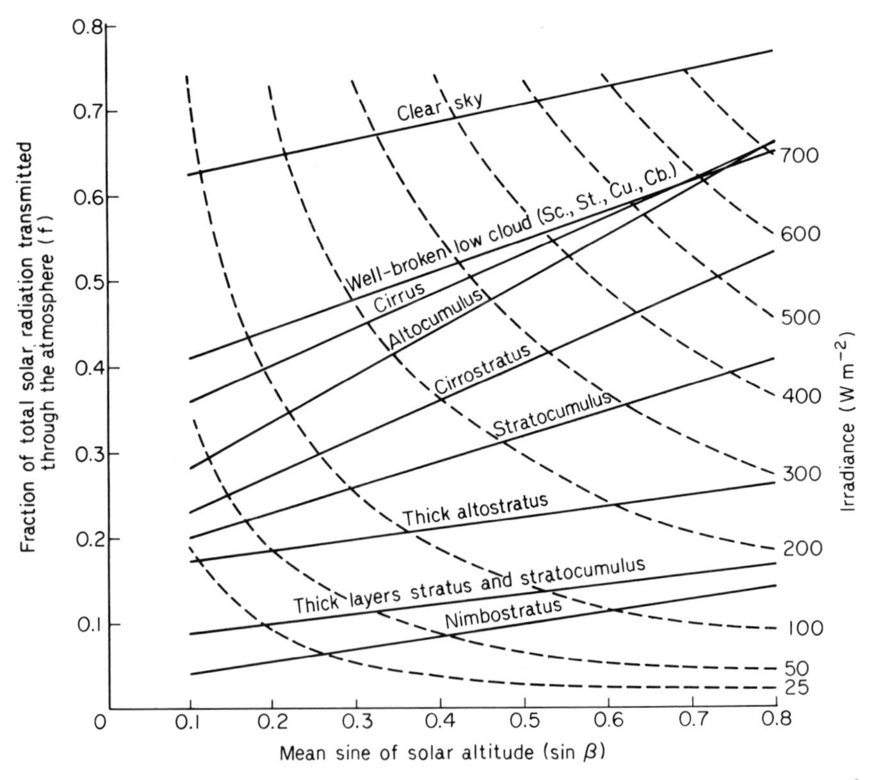

Fig. 4.6 The straight lines represent the empirical relations between solar radiation and solar angle for different cloud types from measurements in the N. Atlantic (52°N, 20°W). The curves are isopleths of irradiance (from Lumb, 1964). Sc, stratocumulus; St, stratus; Cu, cumulus; Cb, cumulonimbus.

The formation of a small amount of cloud in an otherwise clear sky always increases the diffuse flux but the direct component remains unchanged provided neither the sun's disc nor its aureole are obscured. With a few isolated cumulus, total irradiation can therefore exceed the flux beneath a cloudless sky by 5 to 10%. On a day of broken cloud (Fig. 4.7), the distribution of radiation is strongly bimodal: the irradiance is very weak when the sun is completely occluded and strong when it is exposed. For a few minutes before and after occlusion, the irradiance commonly reaches 1000 W m^{-2} in temperate latitudes and even exceeds the Solar Constant in the tropics. This effect is a consequence of strong forward scattering by the small concentration of water droplets present at the edge of a cloud.

As a consequence of cloud, the average insolation over most of Europe in summer is restricted to between 15 and 25 MJ m^{-2}, about 50 to 80% of the insolation on cloudless days. Comparable figures in the USA range from 23 MJ m^{-2} round the Great Lakes to 31 MJ m^{-2} under the almost cloudless skies of the Sacramento and San Joaquim valleys. Winter values range from 1 to 5 MJ m^{-2} over most of Europe and from 6 MJ m^{-2} in the northern USA to 12 MJ m^{-2} in the south. Australian stations record a range of values similar to those of the USA.

Although the difference of irradiance with and without cloud is roughly an order of magnitude, the radiation to which plants are exposed covers a much wider range. In units of micromoles of photons m^{-2}, full summer sunshine is approximately 2000, shade on the forest floor 20, twilight 1, moonlight 3×10^{-4}, and an overcast sky on a moonless night about 10^{-7} (Smith and Morgan, 1981).

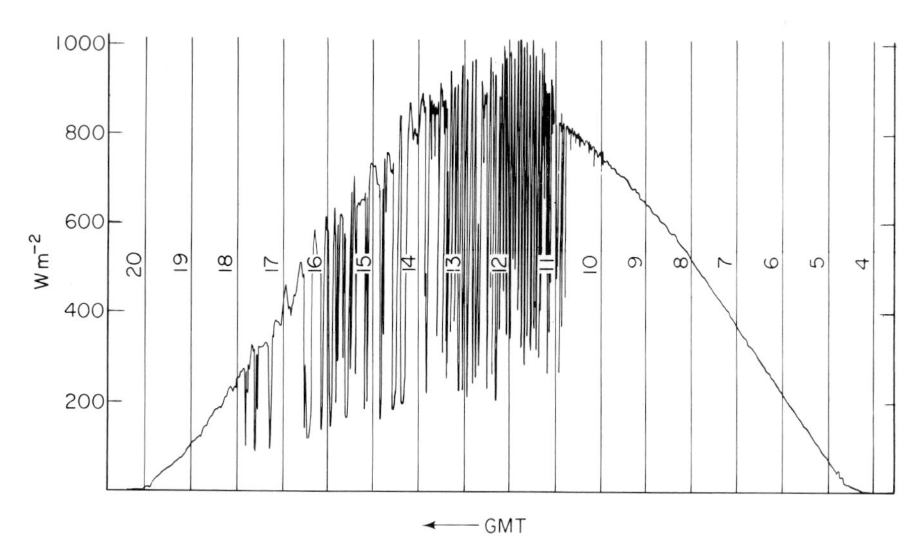

Fig. 4.7 Solar radiation on a day of broken cloud (11 June 1969) at Rothamsted (52°N, 0°W) taken directly from recorder charts. Note very high values of irradiance immediately before and after occlusion of the sun by cloud and the regular succession of minimum values when the sun is completely obscured.

Table 4.2 Short wave radiation balance of atmosphere and surface at Kew Observatory (51.5°N) for 1956–1960 expressed as a percentage of extraterrestrial flux

	Winter (Nov–Jan)	Spring (Feb–April)	Summer (May–July)	Autumn (Aug–Oct)	Year
Extraterrestrial radiation					
three month total $MJ\ m^{-2}$	800	2050	3720	2340	8910
daily mean $MJ\ m^{-2}\ day^{-1}$	8.7	22.3	40.4	25.4	24.4
Losses in atmosphere (per cent)					
(a) Absorption					
water vapour	15	12	13	15	13
cloud	8	9	9	9	9
dust and smoke	15	10	5	8	8
total	38	31	27	32	30
(b) Scattering (away from surface)	37	35	33	34	34
Radiation at surface					
direct	8	14	18	14	15
diffuse	17	20	22	20	21
total	25	34	40	34	36
	100	100	100	100	100
Total as $MJ\ m^{-2}\ day^{-1}$	2.2	7.6	16.2	8.7	8.8

At any station, the way in which insolation changes during the year depends in a complex way on seasonal changes in the water vapour and aerosol content of the atmosphere and on the seasonal distribution of cloud. Table 4.2 shows the main components of attenuation for four 'seasons' at Kew Observatory, a suburban site 10 miles (16 km) west of the centre of London. For the annual average, roughly a third of the radiation received outside the atmosphere is scattered back to space, a third is absorbed and a third is transmitted to the surface. The flux at the surface is 20 to 25% less than it would be in a perfectly clear atmosphere. Because the climate is relatively cloudy, the diffuse component is larger than the direct component throughout the year.

The spectrum of total solar radiation depends, in principle, on solar zenith angle, cloudiness and turbidity and the interaction of these three factors limits the usefulness of generalizations. As zenith angle increases beyond 60°, so does the proportion of scattered radiation and therefore the ratio of visible to total radiation. In one record from Cambridge, England, this ratio increased from about 0.49 at $\psi = 60°$ to 0.52 at $\psi = 10°$ (Szeicz, 1974). Cloud droplets absorb radiation in the infra-red spectrum so with increasing cloud the fraction of visible radiation should increase. Again at Cambridge, the range was between 0.48 in summer and 0.50 in winter. Finally, with increasing turbidity, shorter wavelengths are scattered preferentially, depleting the direct beam but contributing to the diffuse flux so that the net change is relatively small. In another set of measurements at a site close to Cambridge, the visible : total ratio decreased from 0.53 at $\tau = 0.1$ to 0.48 at $\tau = 0.6$ (McCartney, 1978). Elsewhere, a smaller value of 0.44 with little seasonal variation was reported for a Californian site by Howell *et al.* (1983) and, in the tropics, Stigter and Musabilha (1982) found that the ratio increased from 0.51 with clear skies to 0.63 with overcast. It is probable that some of the apparent differences between sites reflect differences or errors in instrumentation and technique rather than in the behaviour of the atmosphere.

At a given site, the relation between quantum content and energy appears to be a conservative quantity and McCartney (1978) obtained a value of 4.56 ± 0.05 μmol per J (PAR) in the English Midlands. This was equivalent to about 2.3 mol per J *total* radiation but other values reported for this quantity range from 2.1 μmol J^{-1} in California to 2.9 μmol J^{-1} in Texas (Howell *et al.*, 1983).

The ratio of energy per unit wavelength in red and far-red wavebands is also conservative and has a value of about 1.1 during the day when solar elevation exceeds 10° (Fig. 4.8). As the sun approaches the horizon, the ratio decreases because red light is scattered more than far-red and because only a small fraction of the forward-scattered light reaches the ground. There is some experimental evidence that the ratio starts to increase and returns to

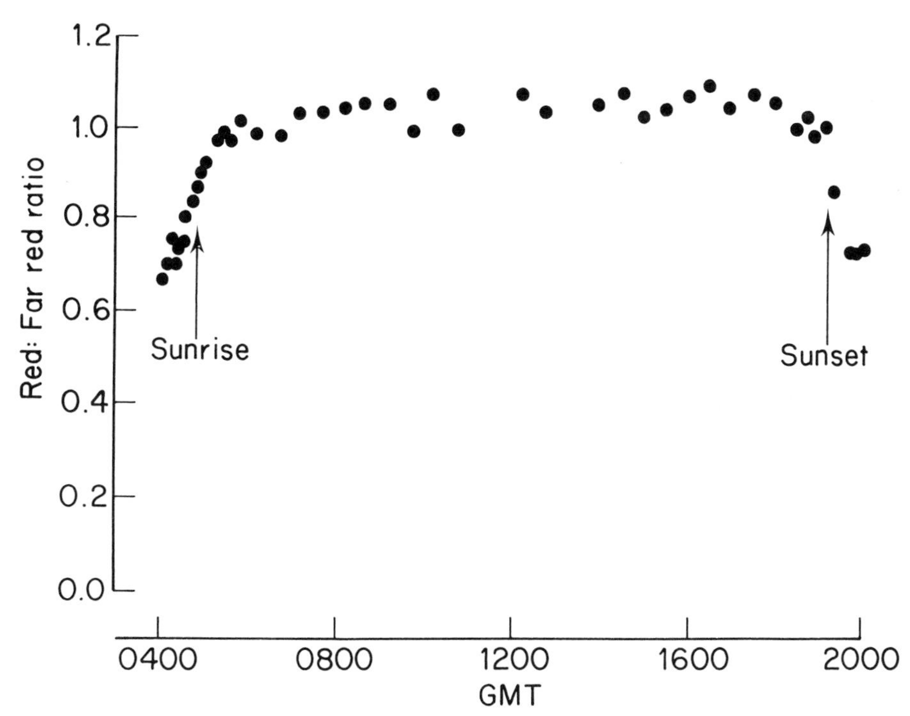

Fig. 4.8 Ratio of spectral irradiance at 660 nm (red) to irradiance at 730 nm (far red) on an overcast day (25 August 1980) near Leicester in the English Midlands from Smith and Morgan (1981).

values greater than 1 when the solar disc falls below the horizon, presumably because skylight alone is relatively rich in shorter wavelengths (Fig. 4.2).

TERRESTRIAL RADIATION

Most natural surfaces can be treated as 'full' radiators which emit 'terrestrial' or long-wave radiation in contrast to the solar or short-wave radiation emitted by the sun. At a surface temperature of 288 K, the energy per unit wavelength of terrestrial radiation reaches a maximum at 2897/288 or 10 μm (Fig. 3.5) and arbitrary limits of 3 and 100 μm are usually taken to define the long-wave spectrum.

Most of the radiation emitted by the earth's surface is absorbed in specific wavebands by atmospheric gases, mainly water vapour and carbon dioxide. These gases have an emission spectrum similar to their absorption spectrum (Kirchhoff's principle, p. 24) and Fig. 4.9 shows the approximate spectral distribution of the downward flux of atmospheric radiation at the earth's surface. Part of the radiation emitted by the atmosphere is lost to outer space; to satisfy the First Law of Thermodynamics for the earth as a planet, the average annual loss of energy must be equal to the average net gain from solar radiation.

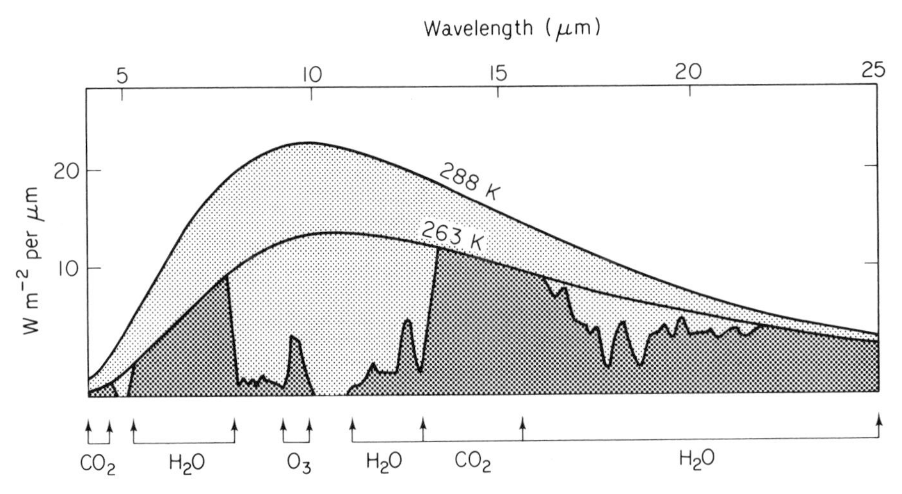

Fig. 4.9 Spectral distribution of long wave radiation for black bodies at 288 K and 263 K. Dark grey areas show the emission from atmospheric gases at 263 K. The light grey area therefore shows the net loss of radiation from a surface at 288 K to a cloudless atmosphere at a uniform temperature of 263 K (after Gates, 1980).

Analysis of the exchange and transfer of long-wave radiation throughout the atmosphere is one of the main problems of physical meteorology but micrometeorologists are concerned primarily with the simpler problem of measuring or estimating fluxes at the surface. The upward flux from a surface $\mathbf{L}_u$ can be measured with a radiometer or from a knowledge of the surface temperature and emissivity. The downward flux from the atmosphere $\mathbf{L}_d$ can also be measured radiometrically, or calculated from a knowledge of the temperature and water vapour distribution in the atmosphere, or estimated from empirical formulae.

Cloudless skies

It is easy to demonstrate with a radiometer that the radiance of a cloudless sky in the long-wave spectrum (or the effective radiative temperature) is least at the zenith and greatest near the horizon. This variation is a direct consequence of the increase in the pathlength of water vapour and carbon dioxide, the main emitting gases. In general, more than half the radiant flux received at the ground from the atmosphere comes from gases in the lowest 100 m and roughly 90% from the lowest kilometre. The magnitude of the flux is therefore strongly determined by temperature gradients near the ground.

It is convenient to define the apparent emissivity of the atmosphere ε_a as the flux density of downward radiation divided by full radiation at screen temperature T_a, i.e.

$$\mathbf{L}_d = \varepsilon_a \sigma T_a^4 \tag{4.11}$$

Similarly, the apparent emissivity at a zenith angle ψ or $\varepsilon_a(\psi)$ can be taken as the flux density of downward radiation at ψ divided by σT_a^4. Many measurements show that the dependence of $\varepsilon_a(\psi)$ on ψ over short periods can be expressed as

$$\varepsilon_a(\psi) = a + b \ln(u \sec \psi) \tag{4.12}$$

where u is precipitable water (corrected for the pressure dependence of radiative emission) and a and b are constants, changing with the vertical gradient of temperature and with the distribution of aerosol (Unsworth and Monteith, 1975). Integration of this equation over a hemisphere using equation (3.24) gives the effective (hemispherical) emissivity as

$$\varepsilon_a = a + b(\ln u + 0.5) \tag{4.13}$$

Comparing equations (4.12) and (4.13) shows that the hemispherical emissivity is identical to the emissivity at a representative angle ψ' such that

$$\ln(u \sec \psi') = \ln u + \ln \sec \psi' = \ln u + 0.5$$

It follows that $\ln \sec \psi' = 0.5$, giving $\psi' = 52.5°$ irrespective of the values of a and b.

Because of the difficulty of choosing appropriate values of a and b *a priori*, $\mathbf{L}_d$ is often estimated from empirical formulae as a function of temperature and/or vapour pressure at screen height. One of the most simple formulae is

$$\mathbf{L}_d = c + d\sigma T_a^4 \tag{4.14}$$

and for measurements in the English Midlands (Unsworth and Monteith, 1975) which covered a temperature range from −6 to 26°C, $c = -119 \pm 16$ W m^{-2} and $d = 1.06 \pm 0.04$. The uncertainty of a single estimate of $\mathbf{L}_d$ was ± 30 W m^{-2}. Measurements in Australia (Swinbank, 1963) gave similar values of c and d but with much less scatter.

Using a linear approximation to the dependence of full radiation on temperature above a base of 283 K allows equation (4.14) to be written in the form

$$\mathbf{L}_d = 213 + 5.5T_a \qquad (4.15)$$

where T_a is now in °C. Outward long-wave radiation assumed to be σT_a^4 is given by a similar approximation as

$$\mathbf{L}_u = 320 + 5.2T_a \qquad (4.16)$$

The net loss of long-wave radiation is therefore

$$\mathbf{L}_u - \mathbf{L}_d = 107 - 0.3T_a \qquad (4.17)$$

implying that 100 W m^{-2} is a good average figure for the net loss to a clear sky (see Fig. 4.10).

An expression for the effective radiative temperature of the atmosphere T can be obtained by writing

$$\mathbf{L}_d = 320 + 5.2T_b = 213 + 5.5T_a$$

so that

$$T_b = (5.5T_a - 107)/5.2 = (T_a - 21) + 0.06T_a$$

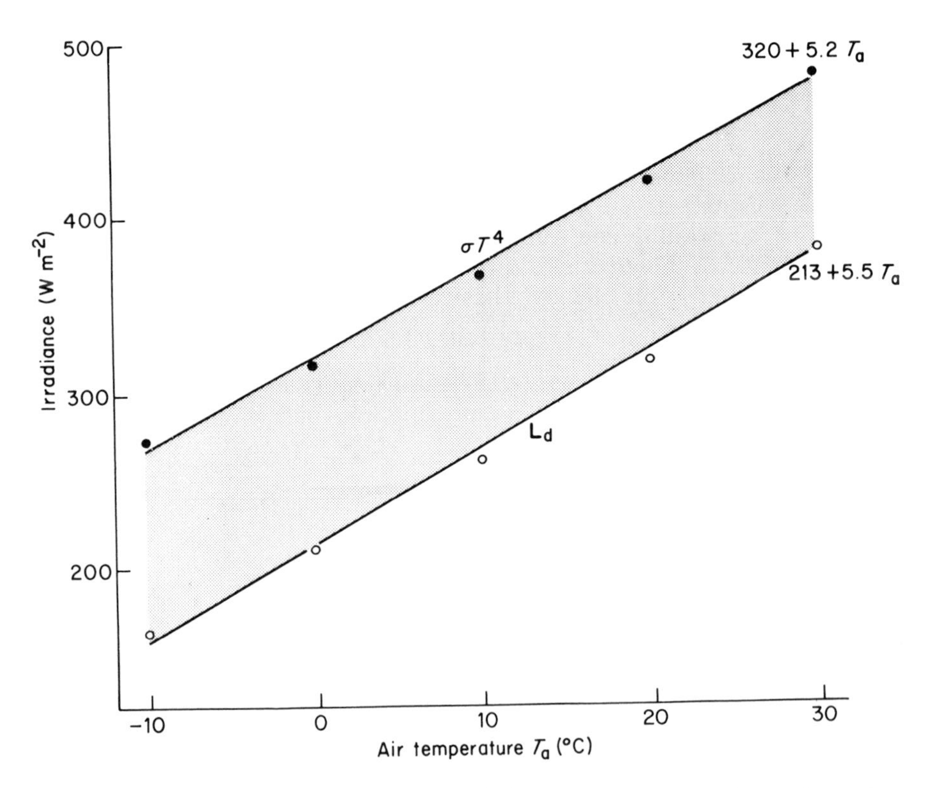

Fig. 4.10 Black body radiation at T_a (full circles) and long wave radiation from clear sky (open circles) from equation (4.14). Straight lines are approximations from equations (4.16) and (4.15) respectively.

This relation shows that when T_a is between 0 and 20°C, the mean effective radiative temperature of a cloudless sky is usually about 21 to 19°C below the mean screen temperature.

Much more complex formulae have been derived to fit observations of $\mathbf{L}_d$ under cloudless skies and the most accurate of these incorporate vapour pressure at screen height as well as temperature in order to allow for differences in precipitable water (Hatfield, 1983). However, these formulae have little additional merit for climatological work where the main uncertainty lies in the influence of cloud—the next topic for discussion.

Cloudy skies

Clouds dense enough to cast a shadow on the ground emit like full radiators at the temperature of the water droplets or ice crystals from which they are formed. The presence of cloud increases the flux of atmospheric radiation received at the surface because the radiation from water vapour and carbon dioxide in the lower atmosphere is supplemented by emission from clouds in the waveband which the gaseous emission lacks, i.e. from 8 to 13 μm (see Fig. 4.9). Because most atmospheric radiation originates below the base of clouds, the gaseous component of the downward flux can be treated as if the sky was cloudless with an apparent emissivity ε_a. From Kirchhoff's principle, the transmission of radiation beneath the layer beneath cloud base is $1-\varepsilon_a$ and if the base temperature is T_c the downward radiation from an overcast sky will be

$$\begin{aligned} \mathbf{L}_d &= \varepsilon_a \sigma T_a^4 + (1-\varepsilon_a)\sigma T_c^4 \\ &= \sigma T_a^4\{1-(1-\varepsilon_a)4\delta T/T_a\} \end{aligned} \tag{4.18}$$

using the linear approximation equation (3.20) with $\delta T = T_a - T_c$.

The analysis of a series of measurements near Oxford (Unsworth and Monteith, 1975) gave an annual mean of $\delta T = 11$ K with a seasonal variation of ±2 K, figures consistent with a mean cloud base at 1 km, higher in summer than in winter. Taking 283 K as a mean value of T_a gives $4\delta T/T_a = 0.16$, so that the emissivity of an overcast sky at this site is

$$\varepsilon_a(1) = \mathbf{L}_d/\sigma T_a^4 = 1 - 0.16\{1-\varepsilon_a\} = 0.84 + 0.16\varepsilon_a \tag{4.19}$$

For a sky covered with a fraction c of cloud, interpolation gives

$$\begin{aligned} \varepsilon_a(c) &= c\varepsilon_a(1) + (1-c)\varepsilon_a \\ &= (1-0.84c)\varepsilon_a + 0.84c \end{aligned} \tag{4.20}$$

The main limitation to this formula lies in the choice of appropriate values for cloud and for δT which depend on base height and therefore on cloud type.

It is important to remember that the formulae presented in this section are statistical correlations of radiative fluxes with weather variables at particular sites and do not describe direct functional relationships. For prediction, they are most accurate under average conditions, e.g. when the air temperature does not increase or decrease rapidly with height near the surface and when the air is not unusually dry or humid. They are therefore appropriate for climatological studies of radiation balance but are often not accurate enough for micrometeorological analyses over periods of a few hours. In particular, the equations cannot be used to investigate the diurnal variation of $\mathbf{L_d}$. At some sites, the amplitude of $\mathbf{L_d}$ in clear weather is much smaller than the amplitude of $\mathbf{L_u}$, behaviour to be expected if changes of atmospheric temperatures were governed by and followed changes of surface temperature. At other sites, $\mathbf{L_d}$ appears to change more than $\mathbf{L_u}$ but this difference is hard to explain in terms of atmospheric behaviour and it may be a consequence of small calibration errors in radiometers.

NET RADIATION

All surfaces receive short-wave radiation during daylight and exchange long-wave radiation continuously with the atmosphere. The net amount of radiation received by a surface with reflection coefficient ρ is defined by the equation

$$\mathbf{R_n} = (1-\rho)\mathbf{S_t} + \mathbf{L_d} - \mathbf{L_u} \tag{4.21}$$

The *micro*climatological application of this equation is considered in Chapter 6. Here, a discussion of the net receipt of radiation by a standard horizontal surface is needed to round off the chapter although net radiation is not strictly a *macro*climatological quantity: it depends on the temperature, emissivity and reflectivity of a surface. Net radiation is measured routinely at very few climatological stations partly because of the problem of providing a standard surface but also because instruments are difficult to maintain. Examples of annual and daily measurements follow.

Figure 4.11 shows the annual change of components in the radiation balance of a short grass surface at Hamburg from February 1954 to January 1955. Each entry in the graph represents the gain or loss of radiation for a period of 24 hours.

The largest term in the balance is $\mathbf{L_u}$, the long-wave emission from the grass surface, ranging between winter and summer from about 23 to about 37 MJ m^{-2} day (or 270 to 430 W m^{-2}). The minimum values of atmospheric radiation $\mathbf{L_d}$ (230 W m^{-2}) were recorded in spring, presumably in cloudless anticyclonic conditions bringing very cold dry air masses; and maximum values (380 W m^{-2}) were recorded during warm humid weather in the autumn. The net loss of long-wave radiation was about 60 W m^{-2} on average (cf. 100 W m^{-2} for cloudless skies on p. 52) and was almost zero on a few, very foggy days in autumn and winter.

In the lower half of the diagram, the income of short-wave radiation forms a Manhattan skyline with much larger day-to-day changes and a much larger seasonal amplitude than the income of long-wave radiation. The maximum value of $\mathbf{S_t}$ is about 28 MJ m^{-2} day^{-1} (320 W m^{-2}) and $\mathbf{S_t}$ is smaller than $\mathbf{L_d}$ on every day of the year. The reflected radiation is about $0.25\mathbf{S_t}$ except on a few days in January and February when snow increased the reflection coefficient to between 0.6 and 0.8.

The net radiation $\mathbf{R_n}$ given by $(1-\rho)\mathbf{S_t} + \mathbf{L_d} - \mathbf{L_u}$ is shown in the top half of the graph. During summer the ratio $\mathbf{R_n}/\mathbf{S}$ was almost constant from day to day at about 0.57 but the ratio decreased during the autumn and reached zero in November. From November until the beginning of February $\mathbf{R_n}$ was negative on most days. In summer, net radiation and mean air temperature were

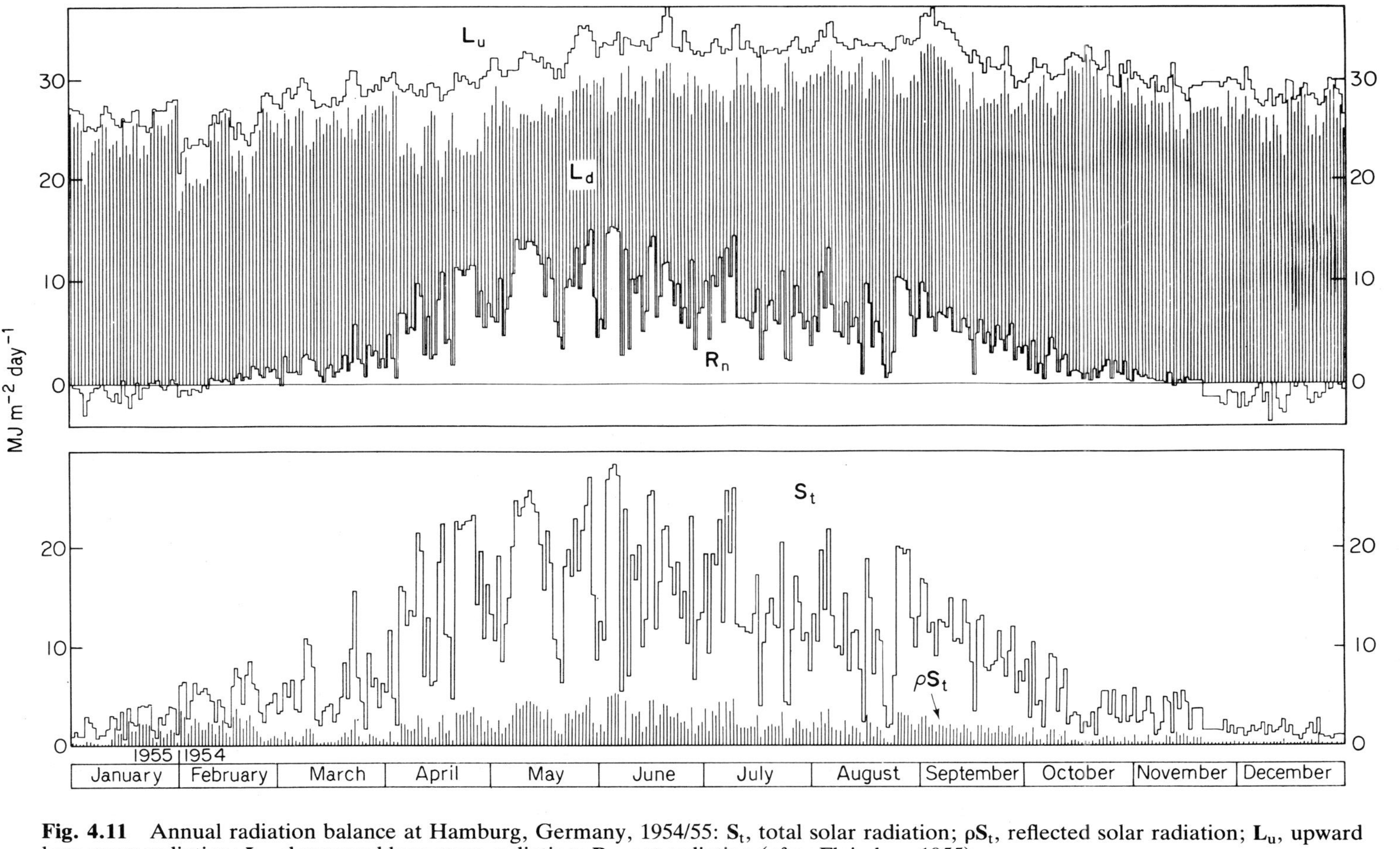

Fig. 4.11 Annual radiation balance at Hamburg, Germany, 1954/55: $\mathbf{S_t}$, total solar radiation; $\rho\mathbf{S_t}$, reflected solar radiation; $\mathbf{L_u}$, upward long wave radiation; $\mathbf{L_d}$, downward long wave radiation; $\mathbf{R_n}$, net radiation (after Fleischer, 1955).

positively correlated with sunshine. In winter the correlation was negative: sunny cloudless days were days of minimum net radiation when the mean air temperature was below average.

Staff at the University of Bergen have maintained records of incoming

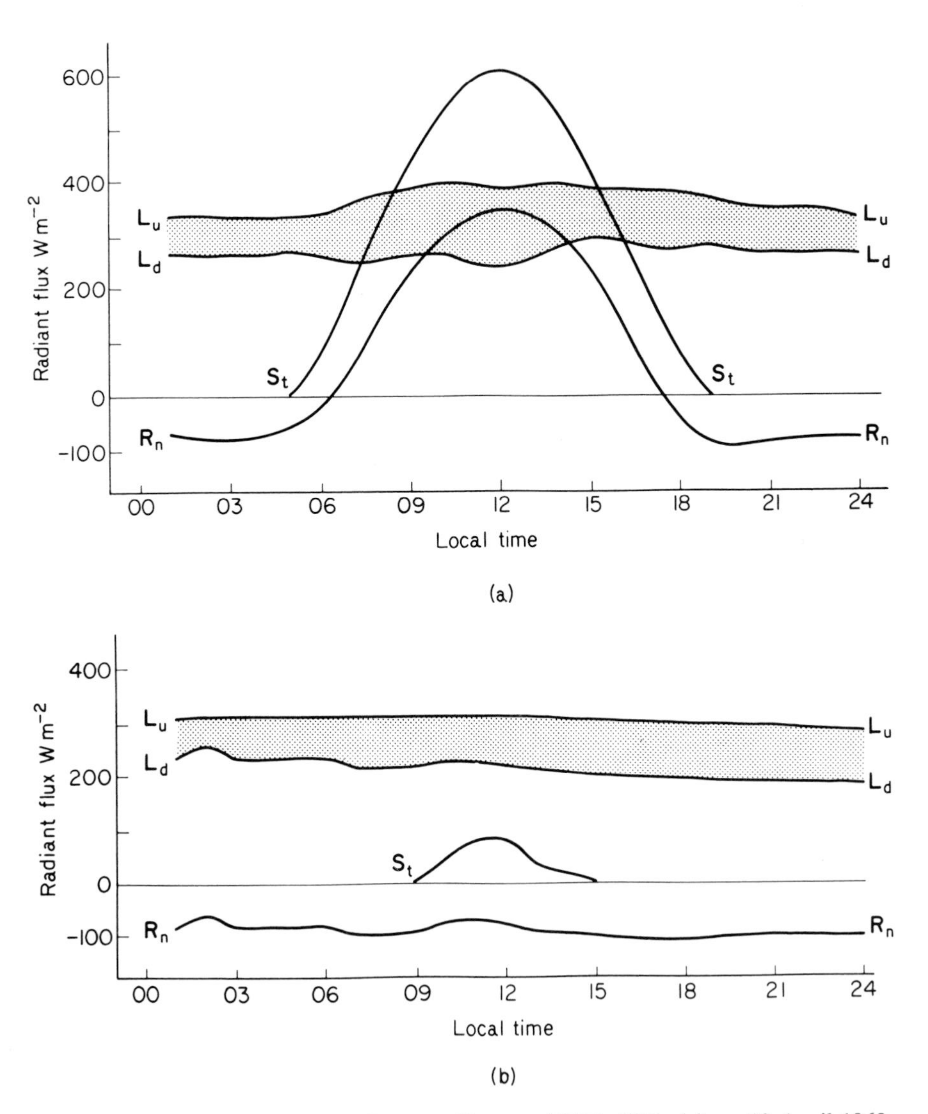

Fig. 4.12 Radiation balance at Bergen, Norway (60°N, 5°E): (a) on 13 April 1968, (b) on 11 January 1968. The grey area shows the net long wave loss and the line $\mathbf{R}_n$ is net radiation. Note that net radiation was calculated from measured fluxes of incoming short and long wave radiation, assuming that the reflectivity of the surface was 0.20 in April (e.g. vegetation) and 0.70 in January (e.g. snow). The radiative temperature of the surface was assumed equal to the measured air temperature.

short- and long-wave radiation for many years and have overcome the problem of maintaining a standard surface by using black-body radiation at air (i.e. screen) temperature for $\mathbf{L}_u$ and by adopting a reflection coefficient of 0.2 (grass) or 0.7 (snow) as appropriate. The shaded part of Fig. 4.12 represents the difference between the fluxes $\mathbf{L}_d$ and $\mathbf{L}_u$, i.e. the net loss of long-wave radiation; the bold line is $\mathbf{R}_n$. In cloudless weather, the diurnal change in the two long-wave components is much smaller than the change of short-wave radiation which follows an almost sinusoidal curve. The curve for net radiation is therefore almost parallel to the $\mathbf{S}_t$ curve during the day, decreases to a minimum value in the early evening and then increases very slowly for the rest of the night (because the lower atmosphere is cooled by radiative exchange with the earth's surface). In summer, the period during which $\mathbf{R}_n$ is positive is usually about 2 or 3 hours shorter than the period during which $\mathbf{S}_t$ is positive. Comparison of the curves for clear spring and winter days (Figs. 4.12a and b) shows that the seasonal change of $\mathbf{S}_t/\mathbf{R}_n$ (Fig. 4.11) is a consequence of (i) the shorter period of daylight in winter and (ii) much smaller maximum values of $\mathbf{S}_t$ in winter, unmatched by an equivalent decrease in the net long-wave loss.

In overcast weather, $\mathbf{L}_d$ becomes almost equal to σT_a^4; $\mathbf{R}_n$ is almost zero at night, and during the day $\mathbf{R}_n \approx (1-\rho)\mathbf{S}_t$.

The device of referring net radiation to a surface at air temperature is convenient in *micro*climatology too (see p. 187) and deserves to be more widely adopted.

5

MICROCLIMATOLOGY OF RADIATION

(i) Interception

In conventional problems of micrometeorology, fluxes of radiation at the earth's surface are measured and specified by the receipt or loss of energy per unit area of a horizontal plane. To estimate the amount of radiation intercepted by the surface of a plant or animal, the horizontal irradiance must be multiplied by a shape factor depending on (i) the geometry of the surface and (ii) the directional properties of the radiation. To make the analysis more manageable, objects such as spheres or cylinders with a relatively simple geometry are often used to represent the more irregular shapes of plants and animals. Appropriate shape factors for these models will be derived.

The radiation intercepted by an organism or its analogue can be expressed as the mean flux per unit area of surface. A bar will be used to distinguish this measure of irradiance from the conventional flux on a horizontal surface, e.g. when solar irradiance is $\mathbf{S}$ (W per m^2 horizontal area), the corresponding irradiance of a sheep or a cylinder will be written $\bar{\mathbf{S}}$ (W per m^2 total area).

DIRECT SOLAR RADIATION

The flux of direct solar radiation is usually measured either on a horizontal plane ($\mathbf{S}_b$) or on a plane perpendicular to the sun's rays ($\mathbf{S}_p$). For any solid object exposed to direct radiation at elevation β, a simple relation between the mean flux $\bar{\mathbf{S}}_b$ and the horizontal flux $\mathbf{S}_b$ can be derived from the relation between the area of shadow A_h cast on a horizontal surface and the area projected in the direction of the beam A_p.

The area projected in the direction of the sun's rays is $A_p = A_h \sin\beta$ (Fig. 5.1) and the intercepted flux is

$$A_p\mathbf{S}_p = (A_h \sin\beta)\mathbf{S}_p = A_h\mathbf{S}_b \qquad (5.1)$$

i.e. the area of shadow on a horizontal plane times the irradiance of that plane. Then if the surface area of the object is A

$$\bar{\mathbf{S}}_b = (A_h/A)\mathbf{S}_b \qquad (5.2)$$

The shape factor (A_h/A) can be calculated from geometrical principles or measured directly from the area of a shadow or from a shadow photograph.

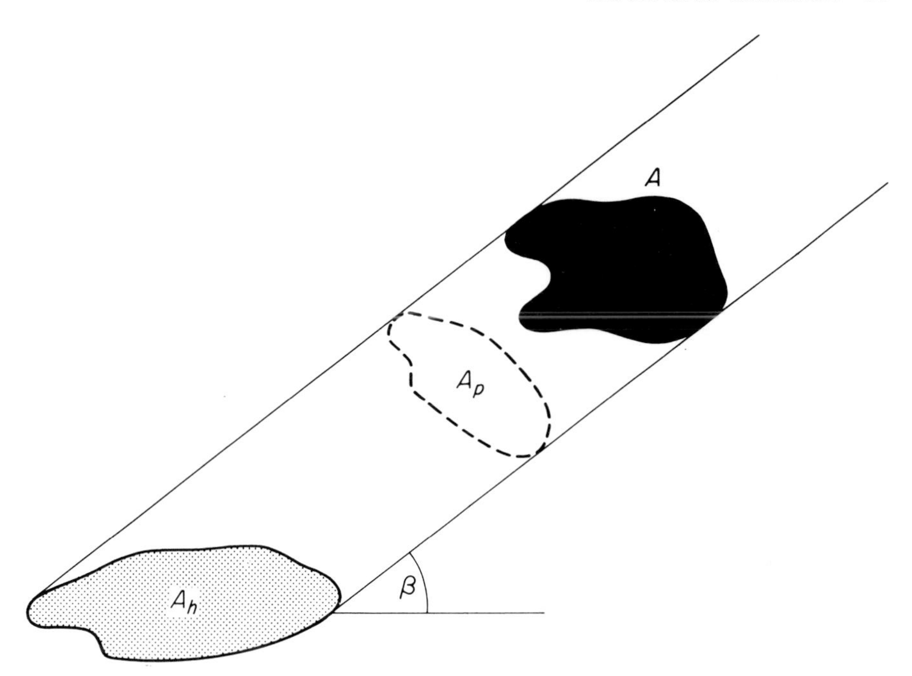

Fig. 5.1 Area A projected on surface at right angles to solar beam (A_p) and on horizontal surface (A_h).

Shape factors

Sphere

The shadow cast by a sphere of radius r has an area of $\pi r^2/\sin\beta$ (Fig. 5.2). The surface area of the sphere is $4\pi r^2$ so

$$\frac{A_h}{A} = \frac{\pi r^2}{4\pi r^2 \sin\beta} = 0.25 \operatorname{cosec}\beta \tag{5.3}$$

The mean irradiance of a sphere is therefore

$$\bar{\mathbf{S}}_b = (0.25 \operatorname{cosec}\beta)\, \mathbf{S}_b = 0.25\mathbf{S}_p \tag{5.4}$$

Ellipsoid

A sphere is a special type of ellipsoid—a solid whose cross-section is elliptical in one direction and circular about an axis of rotation at right-angles to that direction. If the elliptical cross-section through the centre of the ellipsoid has a vertical semi-axis a and a horizontal semi-axis b (Fig. 5.3), it can be shown that a beam at elevation β will cast a horizontal shadow with area

$$A_h = \pi b^2\{1 + a^2/(b^2 \tan^2\beta)\}^{0.5} \tag{5.5}$$

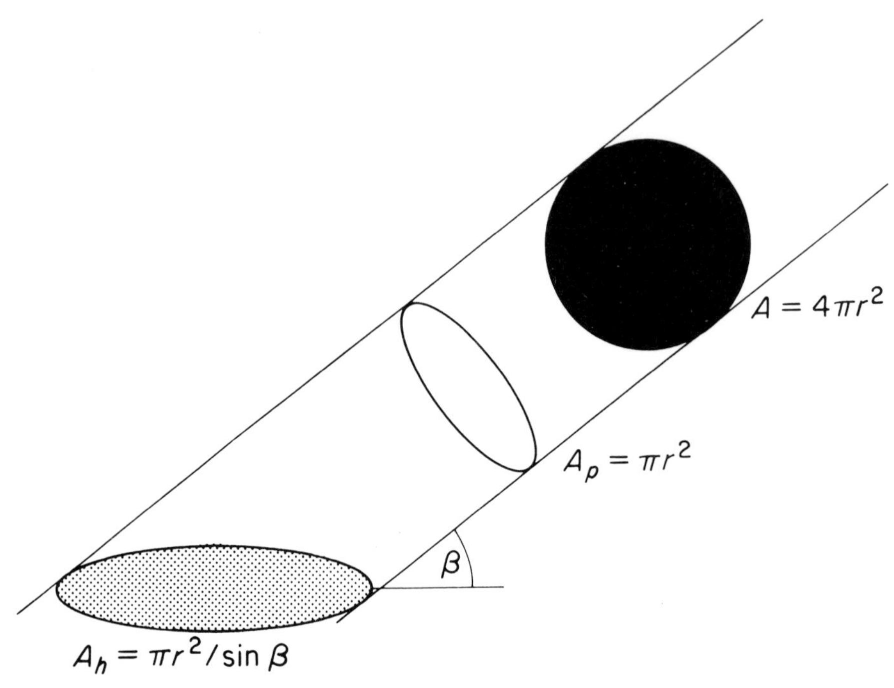

Fig. 5.2 Geometry of sphere projected on horizontal surface.

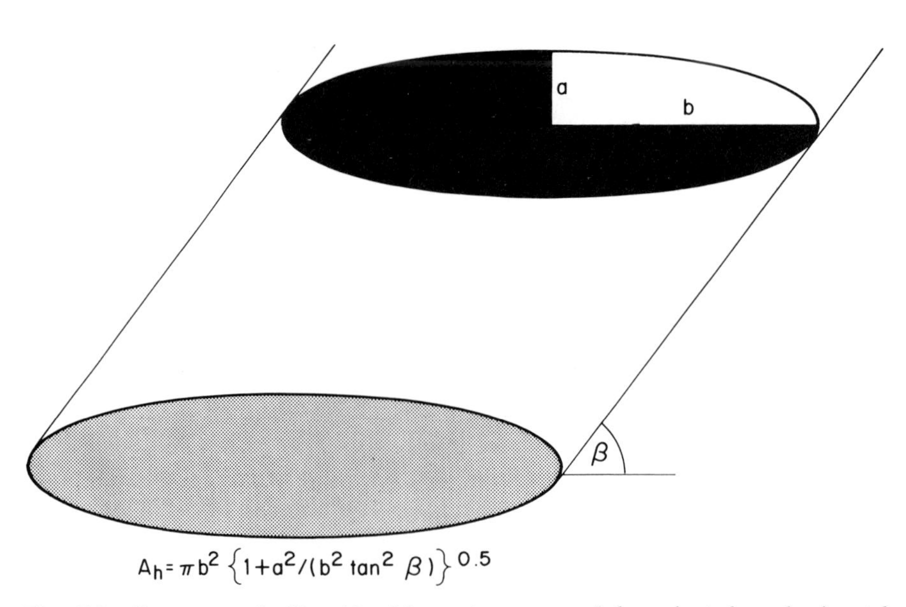

Fig. 5.3 Geometry of ellipsoid with semi-axes a and b projected on horizontal surface.

which reduces to $A_h = \pi b^2$ for a vertical beam and to $A_h = \pi b^2/\sin\beta$ when $a = b$.

The surface area of an *oblate* spheroid ($b > a$) is

$$A = 2\pi b^2[1 + (a^2/2b^2\varepsilon_1) \tag{5.6}$$

where

$$\varepsilon_1 = (1 - a^2/b^2)^{0.5} \tag{5.7}$$

and the area of a *prolate* spheroid ($a > b$) is

$$A = 2\pi b^2\{1 + (a/b\varepsilon_2)\sin^{-1}\varepsilon_2\} \tag{5.8}$$

where

$$\varepsilon_2 = (1 - b^2/a^2)^{0.5} \tag{5.9}$$

The ratio A_h/A can now be found by dividing equation (5.5) by equation (5.6) or (5.8) as appropriate (Fig. 5.3).

Inclined plane

Figure 5.4 shows the end and plan views of a square plane with sides of unit length making an angle α with the horizon XY and exposed to a beam of

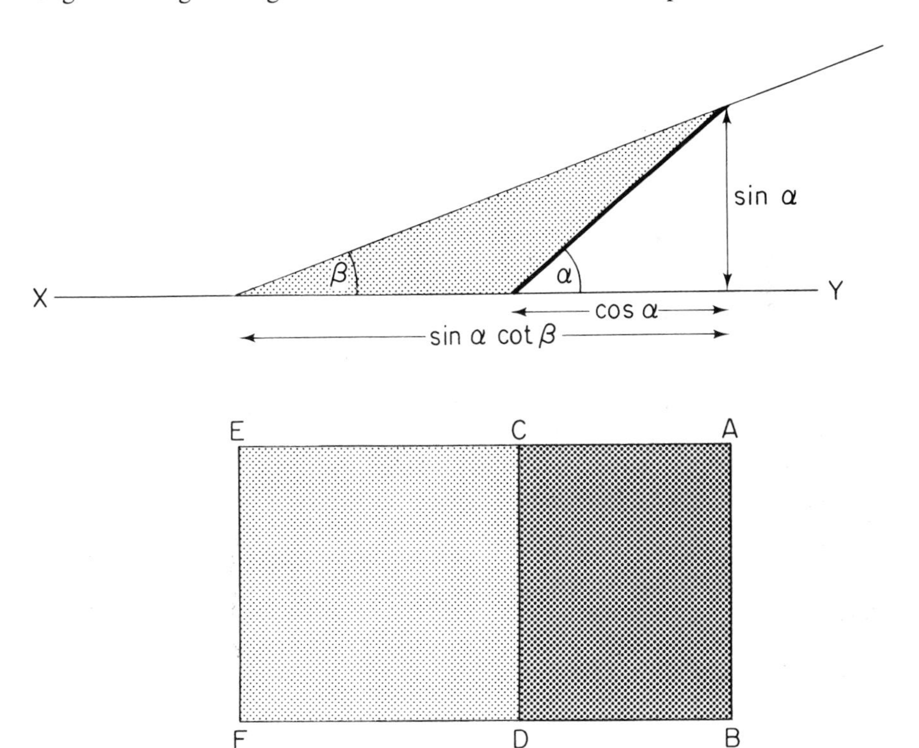

Fig. 5.4 Geometry of rectangle projected on horizontal surface when edge is parallel to solar beam.

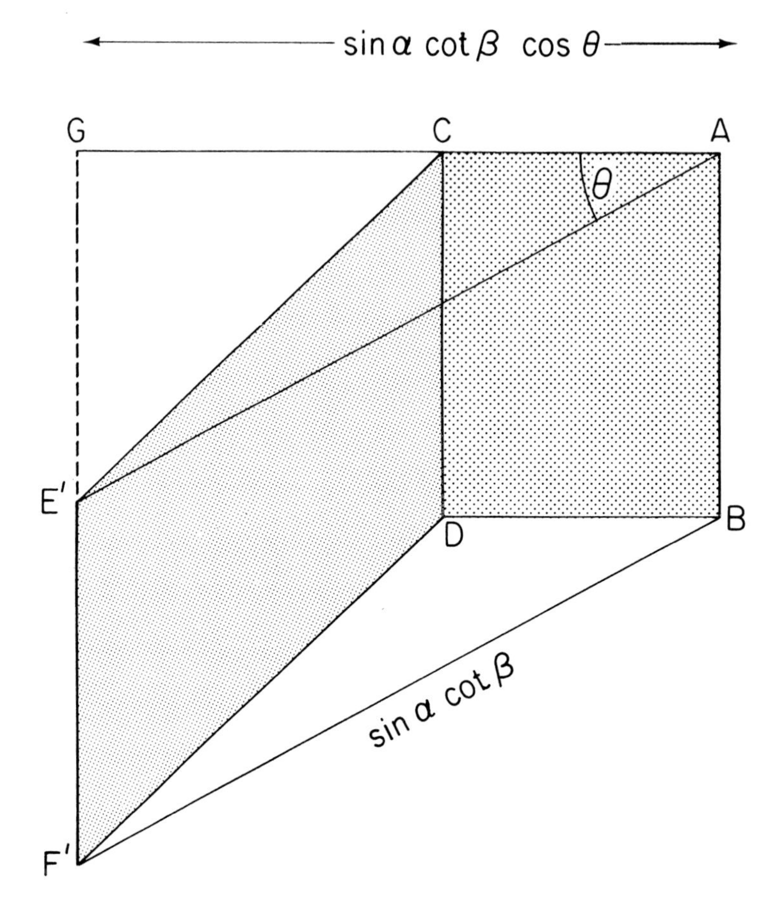

Fig. 5.5 Geometry of rectangle projected on horizontal surface when edge makes angle θ with solar beam.

radiation at an elevation β at right angles to the edge AB. The shadow has a width EF = AB = 1 so A_h is (BF − BD) × 1 or sin α cot β − cos α.

If the beam made an angle θ with respect to the direction AC (Fig. 5.5), the shadow would move to the position indicated by CE′F′D where AE′ = AE = sin α cot β. The shadow becomes a parallelogram with an area CG × CD so A_h is (AE′ cos θ − AC) × 1 or sin α cot β cos θ − cos α. When β > α, the area is {cos α − sin α cot β cos θ} and, for all values of α and β, the projected area can be written |cos α − sin α cot β cos θ|, the positive value of the function. Because any plane with area *A* can be subdivided into a large number of small unit squares, the shape factor is

$$\frac{A_h}{A} = |\cos\alpha - \sin\alpha \cot\beta \cos\theta| \qquad (5.10)$$

This function can be used to estimate the direct radiation on hill slopes or on

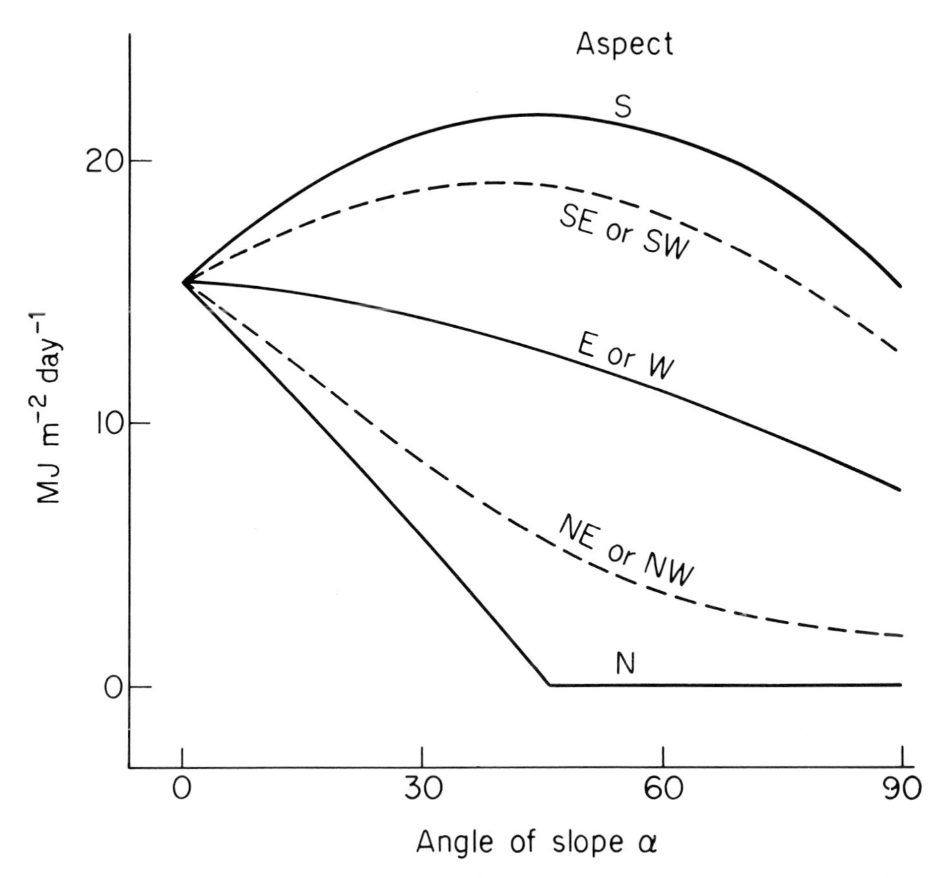

Fig. 5.6 Daily integral of direct solar radiation at the equinoxes for latitude 45°N (from Garnier and Ohmura, 1968).

the walls of houses. Then α depends on the geometry of the system, β depends on solar angle, and θ depends both on geometry and on the position of the sun. Relevant calculations and measurements can be found in a number of texts. The example in Fig. 5.6 emphasizes the large difference of direct irradiance on slopes of different aspect which is often responsible for major differences of microclimate and plant response.

Vertical cylinder

Figure 5.7 shows a vertical cylinder of height h and radius r suspended above the ground. The shadow has two components: $h \cot \beta \times 2r$ from the curved surface and πr^2 from the top. The total area is $2\pi rh + 2\pi r^2$ so

$$\frac{A_h}{A} = \frac{2rh \cot \beta + \pi r^2}{2\pi rh + 2\pi r^2} = \frac{(2x \cot \beta)/\pi + 1}{2x + 2} \tag{5.11}$$

where $x = h/r$.

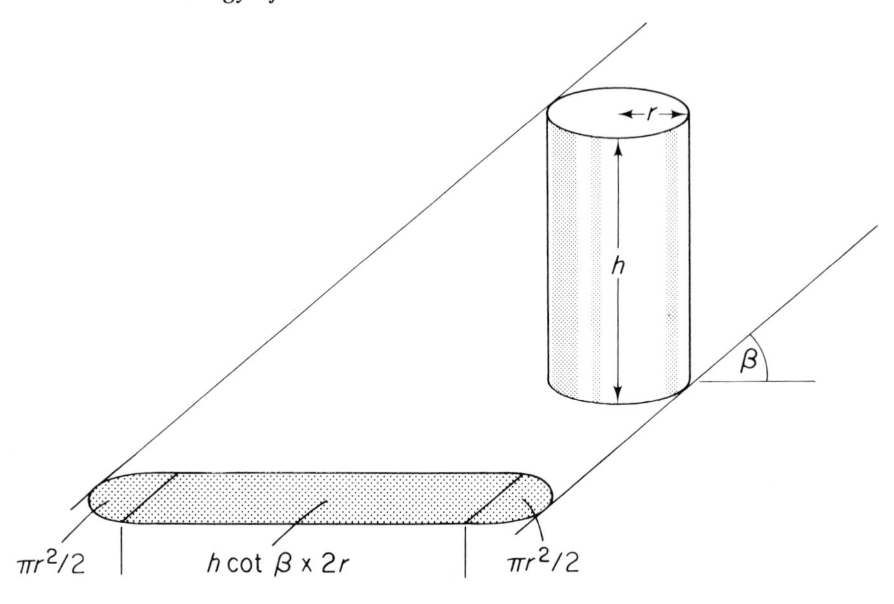

Fig. 5.7 Geometry of vertical cylinder projected on horizontal surface.

Underwood and Ward (1966) measured 25 male and 25 female subjects wearing slips or bathing costumes and photographed them from 19 different angles. (Silhouettes from 8 angles are shown in Fig. 5.8.) The average projected area of all 50 subjects was found by planimetry of each silhouette with proper correction for parallax. When the areas presented at three azimuth angles (0, 45, 90) were bulked, the mean projected body area was very close to the projected area of a cylinder with

$$h = 1.65 \text{ m}, \; r = 0.117 \text{ m}, \; x = 14.1$$

Changes of the projected area with azimuth were taken into account by fitting the measurements to the equation of a cylinder with an elliptical rather than a circular cross-section.

Values of A_h/A for vertical cylinders are plotted in Fig. 5.9 as a function of β and x. At a solar elevation of $\beta = 32.5°$, $\cot \beta/\pi = 0.5$ and $A_h/A = 0.5$ independent of x. When $\beta < 60$ and $x > 10$, A_h/A is almost independent of x, so equation (5.11) with a standard value of $x = 14$ will give a good account of the radiation intercepted by a wide range of human shapes from the ectomorph to the endomorph (e.g. Laurel and Hardy).

Horizontal cylinders

For a horizontal cylinder with dimensions (h,r), A depends on solar azimuth as well as elevation. The azimuth angle θ can be measured from the axis of the cylinder pointing in the direction $\theta = 0$. It can be shown that the length h projected in the direction β,θ is $h(1 - \cos^2 \beta \cos^2 \theta)$ and the projected width is

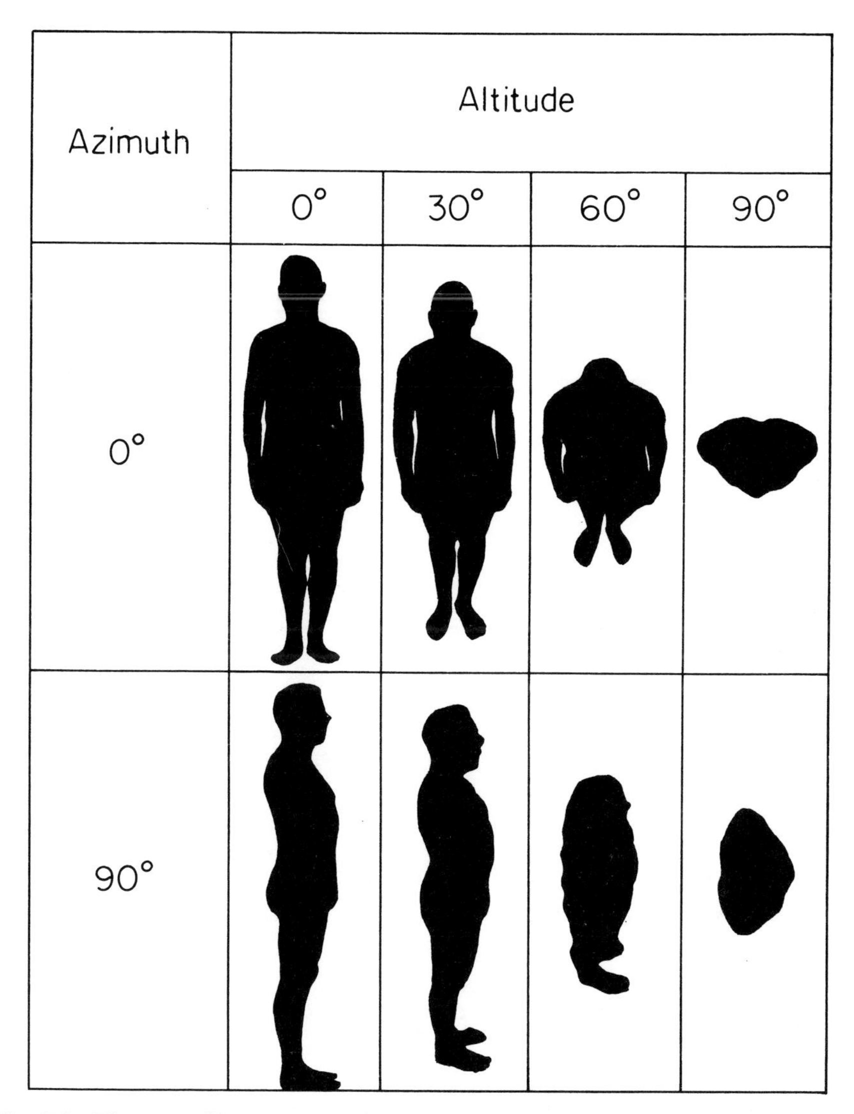

Fig. 5.8 The area of an erect male figure projected in the direction of the solar beam for different values of solar azimuth and altitude. The silhouettes were obtained photographically by Underwood and Ward (1966).

$2r$ independent of β and θ. Thus for the *curved surface* only

$$A_h = A_p \operatorname{cosec} \beta = 2rh \operatorname{cosec} \beta(1 - \cos^2 \beta \cos^2 \theta)^{0.5} \qquad (5.12)$$

The illuminated end of the cylinder can be treated as a vertical plane and if α is taken as $\pi/2$ in equation (5.10)

$$A_h = \pi r^2 \cot \beta \cos \theta \text{ (end only)} \qquad (5.13)$$

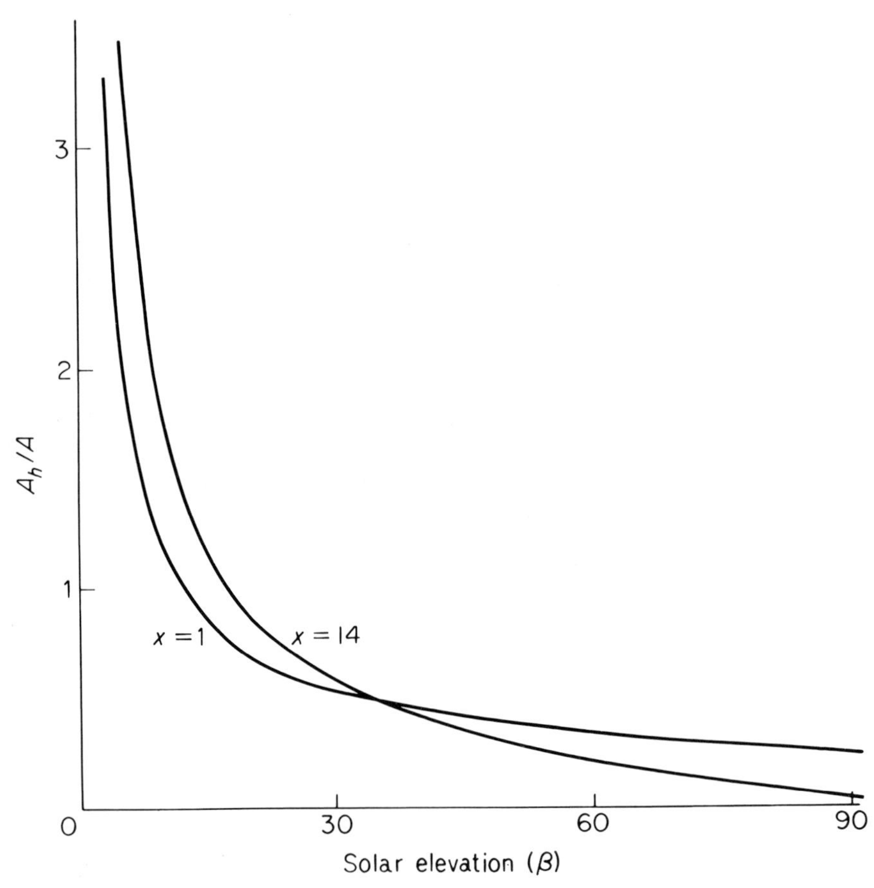

Fig. 5.9 The ratio A_h/A for vertical cylinders with height/radius ratios (x) of 1 and 14.

The total area of the cylinder including the unlit end is $A = 2\pi rh + 2\pi r^2$ giving for the whole cylinder

$$\frac{A_h}{A} = \frac{2rh \operatorname{cosec} \beta (1 - \cos^2 \beta \cos^2 \theta)^{0.5} + \pi r^2 \cot \beta \cos \theta}{2\pi rh + 2\pi r^2}$$

$$= \frac{\operatorname{cosec} \beta \{2\pi^{-1} x (1 - \cos^2 \beta \cos^2 \theta)^{0.5} + \cos \beta \cos \theta\}}{2(x+1)} \tag{5.14}$$

For the special case of a cylinder at right angles to the sun's rays, $\theta = \pi/2$, A_h/A reduces to

$$\frac{A_h}{A} = \frac{x \operatorname{cosec} \beta}{\pi(x+1)} \text{ and } \frac{A_p}{A} = \frac{x}{\pi(x+1)} \tag{5.15}$$

The projected area of sheep was determined photogrammetrically by Clapperton, Joyce and Blaxter (1965) and was compared with the area of equivalent horizontal cylinders whose dimensions were determined from photographs at $\theta = 0$ (end view) and $\beta = \pi/2$ (plan view). One set of values quoted for fleeced sheep was

$$h = 0.91 \text{ m}, r = 0.23 \text{ m}, x = 4.1$$

With the sun at right angles to the axis, the cylindrical model under-estimated the interception of radiation by about 20% when β was less than 60° but with the sheep facing the sun the model over-estimated interception by 10 to 30%. Agreement could probably have been improved by using equation (5.14) instead of calculating h and r from two angles only but, for random orientation, the error in using the dimensions quoted would be very small.

Figure 5.10 shows A_h/A plotted as a function of β for four values of azimuth angle and $x = 4$. When β exceeds 40°, the shape factor is almost independent of θ and is therefore nearly proportional to cosec β. This implies that the radiation intercepted by a cylinder with $x = 4$ will be almost independent of the solar azimuth provided solar elevation exceeds 40°. For $x<4$ the corresponding angle will be less than 30°.

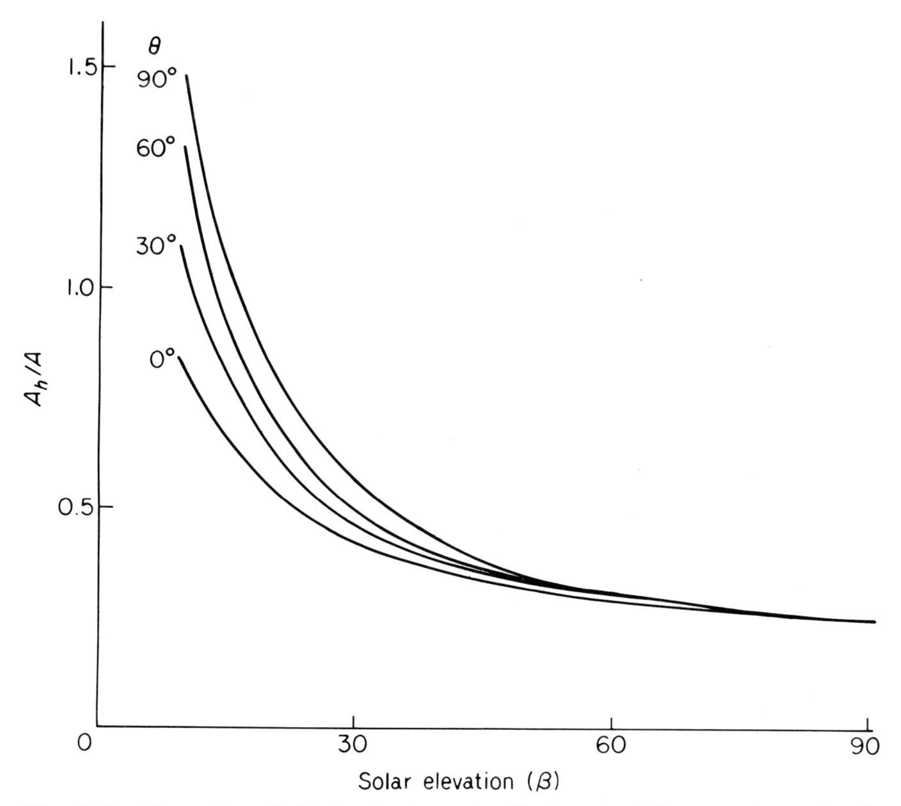

Fig. 5.10 The ratios A_h/A for horizontal cylinders at different values of solar elevation (β) and azimuth (θ) with a length/radius ratio of $x = 4$.

Cone

The interception of radiation by a cone is an interesting problem, relevant to the distribution of radiant energy on a single tree (Fig. 5.11) or on the leaves of a crop when they are randomly distributed with respect to the points of the compass.

The cone in Fig. 5.12 has a slant side of unit length making an angle α with the base, so the perpendicular height is $\sin\alpha$ and the base area is $\pi\cos^2\alpha$. For direct radiation incident at an angle $\beta > x$ (not shown), the walls are fully illuminated and the shadow cast on a horizontal surface by the whole cone is simply the shadow area of the base $A_h = \pi\cos^2\alpha$. As the area of the walls is $A = \pi\cos\alpha$, the shape factor for the walls alone is $A_h/A = \cos\alpha$. When $\beta < \alpha$, the shadow has a more complex form. The walls are now partly in shadow: at the base, CDB is illuminated, BEC is not, and the shadow can be delineated by projecting tangents at B and C to meet at A. The sector of the cone in shadow can now be specified by the angle $\mathrm{AOB} = \mathrm{AOC} = \theta_0$. Now $\mathrm{OB} = \cos\alpha$, AO is the horizontal projection of the axis of the cone, i.e. $\sin\alpha\cot\beta$, and as ABO is a right angle, $\mathrm{AB} = \sin\alpha\cot\beta\sin\theta_0$. The cosine of θ_0 is $\mathrm{OB}/\mathrm{AO} = \cos\alpha/(\sin\alpha\cot\beta)$. It follows that the area of the shadow is

$$\begin{aligned}
\mathrm{ABDC} &= \mathrm{EBDC} + 2\mathrm{ABO} - \mathrm{CEBO} \\
&= \text{circle} + 2\text{ triangles} - \text{sector of circle} \\
&= \pi\cos^2\alpha + \cos\alpha\sin\alpha\cot\beta\sin\theta_0 - \theta_0\cos^2\alpha \\
&= \cos\alpha\{(\pi - \theta_0)\cos\alpha + \sin\alpha\cot\beta\sin\theta_0\}
\end{aligned}$$

Fig. 5.11 Clipped yew trees (*Taxus baccata*) in the grounds of Hampton Court Palace. The topiarist has produced almost perfect cones casting shadows on the ground which closely resemble the outline of the lower part of Fig. 5.12 (p. 69). The amount of direct solar radiation intercepted by the trees (when they are not shading each other) could therefore be calculated from equation (5.16).

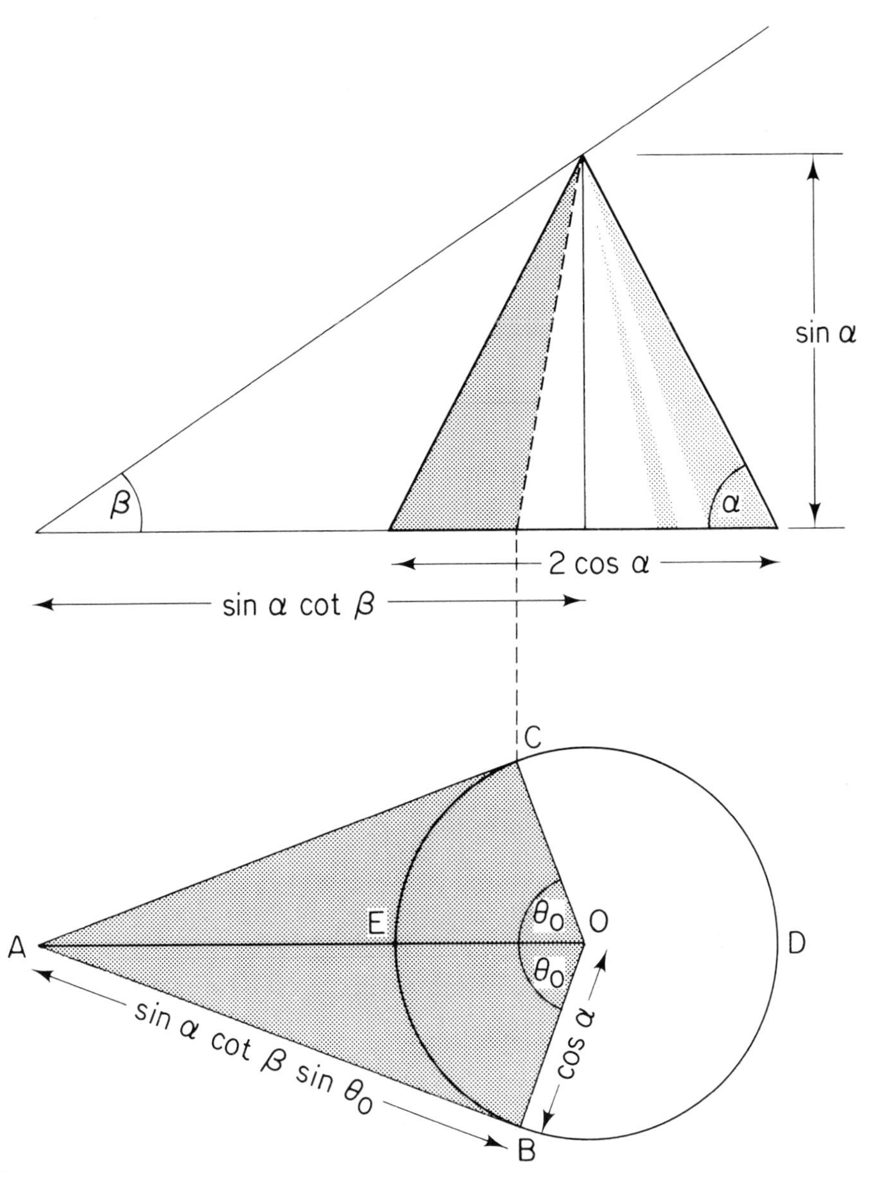

Fig. 5.12 The geometry of a cone projected on a horizontal surface.

As the total area of the cone is $\pi \cos \alpha(1+\cos \alpha)$ the shape factor is

$$\frac{A_h}{A} = \frac{(\pi - \theta_0)\cos \alpha + \sin \alpha \cot \beta \sin \theta_0}{\pi(1+\cos \alpha)} \tag{5.16}$$

where $\theta_0 = \cos^{-1}(\tan \beta \cot \alpha)$.

DIFFUSE RADIATION

Natural objects are exposed to four discrete streams of diffuse radiation with different directional properties.

(1) Incoming short-wave radiation.
The spatial distribution of this flux depends on the elevation and azimuth of the sun and on the degree of cloud cover (pp. 42–43).

(2) Incoming long-wave radiation.
When the sky is cloudless, the intensity of atmospheric radiation decreases by about 20 to 30% from the horizon to the zenith. Under an overcast sky, the flux is nearly uniform in all directions (p. 53).

(3) Reflected solar radiation.
The amount of radiation received by reflection from an underlying surface depends on the reflection coefficient of the surface and the angular distribution of the flux is determined by the structure of the surface. Natural vegetation and farm crops are often composed of vertical elements which shade each other and more radiation is reflected from sunlit than from shaded areas.

(4) Long-wave radiation emitted by the underlying surface.
Like the reflected component of diffuse radiation, the spatial distribution of this flux depends on the disposition of sunlit (relatively warm) and shaded (relatively cool) areas.

The different angular variations of the four diffuse components are difficult to handle analytically but for the purposes of establishing the radiation balance of a leaf or animal they can often be ignored. The following treatment deals with the interception of isotropic radiation, i.e. the intensities of the diffuse fluxes are assumed independent of angle. Additional formulae for long-wave radiation were derived by Unsworth (1975) for slopes and regular solids and by Johnson and Watson (1985) for complex shapes.

Shape factors

Plane surfaces

A horizontal flat plate facing upwards receives diffuse fluxes of $\mathbf{S}_d$ and $\mathbf{L}_d$ in the short- and long-wave regions of the spectrum. The corresponding fluxes on a surface facing horizontally downwards are $\rho\mathbf{S}_t$ and $\mathbf{L}_u$.

A plate at any angle α to the horizon receives radiation from all four sources on both its surfaces. To find how the irradiance from each of these sources depends on α, the atmosphere and the ground can be replaced by two hemispheres ACO and ACO′ (Fig. 5.13). The plane of the horizon is ABCD and the plane of the flat plate is DEBF, making an angle α with the horizontal plane. The irradiance of the plate from the sector of the sky represented by DCBE could be calculated by dividing this sector into a large number of small elements dA and integrating dA times the cosine of the angle between each element and the normal to the surface of the plate. This process can be

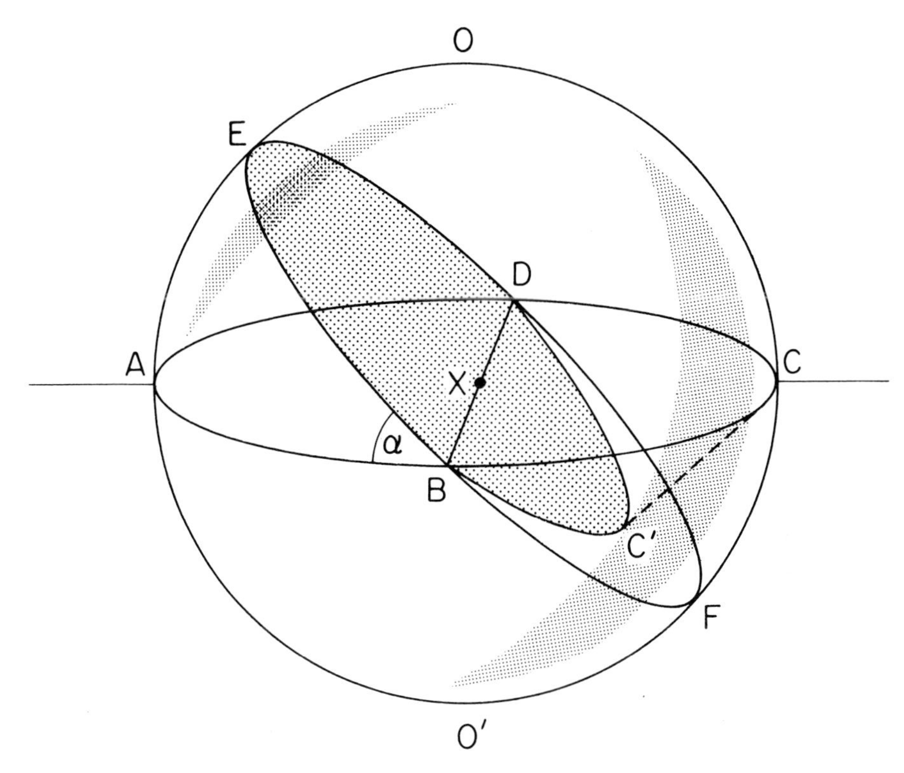

Fig. 5.13 Diagram for calculating the diffuse irradiance at a point X in the centre of a sphere from a sector of the sphere subtending an angle α at the equatorial plane.

avoided by the device described on p. 31, i.e. the sector DEBC is projected on to the plane of the plate.

The area of this projection is the semi-circle DEBX plus the semi-ellipse DC′BX, i.e. $\pi/2 + (\pi \cos \alpha)/2$ if the hemisphere has unit radius. If **N** is the (uniform) radiance emitted by elements on the hemisphere, the irradiance from the sector will be $(\pi/2)(1 + \cos \alpha)\mathbf{N}$. For a horizontal surface, $\alpha = 0$ and the irradiance is $\pi\mathbf{N}$ (p. 31). The irradiance of a surface at angle α is therefore $(1 + \cos \alpha)/2 = \cos^2(\alpha/2)$ times the irradiance of a horizontal surface. For an inclined plane the factor $(1 + \cos \alpha)/2$ for diffuse radiation is equivalent to the factor A_h/A derived for direct radiation.

If a flat leaf or other plane surface is exposed above the ground at an angle α, both its surfaces will receive short- and long-wave radiation from the sky and from the ground. When the four fluxes of radiation are isotropic, they can be written as follows:

	Short-wave	*Long-wave*
Upper surface	$\cos^2(\alpha/2)\mathbf{S}_d + \sin^2(\alpha/2)\rho\mathbf{S}_t$	$\cos^2(\alpha/2)\mathbf{L}_d + \sin^2(\alpha/2)\mathbf{L}_u$
Lower surface	$\sin^2(\alpha/2)\mathbf{S}_d + \cos^2(\alpha/2)\rho\mathbf{S}_t$	$\sin^2(\alpha/2)\mathbf{L}_d + \cos^2(\alpha/2)\mathbf{L}_u$

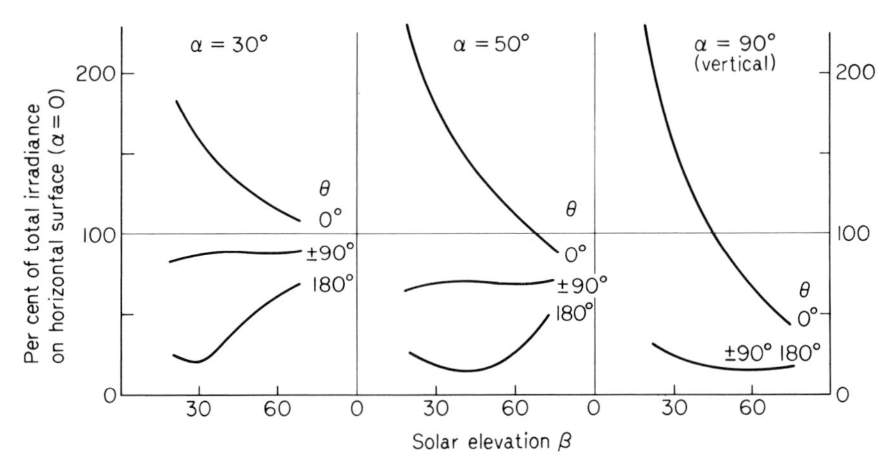

Fig. 5.14 Irradiance of planes (direct and diffuse solar radiation) at latitude 45°N as a function of solar elevation β, elevation of the plane α, and azimuth angle θ between the solar beam and the normal to the plane. From measurements by Kondratyev and Manolova (1960).

The sum of all eight components is simply $(\mathbf{S}_d + \rho\mathbf{S}_t + \mathbf{L}_d + \mathbf{L}_u)$. The condition that the upward fluxes of radiation are approximately isotropic requires that the height of the plane above the ground should be large compared with its dimensions so that its shadow can be neglected.

The total solar radiation received by plane surfaces with different slopes and aspects can be calculated by summing the direct and diffuse components. The curves in Fig. 5.14 were derived by Kondratyev and Manolova (1960) who found that it was essential to allow for the spatial distribution of radiation from the blue sky (p. 43) in order to describe the diurnal change of irradiance on slopes with α exceeding 30°. However, they also found that daily totals of radiation on slopes could be calculated accurately by assuming that the diffuse flux was isotropic so that the daily isolation was the sum of the individual hourly values of the direct radiation on the slope, the diffuse flux from the sky, $\cos^2(\alpha/2)\mathbf{S}_d$; and the diffuse flux from the surrounding terrain, $\sin^2(\alpha/2)\rho\mathbf{S}_t$.

Cone

The diffuse irradiance on the walls of a cone with base angle α is equal to the irradiance of the upper surface of a plate at an elevation of α.

Vertical cylinder

For a vertical surface, $\cos \alpha = 0$, so the receipt of short-wave radiation is $(\mathbf{S}_d + \rho\mathbf{S}_t)/2$ and the receipt of long-wave radiation is $(\mathbf{L}_d + \mathbf{L}_u)/2$.

Horizontal cylinder

The components of irradiance for the upper and lower surfaces of a horizontal cylinder can be found by integrating the factors $(1 + \cos \alpha)/2$ and $(1 - \cos \alpha)/2$.

The integration yields factors of $0.5+\pi^{-1}=0.82$ and $0.5-\pi^{-1}=0.18$. With these approximations, the components are:

	Short-wave	*Long-wave*
Upper half surface	$0.82\mathbf{S}_d+0.18\rho\mathbf{S}_t$	$0.82\mathbf{L}_d+0.18\mathbf{L}_u$
Lower half surface	$0.18\mathbf{S}_d+0.82\rho\mathbf{S}_t$	$0.18\mathbf{L}_d+0.82\mathbf{L}_u$

The components for each half surface are the same as they would be for the upper and lower surfaces of a plane with $\alpha\approx 50°$ and the sum of all the components is simply $(\mathbf{S}_d+\rho\mathbf{S}_t)+(\mathbf{L}_d+\mathbf{L}_u)$ which is the sum for any plane surface.

CANOPIES OF BLACK LEAVES

The principles of radiation geometry can now be applied to estimate the distribution of radiant energy within horizontally uniform foliage. The amount of foliage is conventionally specified as a leaf area index L, the area of leaves per unit area of ground taking one side of each leaf into account. To avoid the complications of scattering for the time being, the leaves are assumed to be black (close to reality for the visible spectrum at least—see p. 86).

Suppose that a thin horizontal layer of black leaves, exposed to direct solar radiation, contains a small leaf area index dL. The amount of energy intercepted by dL is the area of shadow cast by the leaves on a horizontal plane times the horizontal irradiance (p. 59). The required shadow area is dL times the shadow cast by unit area of leaf which is (A_h/A). The product $(A_h/A)dL$ is the area of horizontal shadow per unit ground area (a shadow area index) and the intercepted radiation can now be expressed as

$$\begin{aligned} d\mathbf{S}_b &= -(A_h/A)\mathbf{S}_b dL \\ &= -\mathcal{K}_s\mathbf{S}_b dL \end{aligned} \tag{5.17}$$

where

$$\mathcal{K}_s = A_h/A$$

Equation (5.17) implies that the mean irradiance of a leaf is $\mathcal{K}_s\mathbf{S}_b$. The minus sign is needed if L is measured downwards from the top of the canopy. Integration of this equation gives

$$\mathbf{S}_b(L) = \mathbf{S}_b(0)\exp(-\mathcal{K}_s L)$$

where $\mathbf{S}_b(L)$ is the direct solar irradiance measured on a horizontal plane below a leaf area index L measured from the top of the canopy. This is a special case of Beer's Law (see p. 23).

The ratio $\mathbf{S}_b(L)/\mathbf{S}_b(0)$, which is the relative solar irradiance at L, is also the fractional area of sunflecks on a horizontal plane below L. The area of sunlit foliage between L and $(L+dL)$ is therefore $\{\mathbf{S}_b(L)/\mathbf{S}_b(0)\}dL$ and in a stand with a total leaf area of L_t the leaf area index of sunlit foliage is

$$\int_0^{L_t} \{\mathbf{S}_b(L)/\mathbf{S}_b(0)\}dL = \int_0^{L_t} \exp(-\mathcal{K}_s L)dL = \mathcal{K}_s^{-1}[1-\exp(-\mathcal{K}_s L_t)]$$

which has a limiting value of $1/\mathcal{K}_s$ at large values of L_t.

We now consider how values of $\mathcal{K}_s$ can be deduced from the area of shadows cast by cylinders, spheres, and cones as presented on pages 58 to 69.

Vertical leaf distribution

If all the leaves in a canopy hung vertically facing at random with respect to azimuth or compass angle, they could be rearranged in a mosaic pattern on

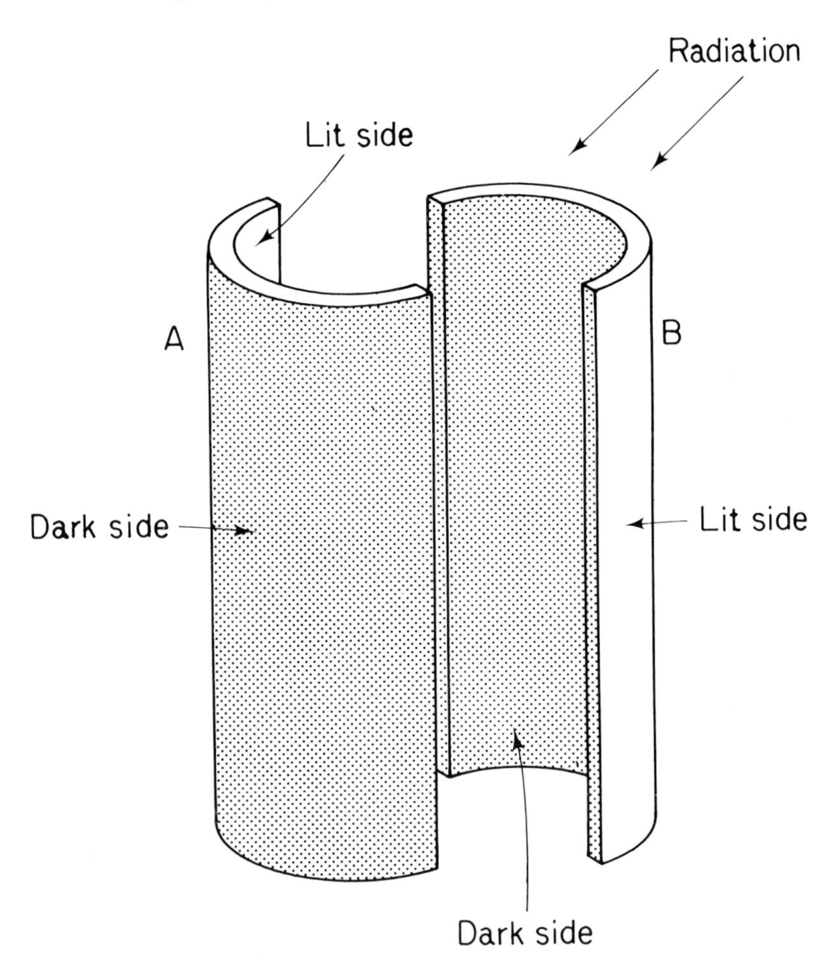

Fig. 5.15 The distribution of radiation over two surfaces of a cylinder representing the irradiance of a large number of vertical leaves.

the curved surface of a vertical cylinder. This cylinder could be split along a central plane at right angles to the sun's rays (Fig. 5.15). The convex half of the cylinder represents leaves illuminated on one side (say the upper surface) and the concave half represents the leaves illuminated on the lower surface. The appropriate value of $\mathcal{K}_s = A_h/A$ is therefore twice the value derived for the curved surface of a solid cylinder, i.e.

$$\mathcal{K}_s = 2(\cot \beta)/\pi \text{ (cf. p. 64)}$$

Spherical and ellipsoid leaf distribution

If the leaves in the canopy were distributed at random with respect to their angles of elevation as well as their azimuth angles, they could be rearranged

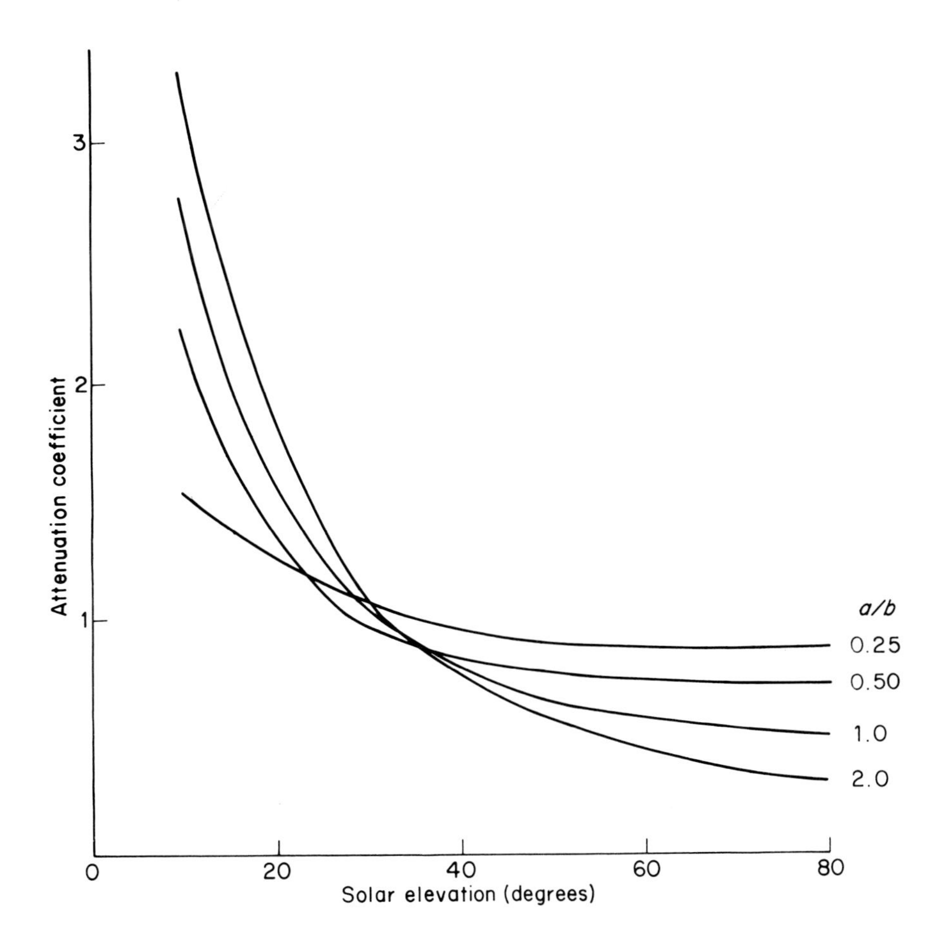

Fig. 5.16 Dependence of attentuation coefficient for direct radiation $\mathcal{K}_s = 2A_h/A$ on solar elevation when leaf angle distribution is ellipsoidal with a/b = ratio of vertical to horizontal radius (see p. 61). The case $a/b = 1$ corresponds to a spherical distribution.

on the surface of a sphere. Splitting the sphere at the equatorial plane normal to the sun's rays gives two hemispheres representing the two sides from which the leaves could be illuminated. The appropriate value of $\mathcal{K}_s$ is twice the value for a sphere, i.e.

$$\mathcal{K}_s = 0.5 \operatorname{cosec} \beta \text{ (cf. p. 59)}$$

More generally, if the distribution of leaves is ellipsoidal, as demonstrated by Campbell (1986) for contrasting types of foliage, the value of $\mathcal{K}_s = 2A_h/A$ can be found from equation (5.6) or (5.8) and Fig. 5.16 shows the extent to which values for oblate ($a<b$) or prolate ($a>b$) spheroids depart from the spherical case $a = b$. Note that when $\beta \approx 30°$, $\mathcal{K}_s \approx 1$, almost independent of β, i.e. ellipsoids behave like planes.

Conical distribution

An assembly of leaves all at an elevation of α but distributed at random with respect to their azimuth angles could be rearranged on the curved surface of a cone with a wall angle of α. If the cone is exposed to a solar beam at elevation β, two cases must be considered:

(i) $\beta>\alpha$

All leaves are illuminated from above, i.e. on their upper (adaxial) surfaces. The whole curved surface of the cone is illuminated so

$$\mathcal{K}_s = \frac{A_h}{A} = \frac{\pi \cos^2 \alpha}{\pi \cos \alpha} = \cos \alpha$$

a value which is independent of solar elevation.

(ii) $\beta<\alpha$

Some of the leaves are illuminated from below, i.e. on their lower (abaxial) surfaces. In accordance with the method used for the sphere and the cylinder, the cone can be split into two parts. The relative shadow area of the convex part representing the abaxial surfaces has been calculated already (p. 68): it is $\cos \alpha\{(\pi - \theta_0) \cos \alpha + \sin \alpha \cot \beta \sin \theta_0\}$. The shadow area of the concave part (adaxial surfaces) is the area ACEB in Fig. 5.12, i.e. the sum of two triangles less the sector of the circle or $\cos \alpha\{\sin \alpha \cot \beta \sin \theta_0 - \theta_0 \cos \alpha\}$. The total shadow area is the sum of these expressions and as $A = \pi \cos \alpha$ for the curved surface alone,

$$\mathcal{K}_s = \frac{A_h}{A} = \pi^{-1}(\pi - 2\theta_0)\cos \alpha + 2\sin \alpha \cot \beta \sin \theta_0.$$

where

$$\theta_0 = \cos^{-1}(\tan \beta \cot \alpha).$$

A similar function, $\mathcal{K}_s \operatorname{cosec} \beta$, which is the relative shadow area on a surface normal to the sun's rays was originally tabulated by Reeve (1960) and presented graphically as a function of α and β by Anderson (1966). Table 5.1 summarizes values of $\mathcal{K}_s$ for idealized leaf distributions and real canopies.

Table 5.1 Transmission coefficients for model and real canopies (from Monteith, 1969)

(a) Idealized leaf distributions	$\mathcal{K}_s$		
	solar elevation β		
	90	60	30
cylindrical	0.00	0.37	1.10
spherical	0.50	0.58	1.00
conical $\alpha = 60$	0.50	0.50	0.58
$\alpha = 30$	0.87	0.87	0.87

(b) Real canopies	$\mathcal{K}_s$
White Clover (*Trifolium repens*)	1.10
Sunflower (*Helianthus annuus*)	0.97
French Bean (*Phaseolus vulgaris*)	0.86
Kale (*Brassica acephala*)	0.94
Maize (*Zea mays*)	0.70
Barley (*Hordeum vulgare*)	0.69
Broad Bean (*Vicia faba*)	0.63
Sorghum (*Sorghum vulgare*)	0.49
Ryegrass (*Lolium perenne*)	0.43
(*Lolium rigidium*)	0.29
Gladiolus	0.20

Irradiance of foliage

To estimate rates of transpiration and photosynthesis for leaves in a canopy, it is essential to calculate the irradiance of individual leaf surfaces as distinct from the irradiance of horizontal surfaces already considered. If $\mathbf{S}_b(L)$ is the direct component of horizontal irradiance below an area index of L, the mean irradiance of foliage at this depth will be $\mathcal{K}_s\mathbf{S}_b(L)$ from equation (5.17). The mean irradiance can also be derived by considering an irradiance of $\mathbf{S}_b$ (W per m^2 field area) distributed over a sunlit leaf area index of $1/\mathcal{K}_s$ (m^2 leaves per m^2 field area) to give a mean irradiance of $\mathcal{K}_s\mathbf{S}_b$ (W per m^2 leaf area).

In the exceptional case when all leaves are facing in the same direction, the irradiance $\mathcal{K}_s\mathbf{S}_b(L)$ will be uniform but in general some leaves will be exposed to a stronger and some to a weaker flux. In the extreme, leaves parallel to the solar beam ($\alpha = \beta$) receive no direct radiation and leaves at right angles to the beam ($\alpha = \beta + \pi/2$) receive $\mathbf{S}_b$ cosec β. This range of irradiance should be taken into account when estimating leaf temperatures precisely and is significant for processes which are not proportional to the irradiance, e.g. photosynthesis rates in strong light.

In addition to the radiation received from the sun at a specific zenith angle, leaves are exposed to three streams of diffuse radiation. Radiation scattered from the atmosphere and from clouds arrives as from a hemisphere and moves downwards through foliage. Radiation scattered by the foliage itself moves both upwards and downwards. Because, in general, $\mathcal{K}$ depends on the geometry of radiation with respect to the architecture of foliage, the value of

$\mathcal{K}$ for direct radiation will usually be different from the values for diffuse fluxes. Equations for handling this complication have been developed from astrophysical theory (Ross, 1975). A simplified treatment for a system in which differences in values of $\mathcal{K}$ for different fluxes may be neglected is presented in the next chapter.

In practice, values of $\mathcal{K}$ are often determined by measuring the attenuation of radiation with tube solarimeters responding over a hemisphere to total solar radiation. Although estimates of $\mathcal{K}$ derived in this way are not strictly comparable with those derived here, they cover the same range, i.e. from about 1 for stands of clover and sunflower with predominantly horizontal leaves to about 0.2 for *Allium* and *Gladiolus* with mainly vertical leaves.

When plants produce their new leaves at the top of a stand, older leaves become progressively shaded and in many species they die when their irradiance falls to a few percent of full sunlight. When production and death are balanced, there is an upper limit L' to the leaf area index. Taking an arbitrary figure of 5% transmission for the level below which leaves die, L' can be related to $\mathcal{K}$ by writing

$$0.05 = \exp(-\mathcal{K}L')$$

so that $L' = -\ln(0.05)/\mathcal{K} = 3/\mathcal{K}$.

This rough calculation is consistent with field experience that stands with predominantly horizontal leaves usually have a maximum leaf area index of between 3 and 5 whereas values up to about 10 have been reported for cereals with a vertical leaf habit (Monteith and Elston, 1983).

6

MICROCLIMATOLOGY OF RADIATION

(ii) Absorption and reflection

RADIATIVE PROPERTIES OF NATURAL MATERIALS

When sunlight is intercepted by soil, by water, or by an object such as a leaf or an animal, energy is absorbed, reflected, and sometimes transmitted. This chapter describes how radiant energy is partitioned and how this process depends on wavelength. It concludes with a discussion of net radiation in terms of the radiative and geometrical characteristics of organisms.

At the short wavelength, high frequency end of the solar spectrum, the radiative behaviour of biological materials is determined mainly by the presence of pigments absorbing radiation at wavelengths associated with specific electron transitions. For radiation between 1 and 3 μm, liquid water is an important constituent of many natural materials because water has strong absorption bands in this region; and even in the visible spectrum where absorption by water is negligible, the reflection and transmission of light by porous materials is often strongly correlated with their water content. In the long-wave spectrum beyond 3 μm, most natural surfaces behave like full radiators with absorptivities close to 100% and reflectivities close to zero.

It is important to distinguish between the **reflectivity** of a surface $\rho(\lambda)$, which is the fraction of incident solar radiation reflected at a specific wavelength, and the **reflection coefficient** which in this context is the average reflectivity over a specific waveband, weighted by the distribution of radiation in the solar spectrum. If this distribution is described by a function $\mathbf{S}(\lambda)$, which is the energy per unit wavelength measured at λ, the energy in the solar spectrum between λ_1 and λ_2 is $\int_{\lambda_1}^{\lambda_2} \mathbf{S}(\lambda)d\lambda$. The reflection coefficient of a surface exposed to solar radiation is therefore

$$\bar{\rho} = \frac{\int_{\lambda_1}^{\lambda_2} \rho(\lambda)\mathbf{S}(\lambda)d\lambda}{\int_{\lambda_1}^{\lambda_2} \mathbf{S}(\lambda)d\lambda} \tag{6.1}$$

For the whole solar spectrum the integrations may be performed from 0.3 to

3 μm. The transmissivity $\tau(\lambda)$ and the transmission coefficient of a surface can be defined in the same way.

For the whole solar spectrum, the reflection coefficient of a natural surface is often called the *albedo*, a term borrowed from astronomy and derived from the Latin for 'whiteness'. However, because whiteness is associated with the visible spectrum, the more general term 'reflection coefficient' is preferable in this context.

Gates (1980) provides a very comprehensive account of the radiative properties of plant and animal surfaces along with tables of reflection coefficient for many species. Representative values for a range of surfaces are in Table 6.1.

Table 6.1 Radiative properties of plant and animal surfaces

(i) Short-wave reflection coefficients ρ (%)

1 Leaves	Upper	Lower	Average
Maize			29
Tobacco			29
Cucumber			31
Tomato			28
Birch	30	33	32
Aspen	32	36	34
Oak	28	33	30
Elm	24	31	28

2 Vegetation at maximum ground cover

(a) Farm crops	Latitude of site	Daily mean
Grass	52	24
Sugar beet	52	26
Barley	52	23
Wheat	52	26
Beans	52	24
Maize	43	22
Tobacco	43	24
Cucumber	43	26
Tomato	43	23
Wheat	43	22
Pasture	32	25
Barley	32	26
Pineapple	22	15
Maize	7	18
Tobacco	7	19
Sorghum	7	20
Sugar cane	7	15
Cotton	7	21
Groundnuts	7	17
(b) Natural vegetation and forest		
Heather	51	14
Bracken	51	24
Gorse	51	18

Maquis, evergreen scrub	32	21	
Natural pasture	32	25	
Derived savanna	7	15	
Guinea savanna	9	19	
(c) Forests and orchards			
Deciduous woodland	51	18	
Coniferous woodland	51	16	
Orange orchard	32	16	
Aleppo pine	32	17	
Eucalyptus	32	19	
Tropical rain forest	7	13	
Swamp forest	7	12	
3 Animal coats			
(a) Mammals	Dorsal	Ventral	Average
Red squirrel	27	22	25
Grey squirrel	22	39	31
Field mouse	11	17	14
Shrew	19	26	23
Mole	19	19	19
Grey fox			34
Zulu cattle			51
Red Sussex cattle			17
Aberdeen Angus cattle			11
Sheep weathered fleece			26
newly shorn fleece			42
Man			
Eurasian			35
Negroid			18
(b) Birds	Wing	Breast	Average
Cardinal	23	40	
Bluebird	27	34	
Tree swallow	24	57	
Magpie	19	46	
Canada goose	15	35	
Mallard duck	24	36	
Mourning dove	30	39	
Starling			34
Glaucous-winged gull			52

(ii) Long-wave Emissivities ε (%)

1 Leaves	Average
Maize	94.4 ± 0.4
Tobacco	97.2 ± 0.6
French bean	93.8 ± 0.8
Cotton	96.4 ± 0.7
Sugar cane	99.5 ± 0.4
Poplar	97.7 ± 0.4
Geranium	99.2 ± 0.2
Cactus	97.7 ± 0.2

2 Animals	Dorsal	Ventral	Average
Red squirrel	95–98	97–100	
Grey squirrel	99	99	
Mole	97	—	
Deer mouse	—	94	
Grey wolf			99
Caribou			100
Snowshoe hare			99
Man			98

Water

When radiation is incident on clear, still water at an angle of incidence ψ less than 45°, the reflection coefficient for solar radiation is almost constant at about 5%. Beyond 45°, the coefficient increases rapidly with increasing ψ, approaching 100% at grazing incidence (Fig. 6.1). The reflectivity of long-wave radiation also increases as ψ increases.

In the visible spectrum, water is usually regarded as transparent but it has a finite absorption coefficient with a minimum in the blue-green between 460 and 490 nm which gives natural bodies of very clean water their characteristic colour, particularly when observed over white sand. Outside this waveband,

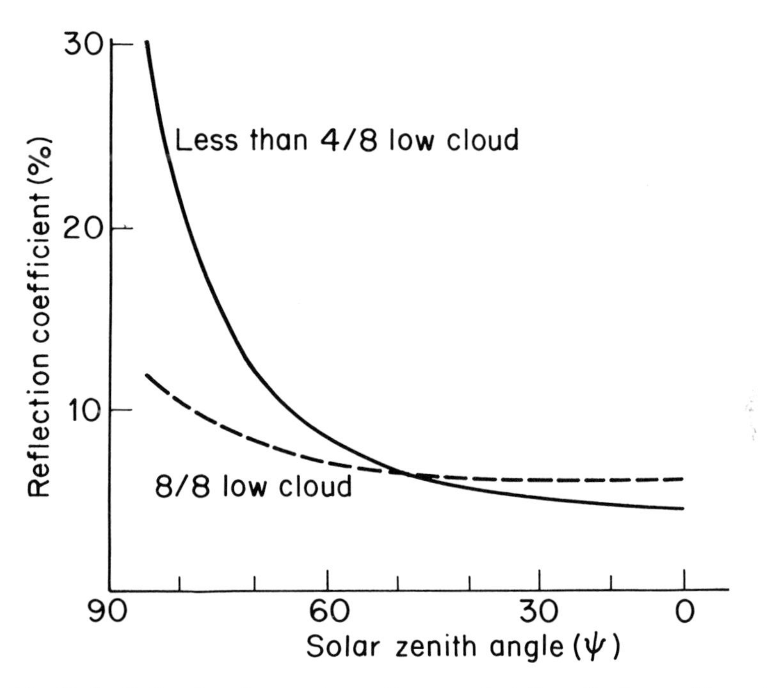

Fig. 6.1 Reflectivity of a plane water surface as a function of solar zenith angle and cloudiness (from Deacon, 1969).

absorption increases in both directions, but particularly towards the red end of the spectrum. Whereas 300 m of pure water are needed to reduce the transmission of blue-green radiation to 1%, the corresponding figure for red light is only 20 m. In natural water, however, the presence of organic matter including chlorophyll depletes the blue end of the spectrum and, in some British lakes, both blue and red wavelengths are attenuated to 1% at depths less than 10 m (Spence, 1976).

In the near infra-red region of the spectrum, water has several absorption bands, readily identified in the transmission and reflection spectra of soil, leaves and animal skins. The centres of the main bands are at 1.45 and 1.95 μm (Fig. 6.2). Beyond 3 μm, the absorptivity and emissivity of water, measured at normal incidence, are about 0.995 but, with increasing angle of incidence, they decrease in a complementary way to the increase of short-wave reflectivity and at 11 μm the emissivity is only about 0.7 when ψ is 80°. The angular dependence of the emissivity of water and other natural surfaces complicates the interpretation of radiometric measurements of skin temperature in human pathology (Clark, 1976) and of ground surface temperature observed from aircraft or satellites (Becker, 1981, Becker *et al.*, 1981).

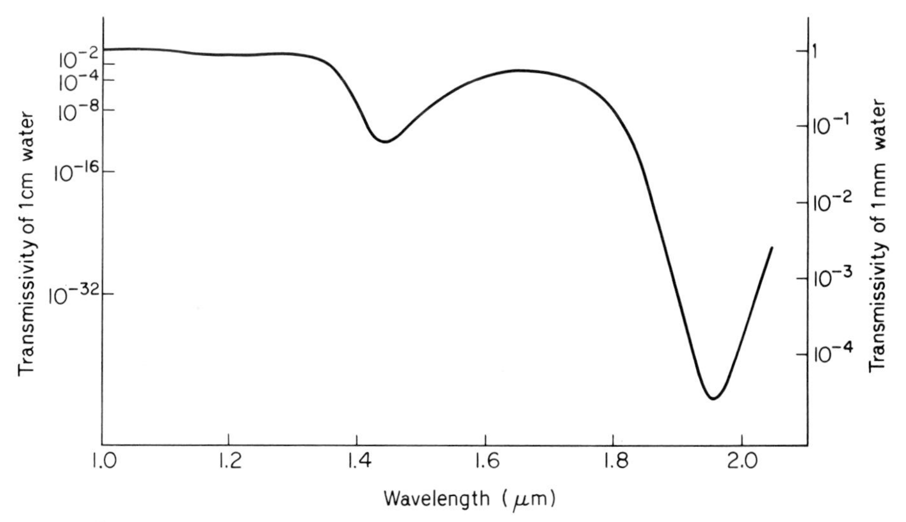

Fig. 6.2 Transmissivity of pure water as a function of wavelength. Note logarithmic scales for 1 cm water (left-hand axis) and 1 mm (right-hand axis).

Soil

The reflectivity of soils depends mainly on their organic matter content, on water content, particle size and angle of incidence. Reflectivity is usually very small at the blue end of the spectrum, increases with wavelength through the visible and near infra-red wavebands and reaches a maximum between 1 and 2 μm. When water is present, its absorption bands are evident at 1.45 and 1.95 μm.

Integrated over the whole solar spectrum, reflection coefficients range from about 10% for soils with a high organic matter content to about 30% for desert sand. Even a very small amount of organic matter can depress the reflectivity of a soil. Oxidizing the organic component of a loam, which was 0.8% by weight, increased its reflectivity by a factor of two over the whole visible spectrum (Bowers and Hanks, 1965).

The reflectivity of clay minerals has been measured as a function of their particle size. Over the spectral range 0.4 to 2 μm, the reflectivity of kaolinite increases rapidly with decreasing particle diameter, e.g. from 56% for 1600 μm particles to 78% for 22 μm particles. Aggregates containing relatively large irregular particles appear to trap radiation by multiple reflection between adjacent faces whereas finely divided powders expose a more uniform surface, trapping less radiation. Particle size also governs the transmission of radiation by soils. Baumgartner (1953) measured the transmission of artificial light by quartz sand. When the particle diameter was 0.2 to 0.5 mm, a depth of 1 to 2 mm of sand was enough to reduce the radiative flux by 95%, but for particles of 4 to 6 mm, a layer 10 mm deep was needed to give the same extinction. In another study, with fine seed compost, irradiance decreased by more than four orders of magnitude in a depth of 3 mm but the corresponding change of red/far-red ratio was small—from 1.2 to 0.8 (Frank-

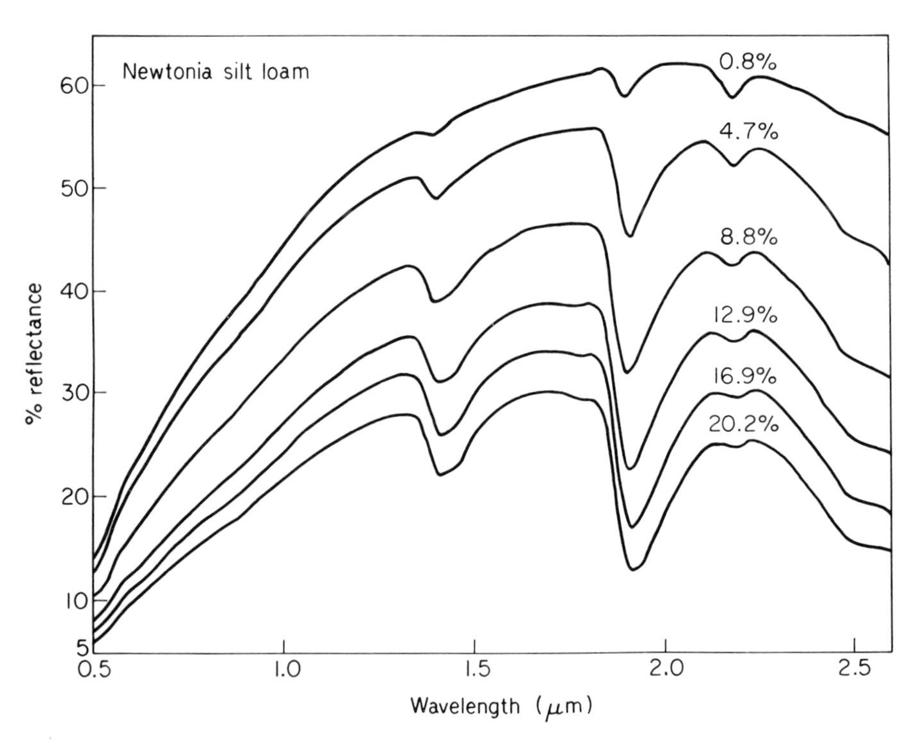

Fig. 6.3 Reflectivity of a loam soil as a function of wavelength and water content (from Bowers and Hanks, 1965).

land, 1981). The transmission of radiation by soils has had little attention from ecologists although the effects of light quality and quantity on seed germination and root development are well established.

The reflectivity of a soil sample decreases as it gets wetter, mainly because radiation is trapped by internal reflection at air-water interfaces formed by the menisci in soil pores. The dependence of reflectivity on water content is evident at all wavelengths but is strongest in the absorption band at 1.95 μm. In an example shown in Fig. 6.3, the reflectivity of a loam at 1.9 μm decreased from 60% at 1% water content to 14% at 20% water. The reflection coefficient of a stable soil can therefore be used to monitor the water content or water potential of the surface layer (Idso *et al.*, 1975; Graser and van Bavel, 1982).

In the long-wave spectrum, most soils have an emissivity between 0.90 and 0.95 and the emissivities of common minerals range from 0.67 for quartz to 0.94 for marble.

Leaves

The fractions of radiation transmitted and reflected by a leaf depend on the angle of incidence ψ. Measurements by Tageeva and Brandt (1961) showed that the reflection coefficient was almost constant for values of ψ between 0 and 50° but as ψ increased from 50 to 90° (grazing incidence), $\rho(\lambda)$ increased sharply as a result of specular reflection. The transmission coefficient was also constant between 0 and 50° but decreased between 50 and 90°. Because changes of $\rho(\lambda)$ and $\tau(\lambda)$ with angle were complementary, the fraction of

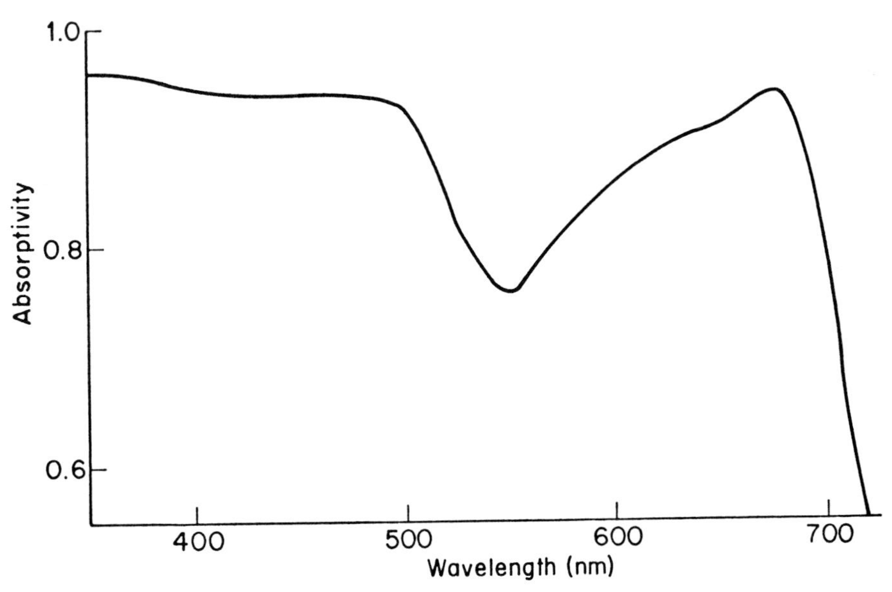

Fig. 6.4 Average absorptivity for leaves of eight field-grown crop species (from McCree, 1972).

radiation absorbed (and available for physiological processes) was almost constant for angles of incidence less than 80°.

For the leaves of many common crop species, the absorptivity of green light at 550 nm is about 0.75 to 0.80, for blue light (400 to 460 nm) it is 0.95 and for red light (600 to 670 nm) it is about 0.85 to 0.95 (Fig. 6.4). The very sharp decrease of absorption as wavelength increases beyond 700 nm is physiologically significant because it implies that within leaf tissue the energy per unit wavelength in an infra-red waveband centred at 730 nm is much larger than the corresponding energy in the red at 660 nm. This ratio plays a major role in determining the state of the pigment phytochrome which governs many developmental processes (Smith and Morgan, 1981). The existence of the complementary 'red edge' in the reflection spectrum has been exploited in remote sensing, as discussed later.

The similarity of the transmission and reflection spectra displayed by many leaves (Fig. 6.5) implies that the radiation within them is scattered in all directions by reflection and refraction at the walls of cells. The only incident radiation not scattered in this way is the component (about 10%) which is reflected from the surface of the cuticle without entering the mesophyll.

To derive approximate values of ρ and τ for the whole solar spectrum, the coefficients can be assumed equal at 0.1 between 0.4 and 0.7 μm and at 0.4 between 0.7 and 3 μm. Because each of these spectral bands contains about half the total radiation, the approximate coefficients for the whole spectrum are

$$\rho = \tau = (0.1 \times 0.5) + (0.4 \times 0.5) = 0.25$$

This value is consistent with measurements on a number of species recorded in Table 6.1. Because the visible spectrum makes a relatively small contribution to the reflection and transmission of solar radiation by leaves, comparisons of colour are largely irrelevant in terms of the (total) reflection coefficient.

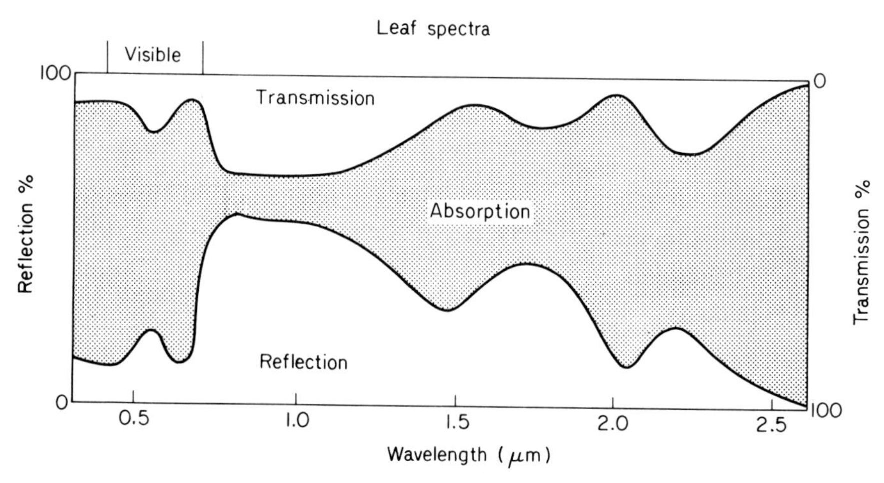

Fig. 6.5 Idealized relation between the reflectivity, transmissivity and absorptivity of a green leaf.

Vegetation

The fractions of incident solar radiation reflected and transmitted by a dense stand of vegetation depend on two main factors: the fraction intercepted by foliage; and the scattering properties of the foliage. (In this context, 'foliage' should include stems, petioles, etc. but their role in interception is usually neglected.) The intercepted fraction depends, in turn, on the area of foliage specified as a **Leaf Area Index** (plan area of leaves per unit ground area) and on the spatial distribution of foliage with respect to the direction of radiation. The scattered fraction depends on the optical properties of leaf cuticles, cell walls and pigments. When the foliage is not dense enough to intercept all the

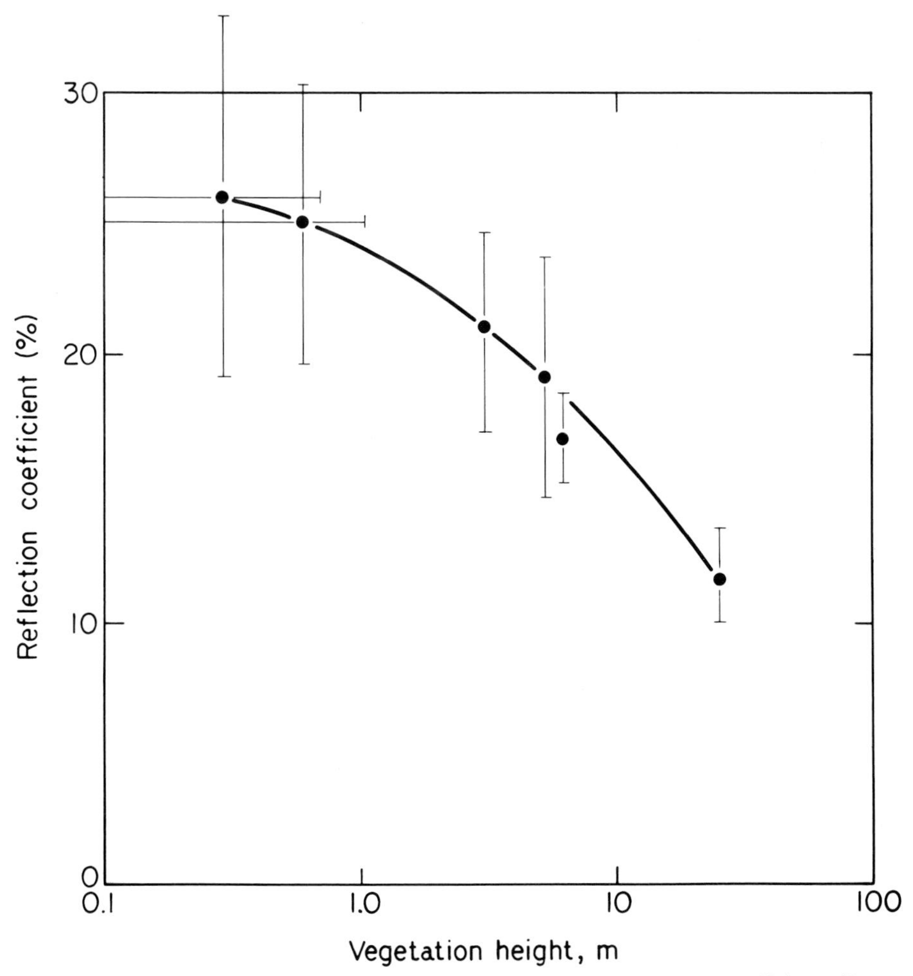

Fig. 6.6 Relation between height of vegetation and reflection coefficient (from Stanhill, 1970).

incident radiation, the reflection coefficient of the stand depends to some extent on reflection from the soil as well as from foliage.

In general, maximum values of ρ (see pp. 80–81) are recorded over relatively smooth surfaces such as closely cut lawns. For crops growing to heights of 50 to 100 cm, ρ is usually between 0.18 and 0.25 when ground cover is complete but values as small as 0.10 have been recorded for forests (Fig. 6.6). These differences can be interpreted in terms of the trapping of radiation by multiple reflection between adjacent leaves and stems. For the same reason, the reflection coefficient for most types of vegetation changes with the angle of the sun. Minimum values of ρ are recorded as the sun approaches its zenith and ρ increases as the sun descends to the horizon because there is less opportunity for multiple scattering between the elements of the canopy. The dependence of ρ on zenith angle may explain why reflection coefficients for vegetation measured in the tropics are usually somewhat smaller than the coefficients for similar surfaces at higher latitudes. Table 6.1 (p. 80) lists values of ρ for canopies of various crops and native species.

Theory and prediction

Assuming that the Kubelka-Munk equations (p. 34) are valid for a deep canopy of foliage and putting representative values of $\rho = \tau = 0.1$ for leaves exposed to radiation in the visible spectrum, the absorption coefficient for leaves is $\alpha_P = 1 - \tau - \rho = 0.8$.

Then from equation (3.31), the reflection coefficient for a deep canopy is given by

$$\rho_c{}^* = (1 - \alpha_P{}^{0.5})/(1 + \alpha_P{}^{0.5}) = 0.06 \qquad (6.2)$$

Corresponding representative figures for the infra-red spectrum are $\rho = \tau = 0.4$ and $\rho_c{}^* = 0.38$. The approximate value of $\rho_c{}^*$ for the whole solar spectrum is therefore

$$\rho_c{}^* = (0.06 \times 0.5) + (0.38 \times 0.5) = 0.22$$

If the fraction of radiation transmitted by a canopy is τ_c and the reflection coefficient is ρ_c, the upward flux of radiation below the canopy will be $\tau_c\rho_s$ when ρ_s is a soil reflection coefficient. The fraction of incident radiation absorbed by the canopy will therefore be, to a good approximation

$$\begin{aligned} \alpha_c &= 1 + \tau_c\rho_s - \tau_c - \rho_c \\ &= 1 - \rho_c - \tau_c(1 - \rho_s) \end{aligned} \qquad (6.3)$$

The quantities α_c, ρ_c and τ_c can be derived from the Kubelka-Munk equations as functions of (i) the corresponding coefficients for leaves; (ii) the leaf area index; (iii) a canopy attenuation coefficient

$$\mathcal{K} = \alpha^{0.5}\mathcal{K}_b \qquad (6.4)$$

(from equation (3.32)) where $\mathcal{K}_b$ is the coefficient for black leaves ($\alpha = 1$) with the same geometry.

Strictly, the theory can be applied only to stands with horizontal leaves so that $\mathcal{K}_b = 1$. This ensures that the intercepting area of all leaves is independent of the direction of incident radiation and therefore a single value of $\mathcal{K}$ is valid for all scattered light as well as for the direct beam. However, Goudriaan (1977) demonstrated that equations (6.2) and (6.4) could be used to describe the behaviour of radiation in canopies where the solar elevation β was larger than the elevation angle for most of the foliage so that the shadow area for an assembly of leaves was independent of β (see p. 76). This class includes canopies with a spherical distribution of leaves provided $\beta > 25°$. It also includes radiation from a uniform overcast sky which is transmitted by a spherical leaf distribution as if all the radiation emanated from an angle of about 40°. For these restricted categories of foliage, the value of $\mathcal{K}_b$ to be inserted in equation (6.4) is given by the relative shadow area of the assembly $\mathcal{K}_s$ as evaluated in Chapter 5.

After integration of the equations (3.27, 3.28) which describe the downward and upward streams of radiation in a canopy, the reflection coefficient is given by

$$\rho_c = \frac{\rho_c^* + f \exp(-2\mathcal{K}L)}{1 + \rho_c^* f \exp(-2\mathcal{K}L)} \tag{6.5}$$

and the transmission coefficient by

$$\tau_c = \frac{\{(\rho_c^{*2} - 1)/(\rho_s \rho_c^* - 1)\} \exp(-\mathcal{K}L)}{1 + \rho_c^* f \exp(-2\mathcal{K}L)} \tag{6.6}$$

where

$$f = (\rho_c^* - \rho_s)/(\rho_c^* \rho_s - 1) \tag{6.7}$$

Neglecting the second-order terms ρ_c^{*2} and $\rho_c^* \rho_s$ gives approximate values of the coefficients as

$$\rho_c = \rho_c^* - (\rho_c^* - \rho_s) \exp(-2\mathcal{K}L) \tag{6.8}$$

and

$$\tau_c = \exp(-\mathcal{K}L) \tag{6.9}$$

where ρ_c clearly has limits of ρ_c^* when L is large and ρ_s when L is small. Now, from equations (6.8) and (6.9), absorption by the canopy is

$$\alpha_c \approx 1 - \{\rho_c^* - (\rho_c^* - \rho_s) \exp(-2\mathcal{K}L)\} - (1 - \rho_s) \exp(-\mathcal{K}L) \tag{6.10}$$

The term $(\rho_c^* - \rho_s) \exp(-2\mathcal{K}L)$ will be much smaller than $(1 - \rho_s) \exp(-\mathcal{K}L)$ when ρ_c^* is small and/or $\mathcal{K}L$ is large. The canopy absorption then becomes

$$\begin{aligned} \alpha_c &\approx 1 - \rho_c^* - (1 - \rho_s)\tau_c \\ &\approx (1 - \rho_c^*)(1 - \tau_c) + \tau_c(\rho_s - \rho_c^*) \end{aligned} \tag{6.11}$$

Some practical implications of these relations will now be considered.

Absorbed and intercepted radiation

In the field, the radiation intercepted by a crop stand is conveniently determined by mounting instruments such as tube solarimeters above and below the canopy. Most tube solarimeters in common use respond uniformly to radiation through the solar spectrum. The fraction of the incident flux density recorded by the lower solarimeter is τ_c and the intercepted fraction is simply $1 - \tau_c$.

Empirically, the amount of total solar radiation intercepted by a canopy is often well correlated with the production of dry matter during periods when the leaf area index is increasing (Russell *et al.*, 1989). Theoretically, however, growth rate should depend on the *absorbed* fraction of radiation. Equation (6.11) provides reassurance that, subject to the validity of the approximations used, and provided the term $\tau_c(\rho_s - \rho_c^*)$ is small, α_c is a nearly constant fraction of $(1-\tau_c)$. The fraction is $1-\rho_c^*$, i.e. about 0.94 for the visible spectrum and 0.77 for total solar radiation using the values derived above.

A further step is needed to obtain absorbed PAR from intercepted *total* radiation. From equations (6.4) and (6.9)

$$\tau_c \approx \exp(-\alpha^{0.5}\mathcal{K}_b L) \tag{6.12}$$

The transmission of PAR, τ_{cP}, can therefore be estimated from the transmission of total radiation, τ_{cT}, using

$$\begin{aligned}\tau_{cP} &= \tau_{cT} \exp\{(\alpha_T{}^{0.5} - \alpha_P{}^{0.5})\mathcal{K}_b L\}\\ &= \tau_{cT} \exp(-0.26\mathcal{K}_b L)\end{aligned} \tag{6.13}$$

if $\alpha_T = 0.4$ and $\alpha_P = 0.8$. The fraction of PAR absorbed by a stand as given by equation (6.11) is

$$\alpha_{cP} = 0.94\{1 - \tau_{cT} \exp(-0.26\mathcal{K}_b L)\} \tag{6.14}$$

assuming $\rho_{cP}^* = 0.06$ and neglecting the small term $\tau_{cP}(\rho_{sP} - \rho_{cP}^*)$. Finally, the fractional absorption of energy (PAR) can be converted to the equivalent fraction of absorbed quanta using a coefficient of 4.6 μmole per joule (p. 49). Figure 6.7 shows how absorbed quanta derived from equation (6.14) are related to intercepted total radiation from equation (6.9).

Equation (6.9) can also be written

$$\tau_{cT} = \exp(-\mathcal{K}' L)$$

where $\mathcal{K}' = \alpha^{0.5}\mathcal{K}_b$ is the attenuation coefficient as conventionally determined by plotting $\ln \tau_{cT}$ against L to find $-\mathcal{K}'$ as a slope.

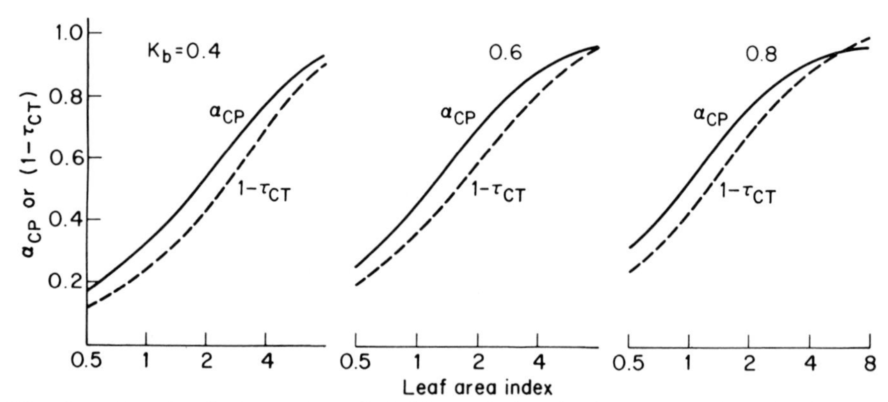

Fig. 6.7 Fractional absorption of visible radiation (or PAR) α_{CP} (full lines), and fractional interception of total radiation $(1-\tau_{cT})$ as functions of leaf area index for three values of the attenuation coefficient for black leaves $\mathcal{K}_b$ (assuming $\rho_P = \tau_P = 0.05$).

The ratio of attenuation coefficients for PAR and total radiation is $(\alpha_P/\alpha_T)^{0.5}$, a conservative quantity because the fraction of total radiation absorbed by a leaf is largely determined by pigments which absorb in the PAR waveband. To demonstrate this point, assume that in the PAR waveband $\rho_P = \tau_P = 0.05$ so that $\alpha_P = 0.9$, that absorptivity in the infra-red waveband is 0.1 and that the solar spectrum contains equal amounts of energy in the two wavebands (p. 48). The absorptivity for total radiation is $\alpha_T = (0.9 + 0.1)/2 = 0.5$ and $\mathcal{K}_P/\mathcal{K}_T = (0.9/0.5)^{0.5} = 1.34$. Doubling ρ_P and τ_P to 0.1 makes $\alpha_P = 0.8$, $\alpha_T = 0.45$ and $\mathcal{K}_P/\mathcal{K}_T = (0.8/0.45)^{0.5} = 1.33$. Green (1984) found that the ratio of $\mathcal{K}$ for quanta (effectively the same as for PAR) and for total radiation was 1.34 for wheat grown with a range of nitrogen applications so that there was a conspicuous range of greenness between treatments.

Remote sensing

Because the spectrum of radiation reflected by foliage has a different shape from the spectrum for all types of soil, the extent to which soil is covered by vegetation can be estimated from the reflection spectrum of the area, as recorded from an aircraft or a satellite. However, interpretation can be very difficult if the cover is not uniform. Most information is obtained by working near the 'red edge' at 700 nm below which leaves are almost black and above which almost all the incident radiation is scattered. In principle, an estimate of cover could be obtained simply from the reflectivity ρ_i in the near infra-red (say between 700 and 900 nm), but in practice, it is often difficult to measure the incident and reflected fluxes simultaneously and error is unavoidable if the incident flux is changing with time. The difficulty is usually overcome by measuring the ratio x_1 of the *absolute* amounts of radiation reflected in the near infra-red and in some part of the visible spectrum, usually in the red. The corresponding ratio x_2 for the spectrum of incident radiation (which changes much more slowly than the absolute flux at any wavelength) can be obtained by recording the spectrum reflected from a standard white surface or by calculation if the measurements are above an atmosphere whose radiative behaviour is known. Then the required ratio of infra and red reflectivities is given by

$$\frac{\rho_i}{\rho_r} = \frac{\text{reflected IR/incident IR}}{\text{reflected R/incident R}}$$

$$= \frac{\text{reflected IR/reflected R}}{\text{incident IR/incident R}} = \frac{x_1}{x_2} \tag{6.15}$$

Many workers use a 'normalized difference vegetation index' (NDVI) defined as

$$\frac{\rho_i - \rho_r}{\rho_i + \rho_r} = \frac{x_1 - x_2}{x_1 + x_2} \tag{6.16}$$

The correlation between fractional ground cover and this quantity is seldom much better than the correlation with the simple ratio x_1/x_2 but the fact that the ratio has limits of -1 and $+1$ makes it convenient to handle when computing.

The main problems in using both types of index to estimate uniform ground cover are (i) changes in the spectral properties of foliage as a consequence of senescence, poor nutrition or disease which alter the dependence of the index on leaf area (Steven, 1983); and (ii) changes in the spectrum of reflected radiation between the ground and a detector as a consequence of scattering in the atmosphere (Steven *et al.*, 1984). Nevertheless, Tucker *et al.* (1985) were able to obtain plausible distributions of ground cover month by month over the whole African continent; and by integration they established a seasonal index of total intercepted radiation and therefore of biomass production. A similar exercise over the North American continent gave biomass estimates consistent with accepted figures for regional classes of vegetation. The basis for this procedure is now outlined.

Following the example of Asrar *et al.* (1984) but using the approximate equations already derived, a relation can be established between the observed value of ρ_i/ρ_r for a canopy and the fraction of PAR which it transmits or absorbs. First, the attenuation coefficient can be written

$$\mathcal{K}_j = \alpha_j^{0.5}\mathcal{K}_b \tag{6.17}$$

where j is replaced by P, r or i for the wavebands of PAR, red or near infra-red radiation.

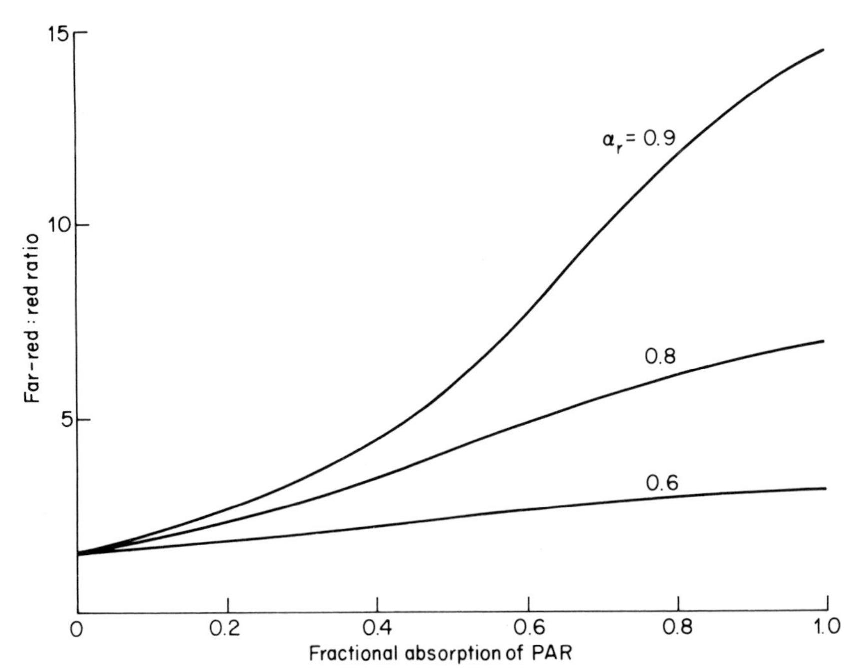

Fig. 6.8 Relation between far-red:red reflectivity ratio for a canopy of vegetation and the amount of PAR absorbed by the canopy. Leaf absorptivity specified.

Then from equations (6.8) and (6.9) and with the same conventions for subscripts

$$\frac{\rho_i}{\rho_r} = \frac{\rho_{ci}{}^* - (\rho_{ci}{}^* - \rho_{si})\tau_P{}^{(2\mathcal{K}_i/\mathcal{K}_P)}}{\rho_{cr}{}^* - (\rho_{cr}{}^* - \rho_{sr})\tau_P{}^{(2\mathcal{K}_r/\mathcal{K}_P)}} \qquad (6.18)$$

It follows that, within the limits of the approximations used, the ratio ρ_i/ρ_r should be a unique function of τ_P, independent of $\mathcal{K}_b$ and therefore of leaf architecture because it involves the ratio of coefficients at different wavelengths. The function also depends on the values of α_i, α_r and α_P which determine the exponents of τ_P and the soil reflectivities ρ_{si} and ρ_{sr}. When the second term in the denominator is small compared with the first term, ρ_i/ρ_r is a function of $\tau_P{}^{(2\mathcal{K}_i/\mathcal{K}_P)}$. For the special case $\alpha_i/\alpha_r = 1/4$ (e.g. $\alpha_i = 0.2$, $\alpha_r = 0.8$), $\mathcal{K}_i/\mathcal{K}_P = \sqrt{1/4} = 1/2$ and ρ_i/ρ_r is a linear function of τ_P. Figure 6.8 shows the extent to which the function departs from linearity because the approximations behind this result are not valid over the whole range of τ_P.

The spectral ratio ρ_i/ρ_r is therefore a valid and extremely useful index of the radiation *intercepted* by and therefore the radiation *absorbed* by a canopy. It can legitimately be used to provide an estimate of growth rate when the relation between growth and intercepted radiation is known. In contrast, the relation between ρ_i/ρ_r and either biomass or leaf area is strongly non-linear and depends on leaf architecture (Fig. 6.7).

Animals

The coats and skins of animals reflect solar radiation both in the visible and in the infra-red regions of the spectrum, and reflectivity is a function of the angle of incidence of the radiation, as for water and leaves. Figure 6.9 shows that the reflectivity of hair coats increases throughout the visible spectrum (like soil); and that intra- and inter-specific differences of reflectivity are much larger than for leaves. The maximum reflectivity of several types of coat is found between 1 and 2 μm and water is responsible for absorption at 1.45 and 1.95 μm. Beyond 3 μm, the absorptivity and emissivity of most types of coat is between 90 and 95%.

The measurements of coat reflectivity plotted in Fig. 6.9 were made in the laboratory on samples of coats exposed to radiation at normal incidence. When Hutchinson *et al.* (1975) used a miniature solarimeter to measure reflection from the coats of live animals standing in the field, they found a marked dependence of reflection coefficient on the local angle of incidence of direct sunlight. For Jersey heifers with relatively smooth coats exposed to sun at an elevation of 48° (Fig. 6.9), a minimum reflection coefficient of 0.25 was recorded when the angle of incidence was minimal but ρ was more than 0.7 at grazing incidence because of the large contribution from specular reflection. The change of reflection with angle was much smaller for Merino sheep with deep, woolly coats (Fig. 6.10). The implication of this work is that laboratory measurements of reflection, usually made at normal incidence, may be an unreliable guide to the effective reflection coefficient for an animal in the field, particularly for species with relatively smooth coats. Table 6.1 summa-

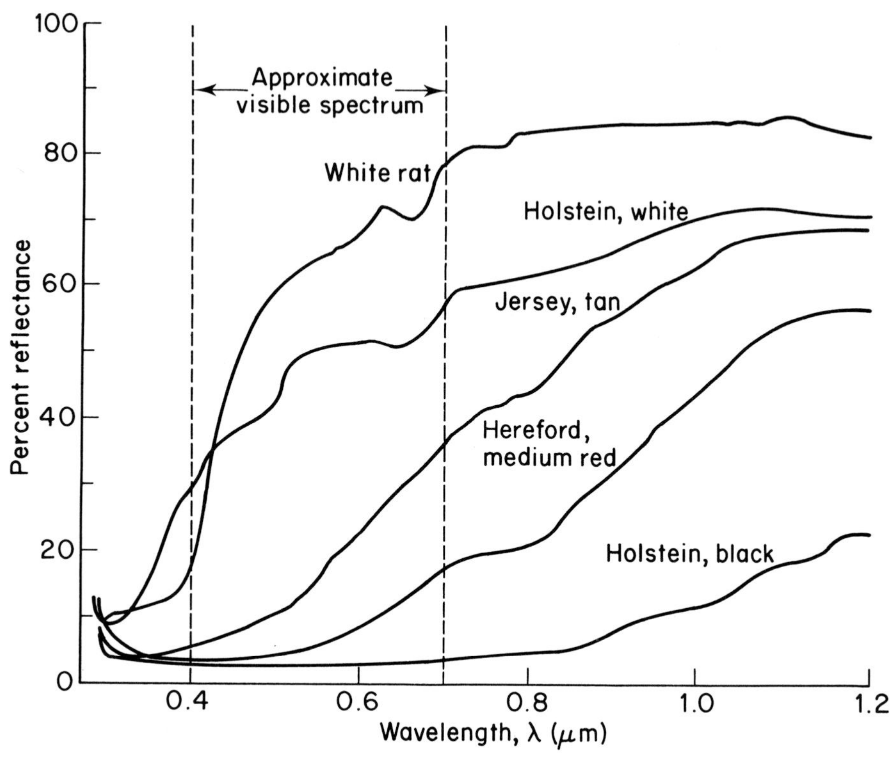

Fig. 6.9 Reflectivity of animal coats (from Mount, 1968).

rizes some measurements of the reflection coefficients of animal and bird coats.

The transmission of radiation within coats was examined by Cena and Monteith (1975a) who applied the Kubelka-Munk equations (p. 34) to measurements made when radiation was incident normally on samples of pelt. The parameters of their analysis were depth of coat, the probability p of a ray striking a hair within unit depth, the fraction of intercepted radiation reflected (ρ), transmitted (τ) or absorbed (α), and the absorptivity of skin. In contrast to leaves, τ was larger than ρ, especially in white coats for which α was extremely small (Table 6.1). The fact that forward scattering exceeds back scattering is probably a consequence of specular reflection when hair is nearly parallel to an incident beam of radiation. The fraction of incident radiation absorbed by animals with light coats (Fig. 6.11) is therefore strongly dependent on the absorptivity of skin. In principle the thermal load on an animal with a white coat and dark skin could be larger than for a dark coat and light skin. The relevant equations have been developed and explored by Walsberg *et al.* (1978).

Measurements of skin reflectivity reported in the literature refer to an almost hairless animal—Man. In the absence of hair, solar radiation pene-

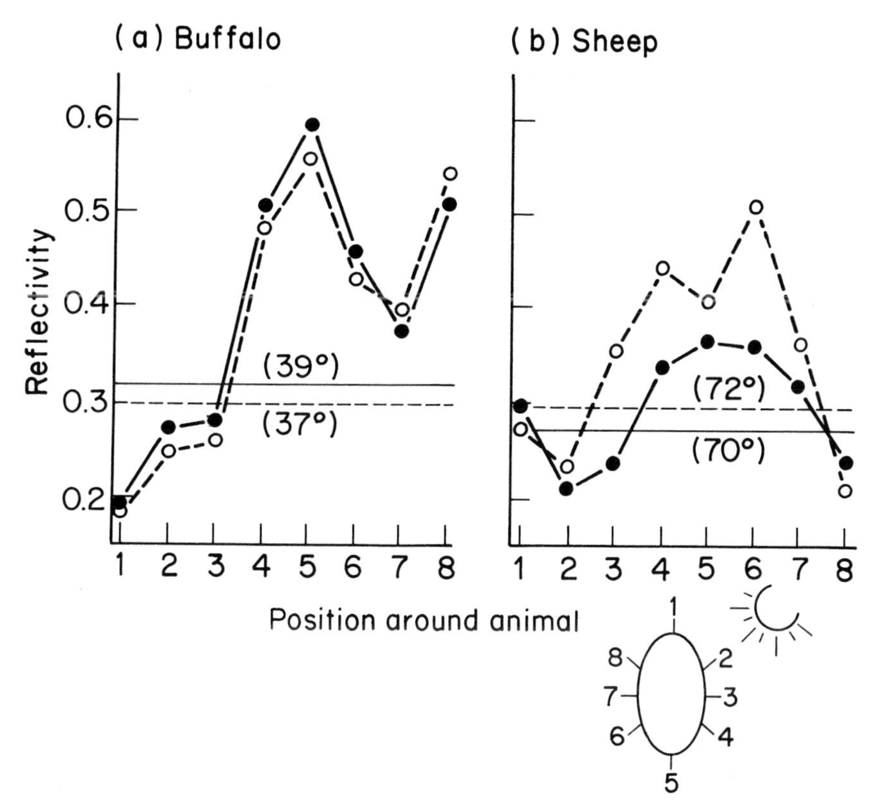

Fig. 6.10 Reflectivity in individual positions and weighted mean reflectance (horizontal lines) (a) for swamp water buffalo calves and (b) for Merino sheep with fleeces 6 cm deep. The animals stood sideways on to the sun. Numbers in brackets represent sun altitudes (from Hutchinson *et al.*, 1975).

trates skin to a depth that depends on pigmentation and in humans the penetration ranges from several millimetres in Caucasian subjects with light skin to a few tenths of a millimetre in negroid subjects with a much greater concentration of melanin in the corneum. Corresponding reflection coefficients range from 20% for dark skins to 40% for light. Figure 6.12 shows that the reflectivity of white skin is maximal at about 0.7 μm and decreases to a few percent at 2 μm.

There have been very few definitive studies of the relation between skin colour, heat stress and live weight gain in livestock. Finch *et al.* (1984) working with Brahman and Shorthorn cattle in Queensland were able to demonstrate a significant positive correlation between reflectivity and live weight gain and the largest response was observed in individuals with the thickest and woolliest coats. In the absence of heat stress, however, several workers have found a negative correlation between reflectivity and live weight gain for reasons which remain obscure.

Fig. 6.11 (a) Farmers at a cattle market in Andhra Pradesh, India. The bareheaded gentleman needs the shade of an umbrella. The two with white turbans do not. (Would a white umbrella be more or less effective? See p. 94). (b) Cattle in a field in Israel. The dark cattle have sought shade, but the white animals are apparently comfortable in full sunshine. (See p. 94).

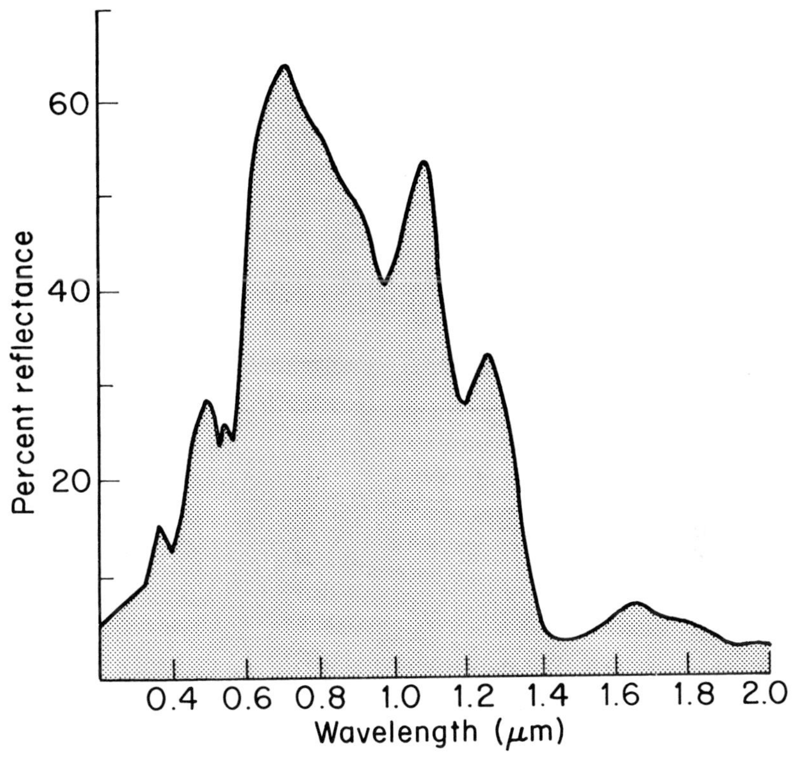

Fig. 6.12 Spectral reflectivity of skin on an author's thumb from a recording by Dr Warren Porter in his laboratory at the University of Wisconsin on 18 April 1969.

NET RADIATION

In Chapter 4, net radiation was presented as a climatological variable with the caveat that its value depends on the temperature and reflectivity of the surface exposed to radiative exchange. Having considered how the interception of radiation by an organism is determined by its geometry and having established characteristic values of reflectivity for natural surfaces, we are now in a position to compare differences in the net flux of radiant energy absorbed by contrasting surfaces exposed to the same radiation environment as specified by a flux of short-wave radiation (received from the sun and sky and reflected from the ground) and a long-wave flux (received from the atmosphere and emitted by the ground).

A suitable general form of the equation describing the radiation budget is

$$\mathbf{R}_n = (1 - \rho)\mathbf{S}_t + \mathbf{L}_d - \mathbf{L}_u \tag{6.19}$$

and applications of this expression are now considered for four contrasting surfaces (Fig. 6.13).

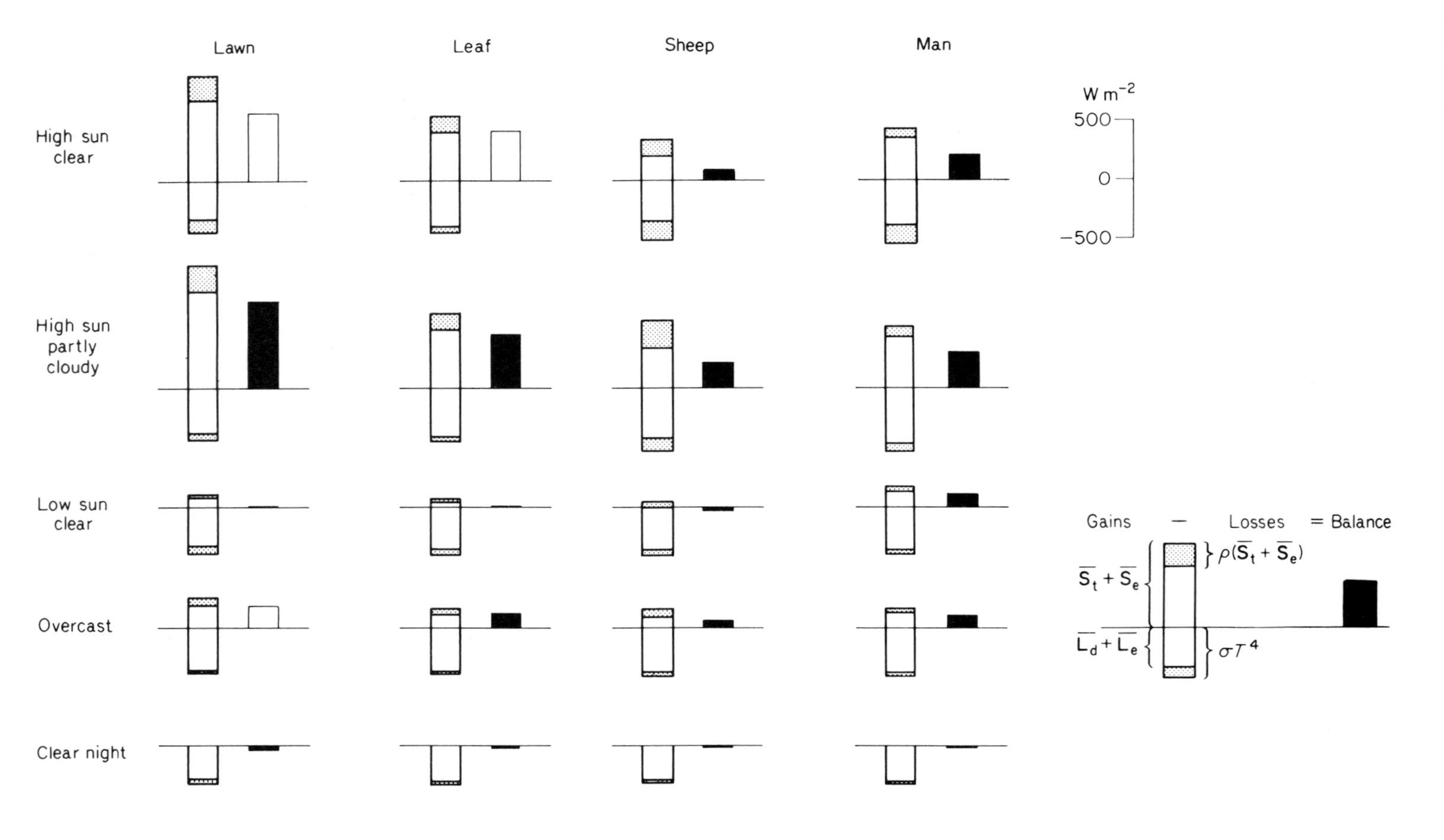

Fig. 6.13 Radiation balance for different surfaces in a range of weather conditions as specified in Table 6.2.

(i) A short grass lawn
For a continuous horizontal surface receiving radiation from above and not from below, the net radiation is simply

$$\mathbf{R}_n = (1-\rho_1)\mathbf{S}_t + \mathbf{L}_d - \sigma T_1^4 \qquad (6.20)$$

where ρ_1, T_1 are the reflection coefficient and radiative temperature of the lawn.

(ii) A horizontal leaf
If the leaf is assumed to be exposed above the lawn, it receives an additional income of short-wave radiation $\mathbf{S}_e = \rho_1\mathbf{S}_t$ and of long-wave radiation $\mathbf{L}_e = \sigma T_1^4$. The net radiation is therefore

$$\mathbf{R}_n = \{(1-\rho_2)(1+\rho_1)\mathbf{S}_t + \mathbf{L}_d + \mathbf{L}_e - 2\sigma T_2^4\}/2 \qquad (6.21)$$

where ρ_2, T_2 are the reflection coefficient and radiative temperature of the leaf. Note that $\mathbf{R}_n$ is the net radiation per unit of *total* leaf area, i.e. twice the plane area or twice the leaf area index.

(iii) A sheep
The sheep is assumed to be standing on the lawn and so receives radiation reflected and emitted by the surface below it. When the area of shadow is ignored the net radiation for the sheep is

$$\mathbf{R}_n = (1-\bar{\rho}_3)(1+\rho_1)\bar{\mathbf{S}}_t + \bar{\mathbf{L}}_d + \bar{\mathbf{L}}_e - \sigma T_3^4 \qquad (6.22)$$

where the bars indicate averaging over the exposed surface. The sheep is assumed to be a horizontal cylinder with its axis at right angles to the sun's rays, reflectivity is ρ_3 and mean surface temperature is T_3.

(iv) A man
For a man standing on the lawn, the radiation balance is formally identical to equation (6.22) with ρ_4, T_4 replacing ρ_3, T_3.

Values assumed for the radiation fluxes and for ρ and T are given in Table 6.2. Long-wave emissivities are assumed to be unity.

It is instructive to compare the net radiation received by different surfaces at the same time, noting the effects of geometry and reflection coefficients; or to compare the same surface at different times to see the influence of solar elevation and cloudiness. Salient features of these comparisons are as follows.

(a) During the day, the lawn absorbs more net radiation than any of the other surfaces including the isolated leaf. The leaf receives less short-wave radiation than the lawn ($(1+\rho_1)\mathbf{S}_t/2$ compared with $\mathbf{S}_t$) and absorbs more long-wave radiation ($(\mathbf{L}_d + \sigma T_1^4)/2$ compared with $\mathbf{L}_d$).
(b) The sheep absorbs less net radiation than the other surfaces. This is partly a consequence of the relatively large reflection coefficient (0.4) and partly a consequence of geometry.
(c) The geometry of the man ensures that $\mathbf{R}_n$ is large in relation to other surfaces when the sun is low.
(d) For all surfaces, the net radiation is greatest when the sun is shining between clouds and is larger under an overcast sky than it is when the sun is near the horizon.

Table 6.2 Conditions assumed for radiation balances, Fig. 6.12

	1 High sun clear	2 High sun partly cloudy	3 Low sun clear	4 Overcast day	5 Clear night
Solar elevation β	60	60	10	—	—
Direct solar radiation S_b (W m^{-2})	800	800	80	—	—
Diffuse solar radiation S_d (W m^{-2})	100	250	30	250	—
Downward long wave radiation L_d (W m^{-2})	320	370	310	380	270
Surface temperature (°C)					
air	20	20	18	15	10
lawn	24	24	15	15	6
leaf	24	25	15	15	4
sheep	33	36	15	20	10
man	38	39	15	20	10
Reflectivities					
lawn	0.23	0.23	0.25	0.23	—
leaf	0.25	0.25	0.35	0.25	—
sheep	0.40	0.40	0.40	0.40	—
man	0.15	0.15	0.15	0.15	—

(e) At night, the leaf, sheep and man receive long-wave radiation from the lawn as well as from the sky so their net loss of long-wave radiation is less than the net loss from the lawn.

Measurement

Measuring the net exchange of radiation at the surface of a uniform plane poses no problems: a net radiometer is exposed with the surface of its thermopile parallel to the plane. For more complex surfaces, the net flux can be derived from an application of Green's theorem which states that the flux of any quantity received within or lost by a defined element of space is the integral of the flux evaluated at right angles to the envelope which defines the space. This principle has been applied to measure the net radiant exchange of apple trees with tubular net radiometers defining the surface of a cylinder and with thermopile surfaces parallel to the surface of the cylinder (Thorpe, 1978). Similarly, Funk (1964) measured the net radiant flux received by a man standing inside a vertical cylindrical framework over which a single net radiometer moved on spiral guides, always pointing towards the axis of the cylinder.

7

MOMENTUM TRANSFER

When plants or animals are exposed to radiation, the energy which they absorb can be used in three ways: for heating, for the evaporation of water, and for photochemical reactions. Heating of the organism itself or of its environment implies a transfer of heat by conduction or by convection; evaporation involves a transfer of water vapour molecules in the system and photosynthesis involves a similar transfer of carbon dioxide molecules. At the surface of an organism, heat and mass transfer are sustained by molecular diffusion through a thin skin of air known as a **boundary layer** in contact with the surface. The behaviour of this layer depends on the viscous properties of air and on the transfer of momentum associated with viscous forces. A discussion of momentum transfer is therefore needed as background to the following three chapters which consider different aspects of exchange between organisms and their environment.

BOUNDARY LAYERS

Figure 7.1 shows the development of a boundary layer over a smooth flat surface immersed in a moving fluid (i.e. a gas or liquid). When the streamlines of flow are almost parallel to the surface the layer is said to be laminar and the flow of momentum across it is affected by the momentum exchange between individual molecules discussed on p. 18. The thickness of a laminar boundary layer cannot increase indefinitely because the flow becomes unstable and breaks down to a chaotic pattern of swirling motions called a turbulent boundary layer. A second laminar layer of restricted depth—the laminar sublayer—forms immediately above the surface and below the turbulent layer.

The transition from laminar to turbulent flow depends on the relative size of inertial forces associated with the horizontal movement of the fluid to viscous forces generated by inter-molecular attraction, sometimes referred to as 'internal friction'. The ratio of inertial to viscous forces is known as the **Reynolds number** (Re) after the physicist who first showed that this ratio determined the onset of turbulence when a liquid flowed through a pipe. When Re is small, viscous forces predominate so that the flow tends to remain laminar but when the ratio increases beyond a critical value Re_c, inertial forces take charge and the flow becomes turbulent.

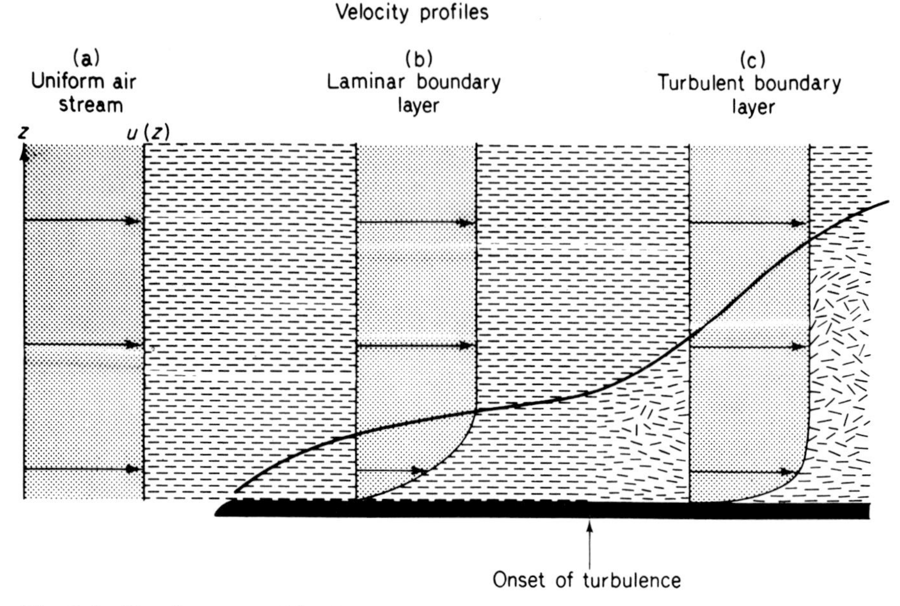

Fig. 7.1 Development of laminar and turbulent boundary layers over a smooth flat plate (the vertical scale is greatly exaggerated).

The general form of Re is Vd/ν where V is the fluid velocity, d is an appropriate dimension of the system, and ν is the coefficient of kinematic viscosity of the fluid. For flow over a flat plate, d is the distance from the leading edge. When a very smooth flat plate is exposed to a parallel flow of air virtually free from turbulence, Re_c is of the order of 10^6 but the engineering literature quotes values of about 2×10^4 observed in less rigorous conditions.

Both in laminar and in turbulent boundary layers, velocity increases from zero at the surface to the free-stream value V at the top of the boundary layer, but definitions of boundary layer depth are necessarily arbitrary. One definition is the streamline where the velocity is $0.99V$, but in problems of momentum transfer it is more convenient to work with an average boundary layer depth.

In Fig. 7.2, the flow of air reaching a flat plate across a vertical cross-section of depth h is proportional to the velocity V times h. At a distance l from the edge of the plate, the velocity profile is represented by the line ABC and the flow through the cross section at C will be less than Vh because the velocity in the boundary layer is less than V. The same reduction in flow would be produced by a layer of completely *still* air with thickness δ (shaded) above which the air moves with a uniform velocity V. The velocity profile in this equivalent system is represented by ADFC. The depth δ is known as the 'displacement boundary layer' and can be regarded as an average depth of the boundary layer between the leading edge of the plate and the cross section at C.

By applying the principle of momentum conservation to flow within a

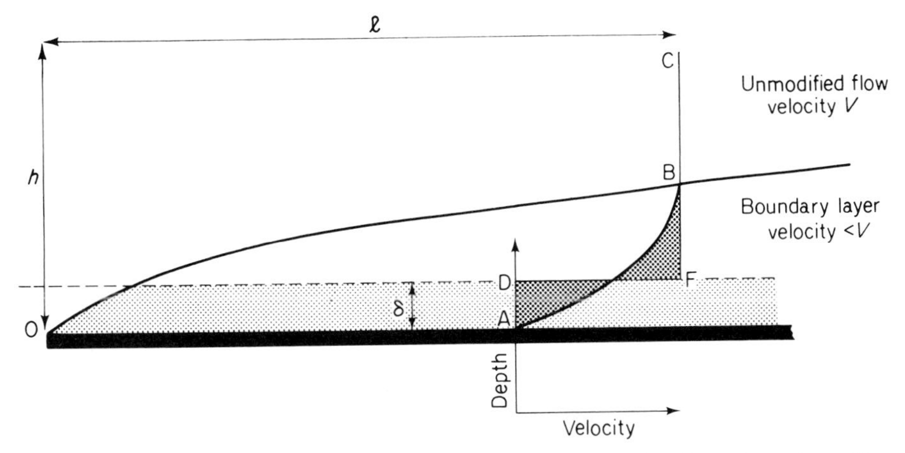

Fig. 7.2 Boundary layer OB, displacement boundary layer (grey) and wind profile (CBA) over a smooth flat plate exposed to an airstream with uniform velocity V.

laminar boundary layer over a smooth plate, it can be shown that the depth of the displacement boundary layer expressed as a fraction of the distance l from the leading edge is

$$\delta/l = 1.72\,(\mathrm{Re})^{-0.5} \tag{7.1}$$

implying that δ increases with $\sqrt{l^{0.5}}$. (For an estimate of the actual boundary layer depth at l as distinct from an average depth over a distance l, the numerical factor in this equation can be replaced by 5.) The depth of a *turbulent* boundary layer increases with $l^{0.8}$.

Skin friction

The force that air exerts on a surface in the direction of the flow is a direct consequence of momentum transfer through the boundary layer and is known as **skin friction**. To establish analogies with heat and mass transfer later, the transfer of momentum is conveniently treated as a process of diffusion (see p. 18). If t is a diffusion path length for momentum transfer from air moving with velocity V to a surface where it is zero, then the frictional force is

$$\tau = \nu\rho V/t = \rho V/r_{\mathrm{M}} \tag{7.2}$$

when $r_{\mathrm{M}} = t/\nu$ is a resistance for momentum transfer. From theoretical analysis for the flow over a smooth plate, the drag per unit surface area is proportional to $V^{3/2}$ and is given by

$$\tau = 0.66\rho V(V\nu/l)^{0.5} \tag{7.3}$$

Comparison of equations (7.2) and (7.3) then gives the resistance as

$$r_{\mathrm{M}} = 1.5(l/V\nu)^{0.5} = 1.5V^{-1}\,\mathrm{Re}^{0.5} \tag{7.4}$$

For example, if $V = 1$ m s^{-1} and $l = 0.05$ m, then $r_{\mathrm{M}} = 90$ s m^{-1}, establishing an order of magnitude relevant to many micrometeorological problems.

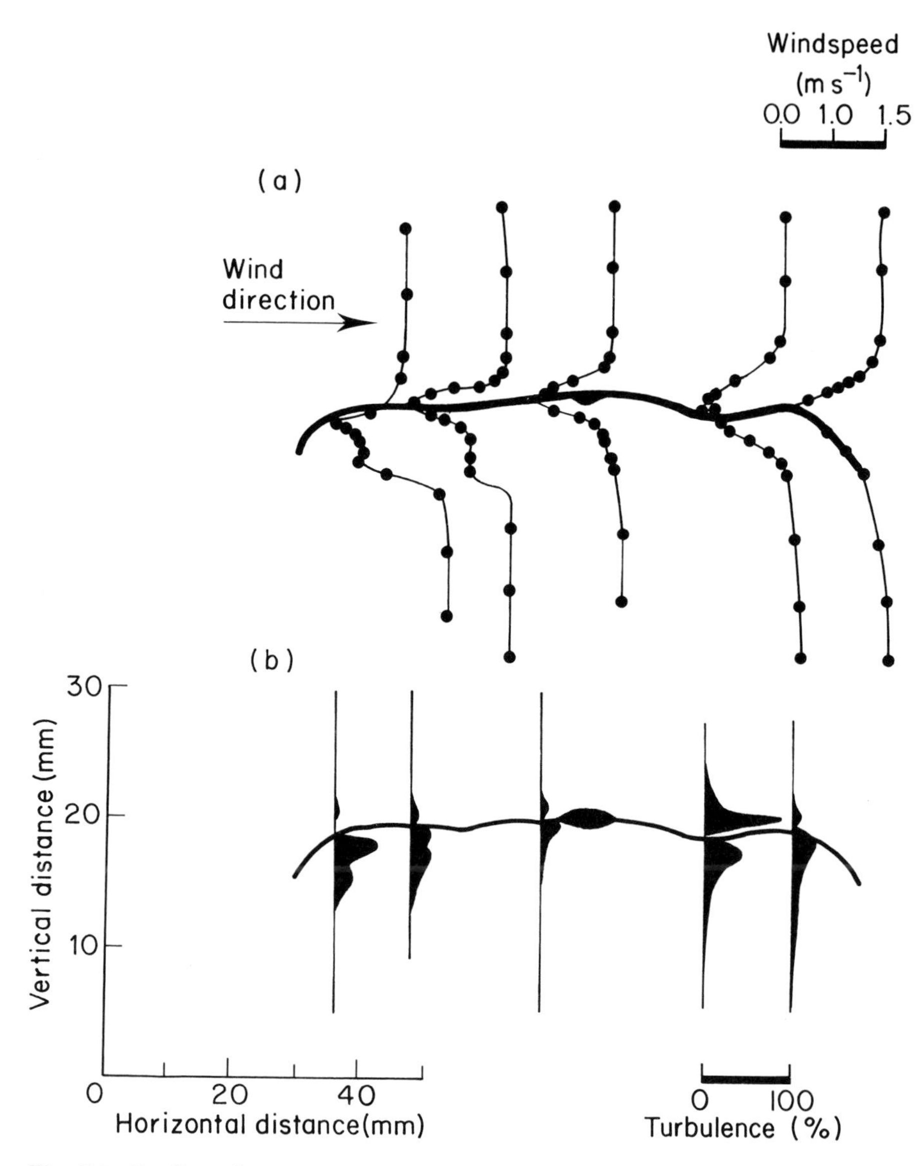

Fig. 7.3 Profiles of mean windspeed (a) and turbulence intensity i (b) around a *Populus* leaf shown in transverse section in a laminar free stream (from Grace and Wilson, 1976).

Over natural surfaces, the flow of air is usually much more complex than Fig. 7.1 suggests, but for a poplar leaf parallel to a laminar flow of air in a wind-tunnel (Fig. 7.3) the profiles of windspeed developed on the upper surface were equivalent to those in Fig. 7.1. At the lower surface, profiles were distorted by curvature of the leading edge which produced shelter but also generated turbulence in its wake. When V was 1.5 m s^{-1} or with minimum turbulence in the airstream, windspeed within a few millimetres of the upper surface was close to the theoretical value for a flat plate upwind of the midrib. The value of Re_c for the onset of turbulence was about 9×10^3. Increasing the velocity or the level of turbulence in the airstream increased the velocity at a fixed point in the boundary layer relative to V and decreased Re_c to 1.9×10^3. These observations are relevant not only to the exchange of momentum and related forces on the surfaces of leaves but also to the forces responsible for detaching fungal spores (Grace and Collins, 1976).

Form drag

In addition to the force exerted by skin friction, a consequence of momentum transfer to a surface across the streamlines of flow, bodies immersed in moving fluids experience a force in the direction of the flow as a result of the deceleration of fluid. This force is known as **form drag** because it depends on the shape and orientation of the body. Maximum form drag is experienced by surfaces at right angles to the fluid flow, and the force can be estimated by assuming that there is a point on the surface where the fluid is instantaneously brought to rest after being uniformly decelerated from a velocity V. If the initial momentum per unit volume of fluid is ρV and the mean velocity during deceleration is $V/2$, the rate at which momentum is lost from the fluid is $\rho V \times V/2 = 0.5\rho V^2$. This is the maximum rate at which momentum can be transferred to unit area on the upstream surface of a bluff body and it is therefore the maximum pressure excess that a fluid can exert in contributing to the total form drag over the body. In practice, fluid tends to slip round the sides of a bluff body so that a force smaller than $0.5\rho V^2$ is exerted on the upstream face. However, in the wake which forms downstream of a bluff body. The total drag force on unit area is conveniently expressed as $c_f.0.5\rho V^2$, where c_f is a drag coefficient.

In most problems, it is appropriate to combine skin friction and form drag to give a total force τ, usually the force on a unit of area projected in the direction of the flow, i.e. $2rl$ for a cylinder of radius r and length l in cross-flow and πr^2 for a sphere. The ratio $\tau/(0.5\rho V^2)$ then defines the total drag coefficient c_d which for spheres and for cylinders at right angles to the flow lies between 0.4 and 1.2 in the range of Reynolds numbers between 10^2 and 10^5. Manufacturers of sleek saloon cars boast drag coefficients of about 0.3.

As background for later discussion of mass and heat transfer, it should be noted that the diffusion of momentum in skin friction is analogous to the diffusion of gas molecules and of heat provided the surface is *parallel* to the airstream. For such a surface, close relations may be expected between r_M, r_H and r_V. For a surface at *right angles* to the air stream however, there is no frictional force in the direction of flow. Friction will operate in all directions at right angles to the flow but the net sum of all these (vector) forces must be zero. In contrast, the net flux of heat or mass, which are scalar quantities, must be finite in the plane of the surface. In this case, r_V and r_H may be similar to each other but will be unrelated to the value of r_M.

Drag on natural surfaces

The atmosphere in motion exerts forces on all natural surfaces—individual leaves, plants, trees, crops, animals, bare soil and open water. Conversely, every object or surface exposed to the force of the wind imposes an equal and opposite force on the atmosphere proportional to the rate of momentum transfer between the air and the surface. Momentum transfer is always associated with wind 'shear': the windspeed is zero at the surface of the object and increases with distance from the surface through a boundary layer of retarded air.

Isolated objects such as single plants or trees tend to have very irregular boundary layers, and disturb the motion of the atmosphere by setting up a train of eddies in their wake rather like the eddies formed downstream from the piers of a bridge. Surfaces such as bare soil and uniform vegetation also generate eddies in the air moving over them because the drag which they exert on the air is incompatible with laminar flow.

Drag on leaves

To avoid the fluctuations of windspeed that are characteristic of the atmosphere the drag on natural objects can be measured in a wind tunnel where the flow is steady and controlled. Thom (1968) studied the force on a replica of a 'leaf' made of thin aluminium sheet. Figure 7.4 shows the dimensions of the replica and Fig. 7.5 shows how the drag coefficient c_d changed with windspeed and direction. Note that the quantity c_d shown in Fig. 7.5 was calculated by dividing the force per unit plan area of the leaf (τ/A) by ρV^2 which is numerically the same as the force per unit of total surface area ($\tau/2A$) divided by $0.5\rho V^2$. This departure from an aerodynamic convention (which uses a projected area) allows the resistance to momentum transfer for the whole leaf to be written as $r_M = 1/(Vc_d)$, equivalent to the combination in parallel of the two resistances of $2/(Vc_d)$ for each surface separately.

When the leaf was oriented in the direction of the airstream ($\varphi = 0$) the drag was minimal and c_d was close to twice the theoretical value derived from equation (7.4) which gives the resistance to momentum transfer for one side of a plate only. When the concave or convex surfaces were facing the airstream ($\varphi = +90$ or $\varphi = -90$ respectively), form drag was much larger than skin friction.

The general form of the curves in Fig. 7.5 is consistent with a combination of form drag proportional to V^2 and skin friction proportional to $V^{1.5}$. The total drag coefficient c_d can be expressed as the sum of a form drag component c_f and a frictional component $nV^{-0.5}$ where n is a constant, i.e.

$$c_d = c_f + nV^{-0.5}$$

The relevance of wind tunnel measurements of leaf drag to momentum transfer in a crop is a matter for debate. In the first place, turbulence is usually suppressed in wind tunnels to achieve laminar flow, whereas the movement of air in canopies is nearly always turbulent. Turbulent eddies of an appropriate size can increase rates of momentum transfer by disturbing flow in the laminar boundary layer, and the disturbance is likely to be

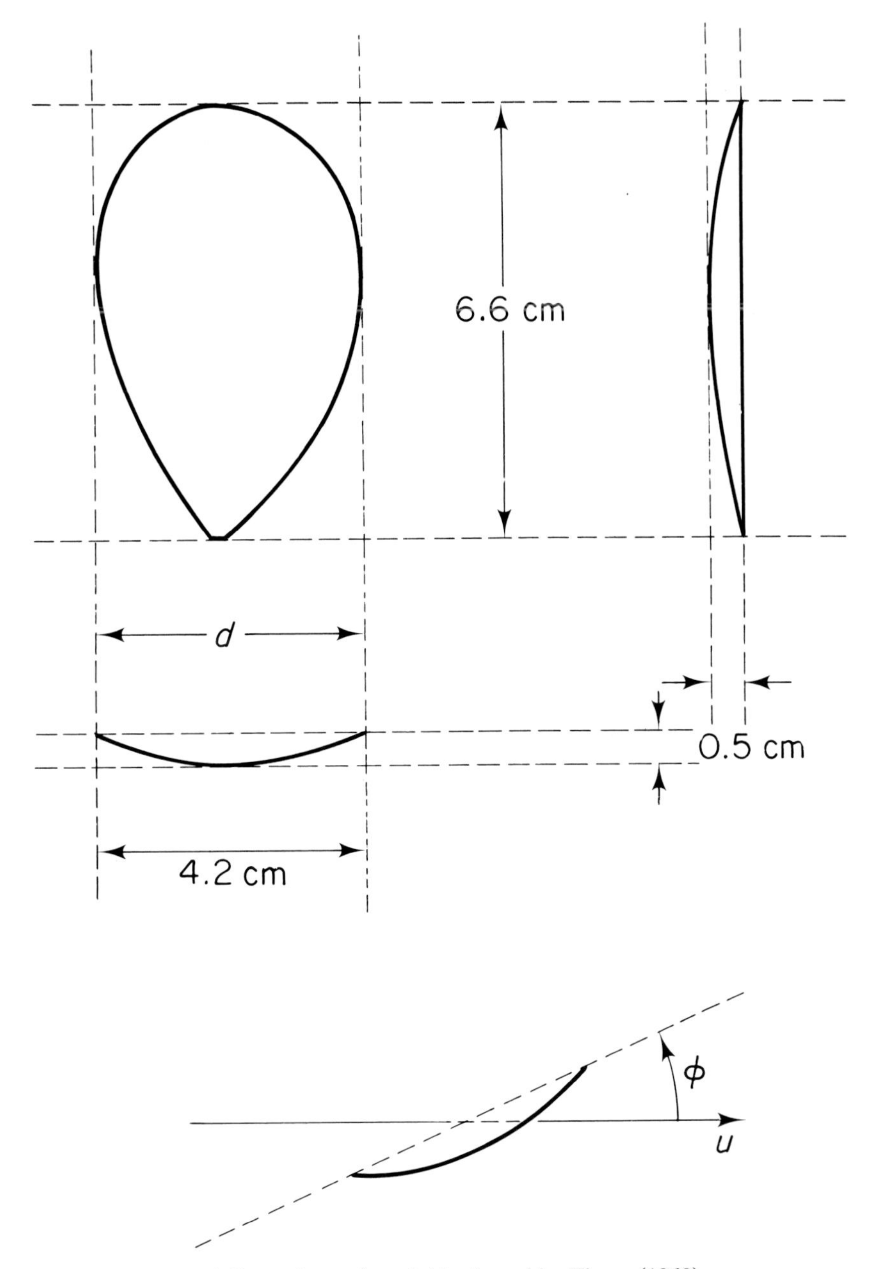

Fig. 7.4 Shape and dimensions of model leaf used by Thom (1968).

accentuated if the turbulence is strong enough to make the leaves flutter. Measurements by Rashke indicate that the characteristic diameter of eddies shed by cylinders of diameter d is about $5d$ when Re exceeds 200. The eddies shed by leaves and stems decay to form smaller and smaller eddies drifting

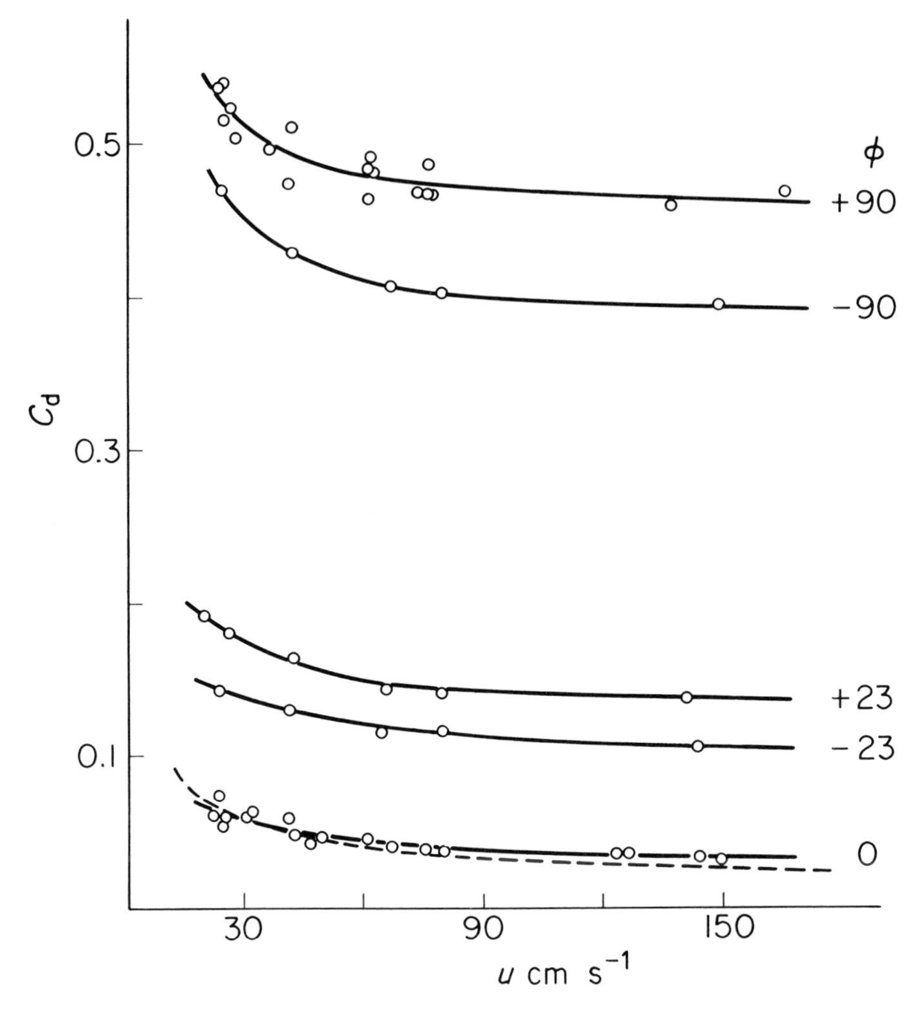

Fig. 7.5 Drag coefficient of a model leaf (full lines) as a function of windspeed u and angle between leaf and airstream (see Fig. 7.4). The broken curve is a theoretical relationship for a thin flat leaf at $\phi = 0$ (from Thom, 1968).

downwind and penetrating the boundary layer of other leaves and stems.

The drag on real leaves may also be increased by the roughness of the cuticle and by the presence of hairs. Sunderland (1968) found that the drag on the aluminium replica of a wheat leaf increased when a real leaf was attached to the metal surface. The increase was about 20% at 1.5 m s^{-1} growing to 50% at 0.5 m s^{-1} because of the increasing importance of skin friction at low windspeeds.

In the real world, leaves rarely exist in aerodynamic isolation and the drag coefficient for foliage therefore depends on its density as well as on windspeed. A convenient way of specifying density in this context is the ratio of the total surface area of laminae, petioles, etc., in a specimen, divided by the plan area facing the wind. In the canopies of arable crops or of a deciduous tree, most leaves are exposed to turbulent air in the wake of their upwind neighbours and in coniferous trees there is similar interference between needles. The extent to which the drag on individual elements of foliage is reduced by the presence of neighbours was expressed by Thom (1971) in terms of a **shelter factor**, the ratio of the actual drag coefficient observed to the coefficient measured (or estimated) for the same element in isolation. Shelter factors depend on foliage density as well as on windspeed. Representative values found in the literature range from 1.2 to 1.5 for shoots of apple trees to 3.5 for a stand of field beans and for a pine forest.

Even without the complications of shelter, the dependence of drag on windspeed is very complex because of the interaction between aerodynamic forces on leaves and elastic forces opposing them. As windspeed increases from zero, fluttering begins, even at a constant windspeed, if the two sets of forces cannot achieve equilibrium and the consequence is a sudden increase of momentum transfer and of drag. With a further increase of windspeed, many types of leaf tend to bend into a streamlined position, reducing drag and eliminating flutter when the windspeed is constant. In nature, however, wind fluctuates continuously both in speed and in direction so that leaves, stems and small branches oscillate even in quite light winds.

Drag on particles

Particles in the atmosphere experience drag forces when there is relative motion between the air and the particle. Such forces are important to allow air movements to detach the pollen of some plants; or to transport spores of disease pathogens from surfaces into the atmosphere; and for determining whether airborne particles are captured at the surfaces of vegetation.

When a particle is in steady motion the drag force exerted on it by the air must balance the other external forces such as gravity, electrical attraction etc. The origins of the drag forces on particles can be conveniently treated as three special cases applied to spherical particles:

(i) Particles with radius r much smaller than the mean free path λ of gas molecules.

In this case particles behave like giant gas molecules, and the drag force is the result of more molecules impinging on the surface of a moving particle from in front than from behind.

Provided that the mass of a particle is much larger than the mass m_g of the gas molecules and assuming perfect reflection in collisions, kinetic theory can be used to show that the drag force on a particle moving at velocity V in a gas is

$$F = \frac{4}{3}\pi n m_g \bar{c} r^2 V \tag{7.5}$$

where n is the number of gas molecules per unit volume, and $\bar{c}$ is their mean velocity. Thus the drag force is proportional to particle velocity and to the surface area (r^2).

(At 20°C, there are about 3×10^{25} molecules m^{-3}, the average value of m_g for air is about 5×10^{-26} kg, and $\bar{c}$ is about 500 m s^{-1}. The mean free path of molecules in air is about 0.1 μm, so that this case applies for very small particles with radii less than 0.01 μm.)

(ii) Particles with radius r larger than λ, but where the Reynolds number Re_p for relative motion between gas and particle is small (i.e. $2rV/\nu$ less than about 0.1).
In this case the viscous forces arising from skin friction between the particle and the gas dominate the drag. Stokes showed that for this case

$$F = 6\pi\rho\nu rV \tag{7.6}$$

Thus the drag force if proportional to particle velocity and to radius. (The kinematic viscosity ν of air is about 15×10^{-6} m^2 s^{-1}, and so Stokes' Law applies to 10 μm radius particles typical of pollen grains when $V \times 2 \times 10 \times 10^{-6}/(15 \times 10^{-6}) < 0.1$, i.e. $V < 0.08$ m s^{-1}.)

(iii) Particles with $r > \lambda$ and $Re_p > 1$
In this case the inertial forces corresponding to the form drag as the gas flows around the particle dominate the drag. The drag force is given by

$$F_m = c_d \cdot 0.5\rho V^2 A \tag{7.7}$$

where A is conventionally taken as the cross-sectional area of the particle and ρ is the density of the gas. The drag coefficient of a sphere as a function of Reynolds number is shown in Fig. 7.6.

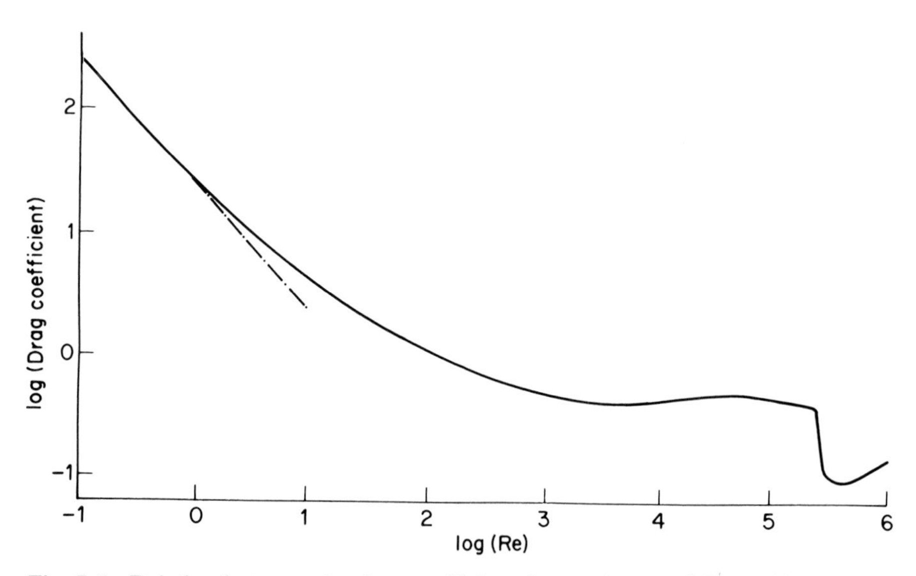

Fig. 7.6 Relation between the drag coefficient for a sphere and Reynolds number. The pecked line is based on Stokes' Law. The discontinuity at $Re \simeq 3 \times 10^5$ corresponds to the transition from a laminar to a turbulent boundary layer (Mercer, 1973).

At small values of $\mathrm{Re_p}$, equating equations (7.6) and (7.7) shows that

$$c_\mathrm{d} = \frac{6\pi\nu\rho r V}{0.5\rho V^2 \pi r^2} = \frac{12\nu}{Vr} = \frac{24}{\mathrm{Re_p}} \quad (7.8)$$

and this reciprocal relation gives the linear section on Fig. 7.6 when $\mathrm{Re_p} < 0.1$. The empirical expression

$$c_\mathrm{d} = (24/\mathrm{Re_p})(1 + 0.17\ \mathrm{Re_p}^{0.66}) \quad (7.9)$$

(Fuchs, 1964) is accurate within a few percent for $1 < \mathrm{Re_p} < 400$; it overestimates c_d by about 7% when $\mathrm{Re_p} \simeq 0.5$. In the fully turbulent region when $\mathrm{Re_p} > 10^3$, c_d is constant and so drag force is proportional to V^2 and to cross-sectional area.

Aylor (1975) used these principles to calculate the force required to detach spores of the pathogen *Helminthosporium maydis* from infected leaves of maize. This fungal pathogen produces roughly cylindrical spores with diameter about 20 μm and projected area about 4×10^{-10} m^2 which grow on stalks projecting 150 μm from the leaf surface. To find the minimum force necessary to detach a spore from its stalk, Aylor observed spores through a miscroscope while dry air was blown on them from a narrow tube. Fig. 7.7 shows that 50% of the spores were removed when the windspeed was about 10 m s^{-1}. The corresponding drag force may be found from equation (7.7) which, taking $\rho = 1.2$ kg m^{-3}, $A = 4 \times 10^{-10}$ m^2 and $c_\mathrm{d} = 4$ for a cylinder at

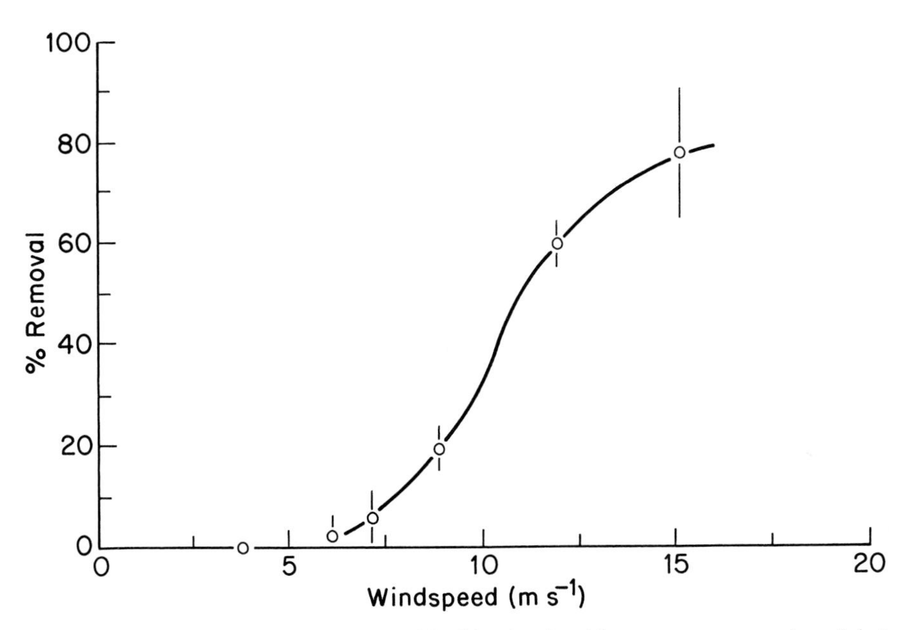

Fig. 7.7 Percentage of spores removed by blowing for 15 s on spores reared on dried plant material (from Aylor, 1975). Vertical bars represent standard deviations.

$\mathrm{Re_p} \simeq 10$ (similar to the value for a sphere, Fig. 7.6), yields $F_m \simeq 1 \times 10^{-7}$ N. Aylor also used a centrifugal method of detaching spores, and this gave excellent agreement with his estimates of F_m. Spores are detached at much lower *mean* windspeeds than this in the field, indicating the importance of brief gusts in breaking down the leaf boundary layer and dispersing pathogens into the atmosphere.

WIND PROFILES AND DRAG ON UNIFORM SURFACES

In Chapter 3, it was shown that the vertical flux of an entity s can be represented by the mean value of the product $\rho w's'$ which is finite when fluctuations of vertical velocity w' are correlated, either positively or negatively, with simultaneous fluctuations of the entity s'. When the entity is horizontal momentum, s' becomes the fluctuation of the horizontal velocity u' and the vertical flux of horizontal momentum is given by $\overline{\rho u'w'}$. Provided the vertical flux is constant with height, this quantity can be identified as the force per unit ground area, otherwise known as the shearing stress (τ). Moreover, if the intensity of turbulence is the same in horizontal and vertical directions, $u' = w'$ so that

$$\tau = \rho u'w' = \rho u_*^2 \tag{7.10}$$

where u_*, known as the **friction velocity**, is given by

$$u_* = \overline{(u'^2)}^{0.5} = \overline{(w'^2)}^{0.5}$$

The shearing stress may also be written as

$$\tau = \rho K_M du/dz \tag{7.11}$$

where K_M is a turbulent transfer coefficient for momentum with dimensions $L^2\,T^{-1}$. From experience, both windspeed and turbulent mixing increase with height, and a simple dimensional argument will now be used to obtain the functional relations.

The simplest assumption for the height dependence of K is

$$K_M = az \tag{7.12}$$

where a is a quantity with the dimensions of velocity. Similarly, the simplest assumption for the wind gradient is

$$du/dz = b/z \tag{7.13}$$

where b is a second constant with dimensions of velocity. Putting equations (7.12) and (7.13) in equation (7.11) gives

$$\tau = \rho ab \tag{7.14}$$

implying that $ab = u_*^2$. Furthermore, because a and b are both velocities, equation (7.14) implies that

$$a = ku_*;\ b = u_*/k$$

where k is a constant. Substitution in equations (7.12) to (7.14) now gives

$$K_m = ku_*z \tag{7.15a}$$

and

$$du/dz = u_*/(kz) \tag{7.15b}$$

$$\tau = \rho u_*^2 \tag{7.15c}$$

Integration of equation (7.15b) yields

$$u = (u_*/k) \ln (z/z_0) \tag{7.16}$$

where z_0 is a constant termed the **roughness length** such that $u = 0$ when $z = z_0$.

This is the logarithmic wind-profile equation, found to be valid over many types of uniform surface provided the conditions listed on p. 233 are satisfied. Measurements confirm that k is a constant—the **von Karman constant** after a famous aerodynamicist—and it is usually assigned a value of 0.41, as determined by experiment.

A more general form of equation (7.16) takes account of the height of the surface elements by assuming the existence of a **zero plane** at a height d such that the distribution of shearing stress over the elements is aerodynamically equivalent to the imposition of the entire stress at height d. The wind profile equation then becomes

$$u = (u_*/k) \ln \{(z-d)/z_0\} \tag{7.17}$$

and equation (7.15a) becomes

$$K_M = ku_*(z-d) \tag{7.18}$$

These equations are valid in the air above the surface elements where the assumption that shearing stress is independent of height is satisfied. Because

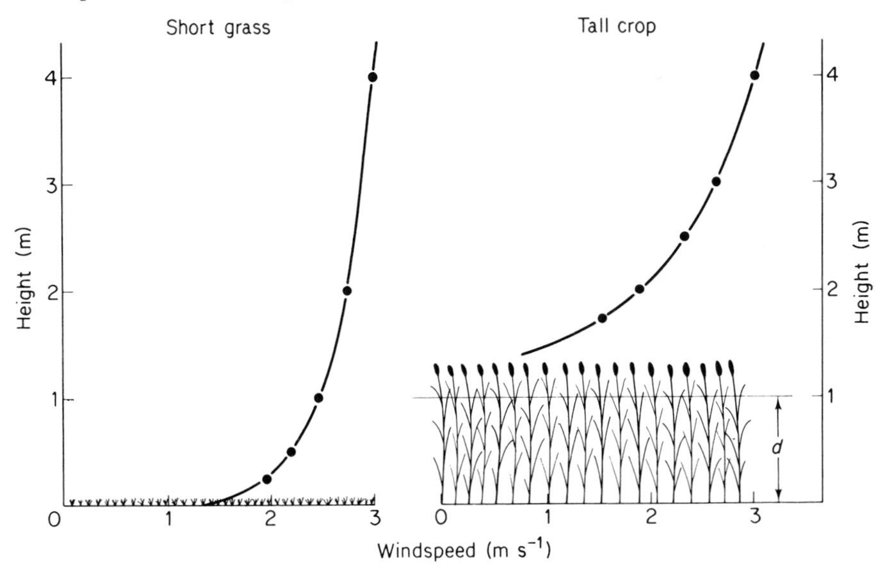

Fig. 7.8 Wind profiles over short grass and a tall crop when the windspeed at 4 m above the ground is 5 m s^{-1}. The filled circles represent hypothetical measurements from sets of anemometers.

of the absorption of momentum below the tops of the elements, the equation is not valid in this region, nor is it valid immediately above very tall vegetation where the relation between flux and gradient defined by equation (7.11) breaks down (see Chapter 14).

It is nevertheless possible to use equation (7.17) to obtain an extrapolated value of u which tends to zero at a height given by $z = z_0 + d$. To recap, d is an equivalent height for the absorption of momentum (a 'centre of pressure') and $(d + z_0)$ is an equivalent height for zero windspeed.

To demonstrate the relation between windspeed and height for contrasting surfaces, Fig. 7.8 contains profiles over short grass ($d = 0.7$ cm, $z_0 = 0.1$ cm) and a tall crop ($d = 95$ cm, $z_0 = 20$ cm) when the windspeed at 4 m is 3 m s^{-1}. Figure 7.9 shows the equivalent logarithmic plots of u as a function of $\ln(z - d)$. Because height appears (by convention) on the vertical axis, the slope is k/u_* giving values of $u_* = 0.15$ and 0.46 m s^{-1} for the grass and tall

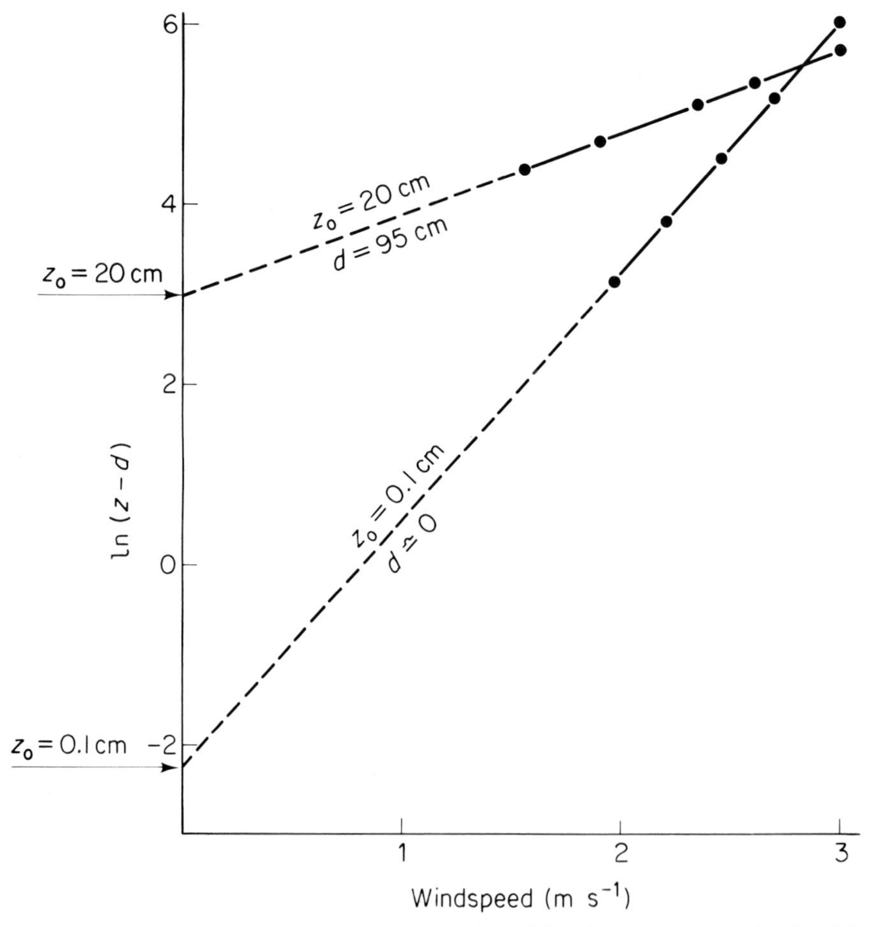

Fig. 7.9 Relations between windspeed and $\ln(z - d)$ for the wind profiles in Fig. 7.8.

crop respectively: $\boldsymbol{\tau} = 0.027$ and $0.25\ \mathrm{N\ m^{-2}}$: and 0.25 and $0.58\ \mathrm{m^2\ s^{-1}}$ for the transfer coefficient K_M at a height of 4 m.

When windspeed is measured over heights much larger than d, this type of analysis requires values u_1 and u_2 at a minimum of two heights z_1 and z_2 so both u_* and z_0 can be eliminated initially to give

$$\ln z_0 = (u_2 \ln z_1 - u_1 \ln z_2)/(u_2 - u_1) \qquad (7.19)$$

When the value of d is significant, at least three heights are needed so that both u_* and z_0 can be eliminated initially to give

$$\frac{u_1 - u_2}{u_1 - u_3} = \frac{\ln(z_1 - d) - \ln(z_2 - d)}{\ln(z_1 - d) - \ln(z_3 - d)} \qquad (7.20)$$

This equation allows d to be found either by iteration in a computer program or graphically by plotting the right hand side as a function of d.

Behaviour of z_0 and d

The dependence of z_0 and d on vegetation height and structure has been

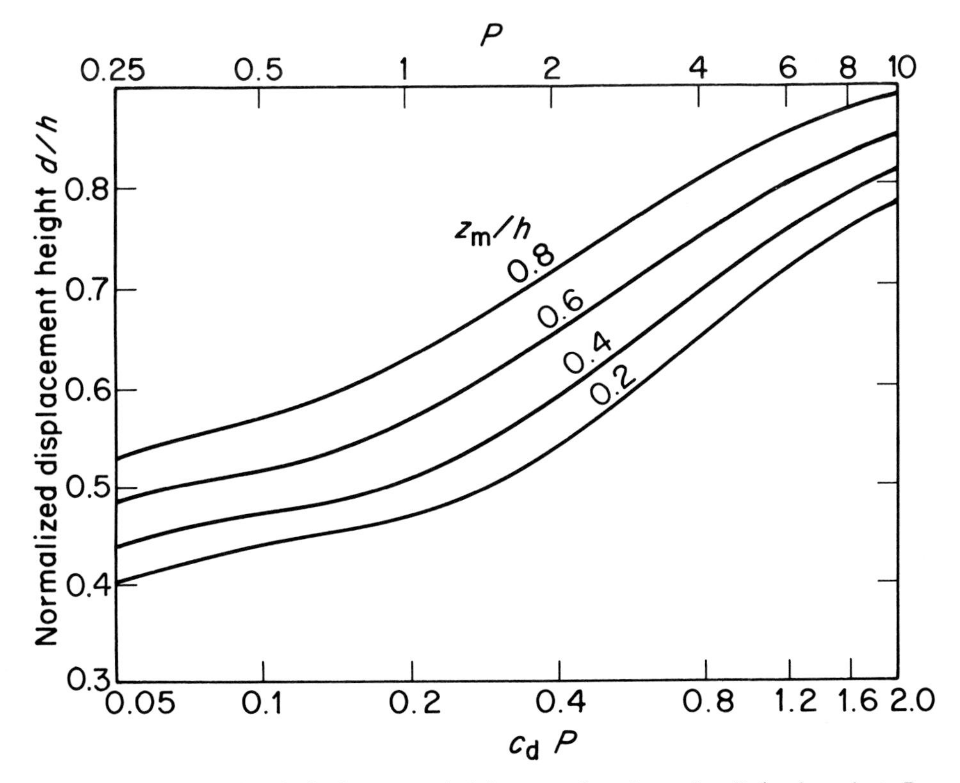

Fig. 7.10 Normalized displacement height as a function of c_dP (and against P assuming that $c_d = 0.2$). Curves are labelled according to the height at which density reaches a maximum (from Shaw and Pereira, 1982).

examined both experimentally and theoretically. Shaw and Pereira (1982) found that attempts to explain the exchange of momentum within canopies in terms of traditional mixing-length models were unsatisfactory, and they developed second-order equations for turbulent mixing. In their numerical model, leaf area index was replaced by a plant area index (P) and the area per unit height was assumed to increase linearly with height from zero at the top of the canopy to a maximum at a height z_m, below which it decreased linearly to the surface. By assuming a uniform drag coefficient c_d within the canopy, and by computing wind profiles, they were able to predict how z_0/h and d/h should depend on the parameter c_dP.

Figure 7.10 shows that, as c_dP increased, the ratio d/h also increased and for representative values of $c_d = 0.5$ and $z_m/h = 0.5$, d/h was between 0.5 and 0.7, consistent with field experience. The dependence of z_0/h on c_dP was more complex. When there are relatively few roughness elements per unit ground

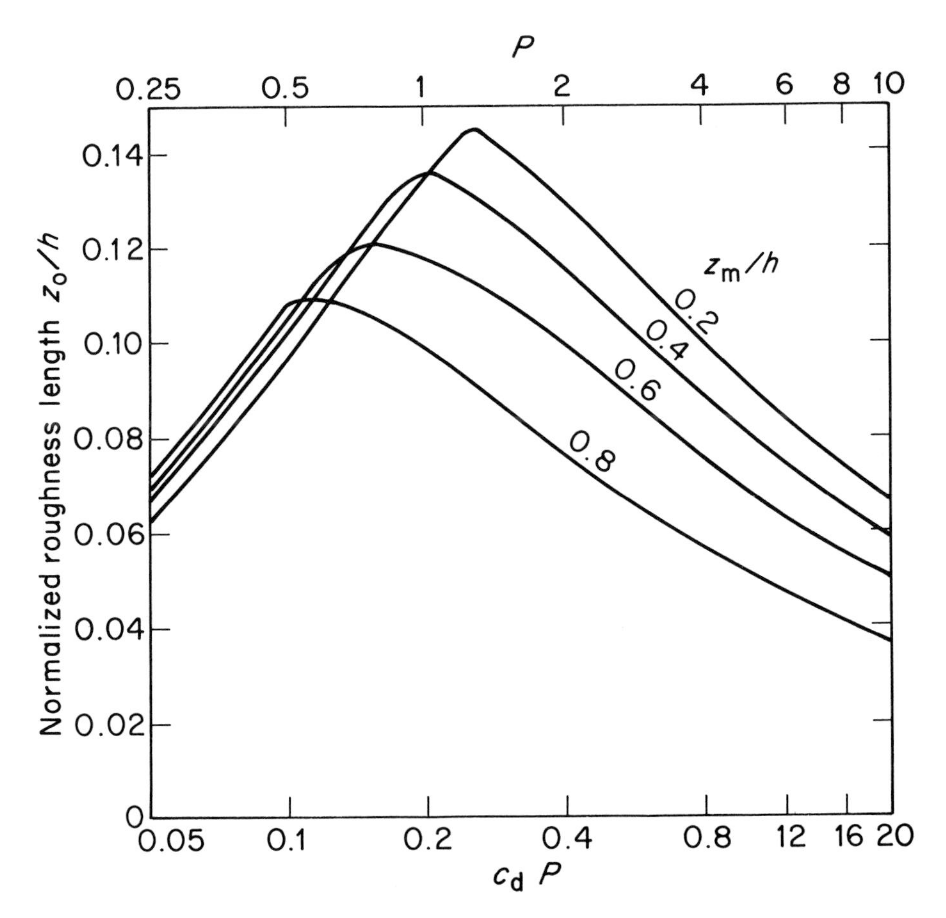

Fig. 7.11 Normalized roughness length as a function of c_dP (and against P assuming that $c_d = 0.2$). Curves are labelled according to the height at which density reaches a maximum (from Shaw and Pereira, 1982).

area, any increase in number increased drag but had relatively little effect on d so z_0/h increased (Fig. 7.11). With a further increase of P, however, a point was reached where an increase in the area of roughness elements tending to increase drag was offset by an increase in the height of the zero plane which reduced the effective depth of the canopy for momentum exchange. Beyond this point, z_0/h decreased as $c_d P$ increased (Fig. 7.11).

As a corollary, when vegetation is *sparse*, drag is greatest when most plant material is near the top of the canopy (z_0/h large) but *dense* vegetation is least rough when z_0/h is large because the canopy then presents a relatively smooth surface to the air passing over it. Below the point of maximum roughness, the value of z_0/h predicted by the model is approximately $0.29\ (1-d/h)$, consistent with measurements, but above the maximum, z_0/h depends on $c_d P$ and on z_m/h as well as on d/h.

Combining field evidence with predictions from the model, when the maximum density of foliage is approximately at half the height of the canopy, z_0/h is expected to be between 0.08 and 0.12 and d/h between 0.6 and 0.7. However, both ratios depend to some extent on windspeed, and for flexible stands of cereals there are many reports that they decrease with windspeed as a consequence of three factors working in the same direction:

(i) decrease in the drag coefficient of individual leaves at a fixed angle to the wind (Fig. 7.5);
(ii) decrease in the drag coefficient of leaves moving into a more streamlined position;
(iii) decrease in the drag coefficient of the whole canopy as stems bend.

Legg *et al.* (1981) found that small changes of d/h and z/h with windspeed were not statistically significant for field beans but, for potatoes at the beginning of the season, d/h decreased with windspeed.

Over surfaces such as sand and water, the force exerted by wind can detach elements which carry momentum upwards into a thin layer of air immediately above the surface before falling back again. In a classic study of the movement of sand in the desert, Bagnold (1941) suggested that the initial vertical velocity of a detached sand grain should be proportional to the friction velocity u_*, a measure of the mean vertical velocity of eddies (p. 112). A grain moving upwards with an initial velocity u_* will come to rest at a height u_*^2/g where g is the gravitational acceleration. It can therefore be argued on dimensional grounds that the depth z_0 of the roughness layer within which horizontal momentum is absorbed should be proportional to u_*^2/g. Chamberlain (1983) pointed out that the relation

$$z_0 = 0.016u_*^2/g \tag{7.21}$$

appears to be valid for sand, snow and sea.

Aerodynamic resistance

The equation expressing momentum flux in terms of the gradient of horizontal momentum per unit volume can be rewritten in the general form of Ohm's Law by introducing an aerodynamic resistance to momentum transfer between heights z_1 and z_2 where windspeeds are u_1 and u_2. Then if

$$\tau = \rho(u_2 - u_1)/r_{aM} \tag{7.22}$$

the resistance can be evaluated as

$$r_{aM} = (u_2 - u_1)/u_*^2 = \ln\{(z_2 - d)/(z_1 - d)\}/ku_* \tag{7.23}$$

The resistance between a single height where the windspeed is $u(z)$ and the level $d + z_0$ where the extrapolated value of u is zero can be written in several equivalent forms, e.g.

$$r_{aM} = \frac{u(z)}{u_*^2} = \frac{\ln\{(z-d)/z_0\}}{ku_*} = \frac{\ln\{(z-d)/z_0\}^2}{k^2u(z)} \tag{7.24}$$

These resistance equations, like all others in the chapter, pertain to momentum transfer in adiabatic conditions, i.e. when the lapse rate of temperature above the ground is equal to the adiabatic lapse rate (p. 7). A discussion of the more general non-adiabatic case is deferred to Chapter 14.

LODGING AND WINDTHROW

Lodging of grass and cereal crops occurs when plant stems are unable to withstand the combined forces of wind and of gravity on the foliage and on the ear. Lodging is most likely to occur when the weight of the upper parts of the plants is increased by the interception of rain or when the lower parts of the stems are weakened by disease or by a heavy application of nitrogenous fertilizer. Very little information is available in the literature about the forces needed to bend the plant stems beyond this elastic limit but Tani (1963) provided figures for rice which indicate the scale of forces which are likely to lodge other cereal crops of similar height, i.e about 1 m.

Tani found that when mature rice plants were exposed to a uniform wind in the laboratory, the stems broke when the forces on the plants produced a moment of about 0.2 N m on the base of the stems. Plants growing to a height of about 0.84 m in the field were expected to lodge when the windspeed exceeded 20 m s^{-1}. At this speed, the moment on the base of the stem had two components: a moment of 0.034 N m induced by the force of the wind (mainly form drag), and a moment of 0.023 N m induced by the force of gravity acting because the top of the stem was displaced by about 40 cm. The total moment needed to break the stem under field conditions was therefore 0.057 N m, about a quarter of the laboratory value. The discrepancy between the figures can be explained by (a) the much larger forces exerted in the field during strong gusts when the instantaneous windspeed can be two or three times larger than the mean, or (b) a resonance set up between the natural period of oscillation of the plants (about 1 s) and the dominant period of turbulent eddies at the top of the canopy, or (c) the effect of disease on field-grown plants.

Drag on trees

Trees are uprooted in gales when the total bending moment exerted by the

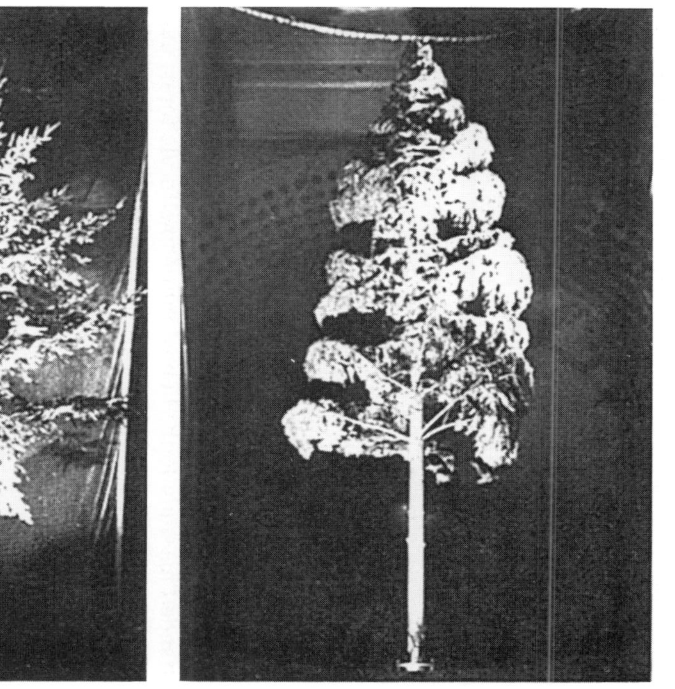

Fig. 7.12 A specimen of Douglas Fir (*Pseudotsuga menziesii*) mounted in a wind tunnel at the National Physical Laboratory. When the windspeed was increased from zero (left-hand) to 26 m s^{-1} (right-hand), there was a marked decrease in the area of foliage presented to the airstream and hence in the drag coefficient (p. 105). (Photographs by A.I. Fraser, reproduced by permission of the Forestry Commission.)

trunk on the root system exceeds the maximum restoring moment that the surrounding soil can exert on the roots. Fraser (1962) exposed fully grown conifers in a wind tunnel with a diameter of 7 m and found that the relation between drag force and windspeed was strongly affected by a decrease in the effective cross section of the crown as the wind got stronger, a result of streamlining by individual leaves as well as by whole branches (Fig. 7.12). Whereas, for rigid objects, the drag at high windspeeds is almost proportional to V^2, the drag on specimens of four species was found to increase almost linearly with velocity, an important result of streamlining.

The drag coefficient c_d was defined as the total drag force divided by $0.5\rho V^2 A_0$ where A_0 was the area of the crown in still air projected in the direction of the airstream. With this convention, c_d at 15 m s^{-1} ranged from 0.57 for spruce (*Picea abies*) to 0.25 for Western hemlock (*Tsuga heterophylla*).

As a test of the validity of the wind tunnel measurements in natural conditions, forest trees were uprooted with a hand winch using a dynamometer to measure the force exerted. Specimens of Douglas fir (*Pseudotsuga menziesii*) growing to sixty feet (18 m) were uprooted by moments of 5 to 8×10^4 N m with respect to the base of the trees. In the wind tunnel, a windspeed of 25 m s^{-1} was needed to produce this moment. Later, similar trees were uprooted by a gale in which the mean gust speed was close to 25 m s^{-1}.

Having established a valid relation between drag and wind velocity for real trees, Fraser extended his study to the behaviour of small scale forest models and the measurement of bending moments on model trees in different patterns of planting.

8

HEAT TRANSFER

Three mechanisms of heat transfer are important in the environment of plants and animals: *radiation*, governed by principles already considered in Chapter 2; *convection*, which is the transfer of heat by moving air; and *conduction* in solids and still gases which depends on the exchange of kinetic energy between molecules. Two types of convection are important in micrometeorology: 'forced' convection or transfer through the boundary layer of a surface exposed to an airstream, proceeding at a rate which depends on the velocity of the flow—a process analogous to skin friction; and 'free' convection which depends on the ascent of warm air above heated surfaces or the descent of cold air beneath cooled surfaces, both a result of differences in air density.

All these mechanisms of transfer are exploited in domestic heating systems. Fan heaters distribute hot air by forced convection; convector heaters and hot water 'radiators' circulate warm air by free convection; underfloor heating depends on the conduction of heat from cables buried below the floor; and the conventional bar-type radiator loses heat both by convection and radiation.

Convection, conduction and systems of mixed heat transfer will be considered in turn.

CONVECTION

The analysis of convection is greatly simplified by using non-dimensional groups of quantities and a short description of these groups is needed to introduce a comparison of the convective heat loss from objects of different size and shape.

Non-dimensional groups

When a surface immersed in a fluid loses heat through a laminar boundary layer of uniform thickness δ, the heat transfer per unit area (**C**) can be written as

$$\mathbf{C} = k(T_s - T)/\delta \qquad (8.1)$$

where k is the thermal conductivity of the fluid, T_s is surface temperature and

T is fluid temperature. The same equation can be used in a purely formal way to describe the heat loss by forced or free convection from any object with a mean surface temperature of T_s surrounded by fluid at T, even though the boundary layer is neither laminar nor uniformly thick. In this case, δ is the thickness of an equivalent rather than a real laminar layer. It is determined by the size and geometry of the surface and by the way in which fluid circulates over it. A more useful form of equation (8.1) can be derived by substituting a characteristic dimension of the body d for the equivalent boundary layer thickness which cannot be measured directly. For a sphere or cylinder, the diameter is a logical choice for d and for a rectangular plate d is the length in the direction of the wind. The equation then becomes

$$\mathbf{C} = \left(\frac{d}{\delta}\right) k(T_s - T)/d \tag{8.2}$$

The ratio d/δ is called the **Nusselt number** after its first exponent and is often written Nu. Just as the Reynolds number is a convenient way of comparing the forces associated with geometrically similar bodies immersed in a moving fluid, the Nusselt number provides a basis for comparing rates of convective heat loss from similar bodies of different scale exposed to different wind-speeds.

The rate of convective heat transfer in air can be written as

$$\mathbf{C} = \rho c_p(T_s - T)/r_H \tag{8.3}$$

where r_H is a thermal diffusion resistance (p. 22). Comparison of equations (8.2) and (8.3) gives

$$r_H = \frac{\rho c_p d}{k\mathrm{Nu}} = \frac{d}{\kappa \mathrm{Nu}} \tag{8.4}$$

where κ is the thermal diffusivity of air.

Forced convection

In forced convection, the Nusselt number depends on the rate of heat transfer through a boundary layer from a surface hotter or cooler than the air passing over it, a process analogous to the transfer of momentum by skin friction. The Nusselt number is therefore expected to be a function of the Reynolds number (specifying the boundary layer thickness for momentum) and the ratio of boundary layer thicknesses for heat (t_H) and for momentum (t_M). This ratio is a function of the **Prandtl number** defined by (ν/κ). Measurements of heat loss by forced convection from planes, cylinders and spheres can be described by the general relation

$$\mathrm{Nu} \propto \mathrm{Re}^n \mathrm{Pr}^m \tag{8.5}$$

where m and n are constants and $t_M/t_H = \mathrm{Pr}^m$.

The ratio of resistances for heat and momentum transfer can be expressed as

$$\frac{r_H}{r_M} = \frac{t_H/\kappa}{t_M/\nu} = \left(\frac{\nu}{\kappa}\right) \Big/ \left(\frac{t_M}{t_H}\right) = \mathrm{Pr}^{1-m} \tag{8.6}$$

For forced convection over plates, $m = 0.33$ and for air $\mathrm{Pr} = 0.71$ so that $r_H/r_M = 0.89$.

Because the Prandtl number is independent of temperature (p. 21) and because micrometeorologists are rarely concerned with heat transfer in any gas except air, Pr^m can be taken as constant in order to simplify equation (8.5) to

$$\mathrm{Nu} = A\mathrm{Re}^n \tag{8.7}$$

Values of the constants A and n for different types of geometry are given in Table A5(a) (p. 269).

Free convection

In free convection, heat transfer depends on the circulation of fluid over and around an object, maintained by gradients of temperature which create gradients of density. In this case the Nusselt number is a function of another non-dimensional group, the **Grashof number** Gr as well as of the Prandtl number Pr. The Grashof number is determined by the temperature difference between the hot or cold object and the surrounding fluid $(T_s - T)$, the characteristic dimension of the object d, the coefficient of thermal expansion of the fluid a, the kinematic viscosity of the fluid ν and the acceleration of gravity g. Physically, the Grashof number is the ratio of a buoyancy force times an inertial force to the square of a viscous force. Numerically, it is calculated from

$$\mathrm{Gr} = agd^3(T_s - T)/\nu^2 \tag{8.8}$$

In a system with a large Grashof number, free convection is vigorous because buoyancy and inertial forces which promote the circulation of air are much larger than the viscous forces which tend to inhibit circulation.

The Nusselt number for free convection in a gas is proportional to $(\mathrm{Gr\,Pr})^m$ and can therefore be written as

$$\mathrm{Nu} = B\mathrm{Gr}^m \tag{8.9}$$

for a specific gas such as air ($\mathrm{Pr} = 0.71$). The numerical constants B and m which depend on geometry are tabulated in Appendix A.5(b) (p. 271). When appropriate values of a and ν for air at 20°C are inserted in equation (8.8), it can be shown that

$$\mathrm{GrPr} = 112d^3(T_s - T) \tag{8.10a}$$

and

$$\mathrm{Gr} = 158d^3(T_s - T) \tag{8.10b}$$

where d is the characteristic dimension in *centimetres*.

Note that for laminar free convection, $m = 1/4$ irrespective of the shape of the object losing heat. In this case, the rate of convective heat loss is proportional to $(T_s - T)^{5/4}$, the so-called five-fourths power law of cooling.

Mixed convection

In many natural systems, convection is a very complex process because of continuous changes in windspeed and direction, often coupled with movement of the surface losing heat. During strong gusts, the loss of heat from a leaf or an animal will usually be determined by forced convection, but during lulls free convection may be dominant. The convection regime may therefore be described as 'mixed' in the sense that both modes of convection are present although their relative importance changes with time. Because this situation is too complex to handle either experimentally or theoretically, heat transfer in the environment is usually calculated from a mean windspeed when forced convection is thought to be dominant or from a temperature difference if the windspeed is very light.

Figure 8.1, based on measurements by Yuge (1960), provides a general

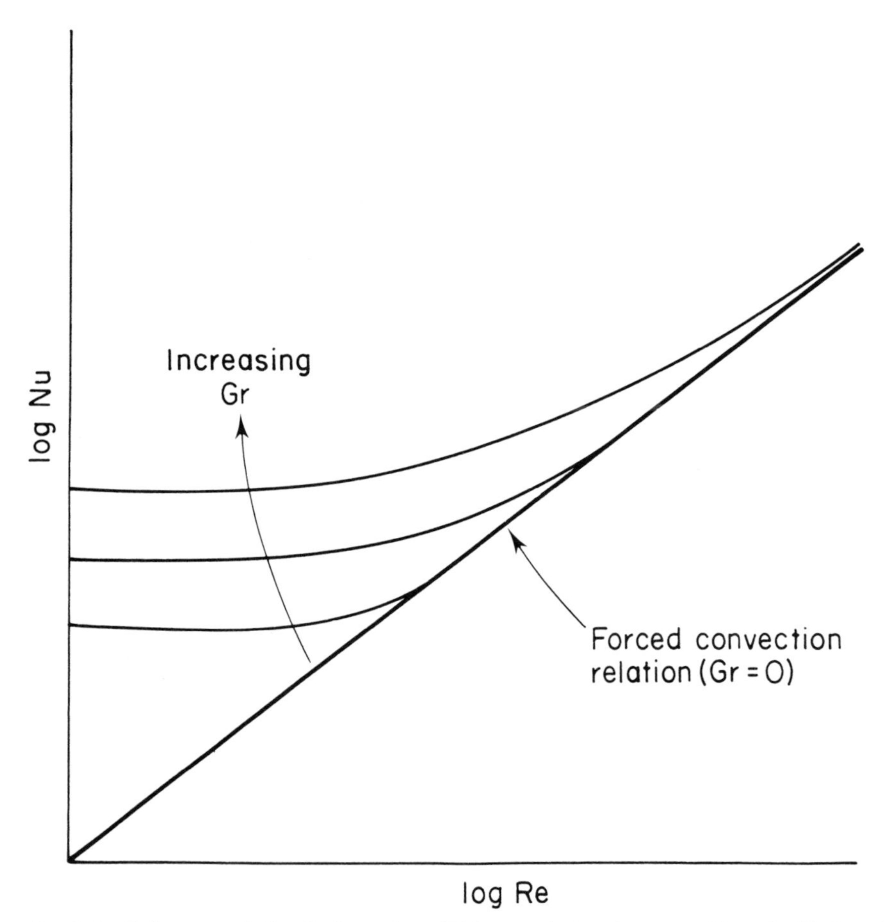

Fig. 8.1 Influence of Grashof number (Gr) on relation between Nusselt (Nu) and Reynolds number (Re) for heated sphere exposed to horizontal wind (after Yuge, 1960).

illustration of the way in which the Nusselt number for heated spheres in a crosswind is determined almost solely by Grashof number when Re is less than about 16 but approaches the value for forced convection as Re increases. Yuge found that when Gr = 400, there was a relatively sharp transition from free to forced convection but with Gr = 1800, there was a wide range of Reynolds numbers over which the Nusselt number had a value substantially greater than the value for forced or free convection separately but substantially less than the sum of the two Nusselt numbers. This is what is usually understood as a regime of 'mixed' convection.

As a rough criterion for distinguishing the two regimes, the Grashof number may be compared with the square of the Reynolds number. As Gr depends on

$$\frac{\text{buoyancy} \times \text{inertial forces}}{(\text{viscous forces})^2}$$

and Re^2 depends on $(\text{inertial forces})^2/(\text{viscous forces})^2$, the ratio $\mathrm{Gr}/\mathrm{Re}^2$ is proportional to the ratio of buoyancy to inertial forces. When Gr is much larger than Re^2, buoyancy forces are much larger than inertial forces and heat transfer is governed by free convection. When Gr is much less than Re^2, buoyancy forces are negligible and forced convection is the dominant mode of heat transfer.

For example, when a leaf with $d = 5$ cm is 5°C warmer than the surrounding air its Grashof number is about 10^5 whereas Re^2 is about $10^7 V^2$ when V is in m s^{-1}. A regime of forced convection is expected when V exceeds 1 m s^{-1} but at windspeeds between 0.1 and 0.5 m s^{-1}, which are often found in crop canopies, both forced and free convection will be active mechanisms of heat transfer.

A cow with $d = 0.5$ m and a surface temperature of 20°C above the ambient air has $\mathrm{Gr} = 4 \times 10^8$ and $\mathrm{Re}^2 = V^2 \times 10^9$ when V is in m s^{-1}. In this case, free convection will be the dominant form of heat transfer when the animal is exposed to a light draught indoors but, at windspeeds of the order 1 m s^{-1} in the field, the convection regime will again be mixed.

Laminar and turbulent flow

Both in forced and in free convection, the size of the Nusselt number depends on the degree of turbulence in the boundary layer. In turn, this depends partly on the turbulence in the airstream and partly on the roughness of the surface which tends to generate turbulence. When a smooth flat plate is exposed to an airstream effectively free from turbulence, the transition from laminar to turbulent flow in the boundary layer occurs at Reynolds number of the order of 10^5, but in a turbulent airstream the critical Reynolds number decreases to an extent that depends partly on the amplitude of the velocity fluctuations and partly on their frequency, and was reported by Grace (1978) to be about 4×10^3 for a poplar leaf. In micrometeorological problems involving leaves or other plant organs, Re is usually between 10^3 and 10^4 but it has never been clearly demonstrated whether the boundary layer of a leaf in a crop canopy, for example, should be regarded as laminar or turbulent. At a

Reynolds number of 10^4, the Nusselt number for laminar forced convection from a flat plate is $0.60 \times (10^4)^{0.5}$ or 60 compared with $0.032 \times (10^4)^{0.8}$ or 51 for a turbulent boundary layer; and at $\mathrm{Re} = 4 \times 10^4$, the corresponding numbers are 120 and 150. Thus for values of Re in the range of micro-meteorological interest, there will usually be little difference between the conventional Nusselt numbers for laminar and turbulent boundary layers. It does not necessarily follow that the same Nusselt numbers will be valid when the airstream itself is turbulent or when the surface is rough. When the air is free from turbulence, elements of surface roughness with a height of less than 1% of the characteristic dimension can increase the Nusselt number for a cylinder by a factor of about 2 and can reduce the critical Reynolds number for the transition to a turbulent boundary layer by an order of magnitude (Achenbach, 1977). A more detailed discussion of these matters was given by Gates (1980).

The onset of turbulence in free convection occurs when the Grashof number exceeds 10^8, an unusual situation in micrometeorology. For example, the mean surface temperature of a sheep or a man would need to be at least 30°C above the temperature of the ambient air to achieve $\mathrm{Gr} = 10^8$. The assumption of laminar flow will therefore be valid in cases of free convection as well as in forced convection.

MEASUREMENTS OF CONVECTION

Plane surfaces

When the boundary layer over a plane surface is laminar, the rate of heat transfer between the surface and the airstream can be calculated from first principles for two discrete cases. First, if the temperature is uniform over the whole surface the Nusselt number is

$$\mathrm{Nu} = 0.66\mathrm{Re}^{0.5}\mathrm{Pr}^{0.33} \tag{8.11}$$

and this relation is quoted in engineering texts which are concerned mainly with the heat transfer from metal surfaces with high thermal conductivity.

In the second case, which is more biologically relevant, the heat flux per unit area is constant over the whole surface. This condition should be valid for poor thermal conductors exposed to a uniform flux of radiation, e.g. leaf laminae in sunshine. The uniformity of heat flux from a leaf surface has not been established experimentally but it is clear from radiometric measurements of leaf temperature (see Fig. 8.3) that it is not legitimate to treat sunlit leaves as isothermal surfaces. According to Parlange *et al.* (1971), the assumption of a uniform heat flux leads to the prediction that the excess leaf temperature $T_s - T$ should increase with the square root of the distance from the leading edge x (cf. the uniform temperature case in which the flux decreases with the square root of x). It is convenient to incorporate x in a local Reynolds number $\mathrm{Re}_x = (Vx/\nu)$. In laminar flow the excess temperature then becomes

$$T_s - T = 2.21\mathbf{C}\frac{x}{k}(\mathrm{Re}_x)^{-0.5}\mathrm{Pr}^{-0.33} \tag{8.12}$$

and the mean temperature excess over a plate of length d is

$$\bar{\theta} = \frac{\int_0^d (T_s - T)dx}{\int_0^d dx} = 1.47\mathbf{C}\frac{d}{k}\mathrm{Re}^{-0.5}\mathrm{Pr}^{-0.33} \tag{8.13}$$

The mean Nusselt number defined as $\overline{\mathrm{Nu}} = \mathbf{C}d/k\bar{\theta}$ is

$$\overline{\mathrm{Nu}} = 0.68\mathrm{Re}^{0.5}\mathrm{Pr}^{0.33} \tag{8.14}$$

which is only a few per cent larger than the Nusselt number for the uniform temperature case.

A Nusselt number for a plate of irregular length, e.g. serrated or compound leaves, can be calculated from an appropriate average length in the direction of the airstream.

If w is the width of a leaf at right angles to the flow, the leaf area can be expressed as $t\int_0^w y dx$ (Fig. 8.2). When the Nusselt number is $A\mathrm{Re}^n$, an effective mean length $\bar{y}$ can be defined by writing the total heat loss from the leaf as

$$\mathbf{C} = A\left(\frac{V\bar{y}}{\nu}\right)^n \frac{k}{\bar{y}} \Delta T \int_0^w y dx \tag{8.15}$$

But the total heat loss can also be written in the form

$$\mathbf{C} = \int_0^w A\left(\frac{Vy}{\nu}\right)^n k\Delta T dx \tag{8.16}$$

and by equating these expressions the mean length is given by

$$\bar{y} = \left. \int_0^w y^n dx \middle/ \int_0^w y dx \right\}^{1/(n-1)} \tag{8.17}$$

For laminar forced convection, $n = 0.5$. Equation (8.17) is also valid for free convection with $n = 0.75$. Parkhurst *et al.* (1968) measured the heat loss and excess temperature from a series of metal leaf replicas with a wide range of shapes. Almost all the Nusselt numbers based on the mean dimension $\bar{y}$ lay between the values 0.60 $\mathrm{Re}^{0.5}$ and 0.80 $\mathrm{Re}^{0.5}$.

Leaves

Many attempts have been made to determine the Nusselt number of leaves or of heated metal replicas of leaves as a function of size, shape, windspeed, intensity of turbulence, degree of fluttering, etc, and a very comprehensive

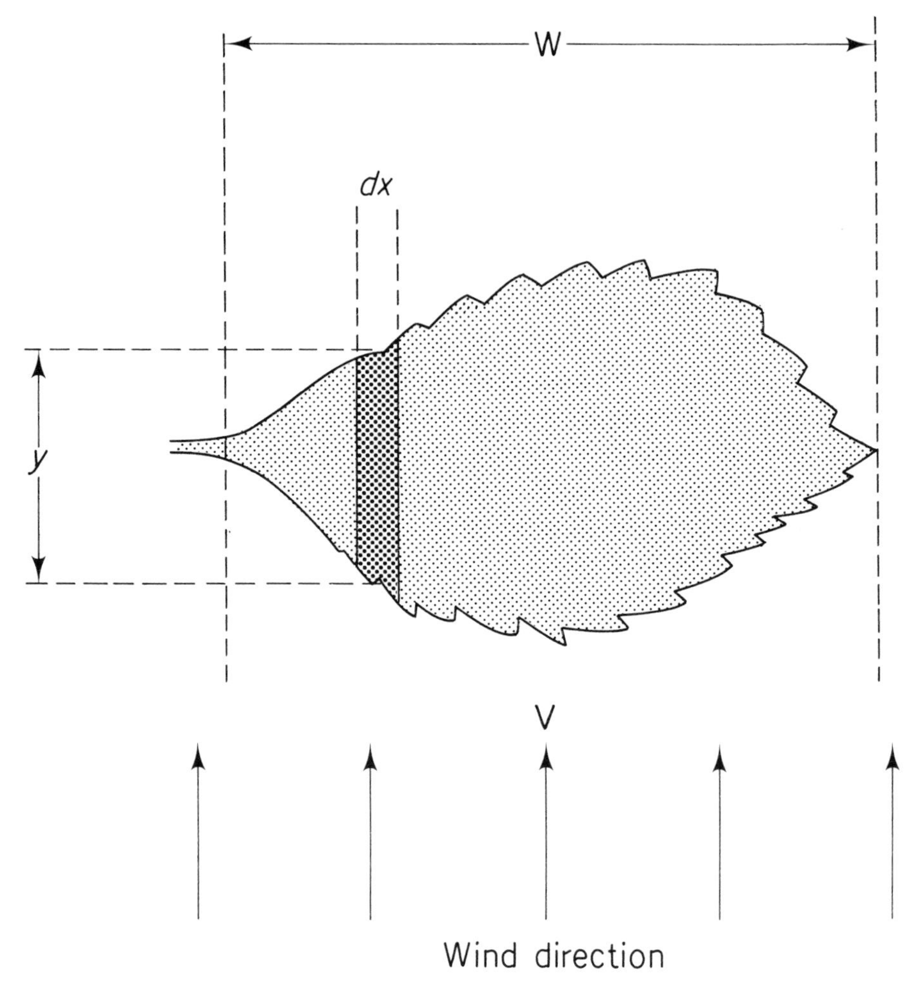

Fig. 8.2 Coordinates for integrating heat loss over the surface of a leaf of irregular shape.

summary was provided by Gates (1980). Values of Nu reported by some workers have been close to the so-called engineering values summarized in Table A.5 but some are larger by a factor between 1 and 2.5. The apparent excess loss of heat from leaves has sometimes been a consequence of undetected free convection at low windspeeds or of turbulence at high windspeeds but the main reason probably lies in the difference between the uniform boundary layer which develops downwind from the edge of a wide, flat plate and the very irregular and unstable boundary layer which must exist over a relatively narrow leaf with irregular edges, sometimes curled, and with protrusions in the form of veins.

Figure 8.3 clearly illustrates the difference in the thermal behaviour of a leaf-shaped plate exposed to a uniform airstream and a real leaf exposed in a comparable environment. Since the imposed heat flux was uniform in both

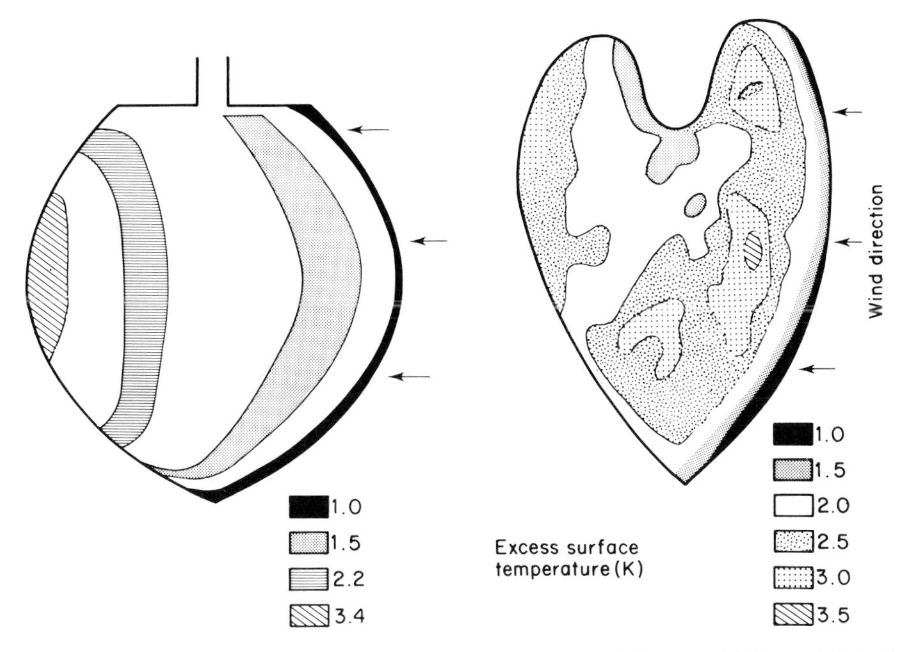

Fig. 8.3 Surface temperature distribution for a model and for a real (*Phaseolus*) leaf in comparable environments measured with a thermal imaging camera. The values in the key indicate the mean temperatures of isotherm bands on the original thermograms. Narrow isotherm bands employed in measurements on the model (left of figure) result in the white areas, intermediate between the adjacent labelled temperatures (from Wigley and Clark, 1974).

cases, the temperature distributions give a striking impression of how the boundary layer thickness changed across the two contrasting laminae. With this contrast in mind, it is surprising that convection from leaves does not differ much more from predictions based on plates. For comparisons, most workers use the ratio β of the observed Nusselt number for a leaf (or group of leaves) to the corresponding value of Nu for a smooth plate at the same windspeed.

For the simple leaf replica shown in Fig. 7.4, heated electrically and exposed in a wind-tunnel, β was about 1.1 and the ratio of resistances for heat and momentum transfer was almost exactly the value predicted from equation (8.6). In a field experiment where heat loss was measured from leaves attached to apple trees growing in rows (Thorpe and Butler, 1977), the relation between Nu and Re was very scattered but the line of best fit gave $\beta \approx 1$, possibly because turbulence compensated for the decrease of windspeed between the alleys (where it was measured) and in the trees (where leaves were exposed). In contrast, values of $\beta = 2.5$ were reported for poplar leaves exposed to laminar flow in a wind tunnel.

Measurements derived from replicas of *Phaseolus* leaves (as in Fig. 8.3) gave $\beta = 1.1$ for laminar flow (intensity of turbulence: $i = 0.01$ to 0.02) and for Re up to 2×10^4 (Wigley and Clark, 1974). For turbulent flow ($i = 0.3$ to 0.4), $A = 0.04$ and $n = 0.84$, Nu exceeded the laminar flow value above (but not below) Re = 10^3, and β was about 2.5 for Re = 10^4.

Relatively little work has been published on free convection from leaves despite its significance for heat transfer in canopies where windspeeds below 0.5 m s^{-1} are common. In one study with leaves of *Acer* and *Quercus*, Nu was close to the prediction from equation (8.9) when GrPr was 10^6 but β increased as GrPr decreased below this figure and was about 2 at GrPr = 10^4.

The dissipation of heat from a set of copper plates exposed to low windspeed was studied by Vogel (1970). All the plates had the same area but their shapes ranged from a circle and a regular 6-point star to replicas of oak leaves with characteristic lobes (Fig. 8.4). The amount of electrical energy needed to keep each plate 15°C warmer than the surrounding air was recorded at windspeeds from 0 to 0.3 m s^{-1} and at different orientations. This energy is proportional to Nu and inversely proportional to the resistance r_H under the conditions of the experiment. The main conclusions were:

(i) increasing airflow from 0 to 0.3 m s^{-1} decreased r_H by 30 to 95%: the decrease was greater for the leaf models than for the stellate shapes;
(ii) the resistances of all the stellate and lobed plates were smaller than the resistance of the round plate and were less sensitive to orientation;
(iii) a deeply lobed model simulating a sun leaf of oak always had a smaller resistance than a shade leaf with smaller lobes;
(iv) the resistance of the leaf models was least when the surface was oblique to the airstream;
(v) serrations about 5 mm deep on the periphery of the circular plate had no perceptible effect on its thermal resistance;
(vi) the measurements could not be correlated using a simple Nusselt number based on a weighted mean width as described above.

These measurements support the hypothesis favoured by some ecologists that the shapes of leaves may represent an adaptation to their thermal environment. Because the natural environment is very variable and because physiological responses to changes of leaf tissue temperatures are complex, conclusive experimental proof is still lacking.

To recap, non-dimensional groups provide a useful way of summarizing and comparing heat loss from objects with similar geometry but different size exposed to different wind regimes. Standard values of these groups should be regarded as a useful yardstick for estimating heat transfer in the field,

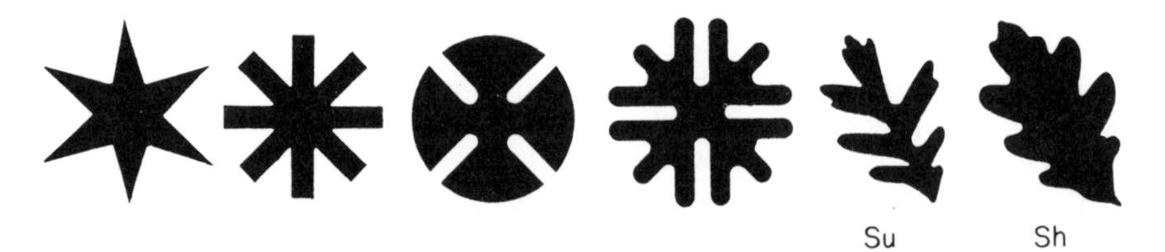

Fig. 8.4 Metal replicas of leaves used by Vogel (1970) to study heat losses in free convection. The shapes Su and Sh represent sun and shade leaves of white oak.

allowing mean surface temperature to be estimated within the precision needed for most ecological studies. When precision is essential, appropriate values of Nu must be obtained experimentally, preferably by measurements on real leaves at the relevant site.

Cylinders and spheres

The flow of air over cylindrical and spherical objects is more complex than the flow over plates because of the separation of the boundary layer that occurs towards the rear, and the wake generated by this separation. The Nusselt number can again be related to the Reynolds number by the expression $\mathrm{Nu} = A\mathrm{Re}^n\mathrm{Pr}^{0.33}$ but, unlike the corresponding constants for flat plates, both A and n change with the value of Re (Table A.5(a)). Both for cylinders and spheres, the obvious quantity to adopt for the characteristic dimension is the diameter but, for the irregular bodies of animals, the cube root of the volume may be more appropriate.

Mammals

McArthur measured the heat loss from a horizontal cylinder electrically heated and with a diameter of 0.33 m to simulate the trunk of a sheep (McArthur and Monteith, 1980a). In a second set of measurements, the cylinder was covered with the fleece of a sheep.

For the bare cylinder, estimates of Nu fitted the relation $\mathrm{Nu} = A\mathrm{Re}^n$ with $A = 0.095$ and $n = 0.68$ in the region $2 \times 10^4 < \mathrm{Re} < 3 \times 10^5$, cf. $A = 0.17$ and $n = 0.62$ in Table A.5(a).

To avoid the difficulty of comparing equations with different exponents, Fig. 8.5 shows the resistance of cylinders and cylindrically-shaped animals obtained from various sources. Consistency is reassuring, and as the aerodynamic resistance is often small compared with the coat resistance of an animal, the degree of uncertainty implied by Fig. 8.5 is immaterial for most calculations.

Nusselt numbers for hair surfaces are harder to establish because of the problem of measuring the temperature at the tips of the hairs. Using a radiation thermometer to measure this quantity, McArthur found $A = 0.112$ and $n = 0.88$. Comparison of the relation between Nu and Re for smooth and hairy cylinders (Fig. 8.6) suggests that in a light wind (<1 m s^{-1}) the effective boundary layer depth was thicker when the surface was composed of hairs, presumably because the effective surface 'seen' by the radiometer was some distance below the physical hair tips. But when Re exceeded 10^5, the boundary layer depth was smaller with hair present, suggesting that wind penetrated the coat, possibly generating turbulence.

When measurements of this kind are used to estimate heat loss from mammals, Nusselt numbers need to be obtained for appendages too: legs and tails can be treated as cylinders and heads as spheres.

Rapp (1970) reviewed several sets of measurements of the loss of heat from nude human subjects in different postures and exposed to a range of environments. Agreement between measured and standard Nusselt numbers was excellent except when the convection regime was mixed. In one

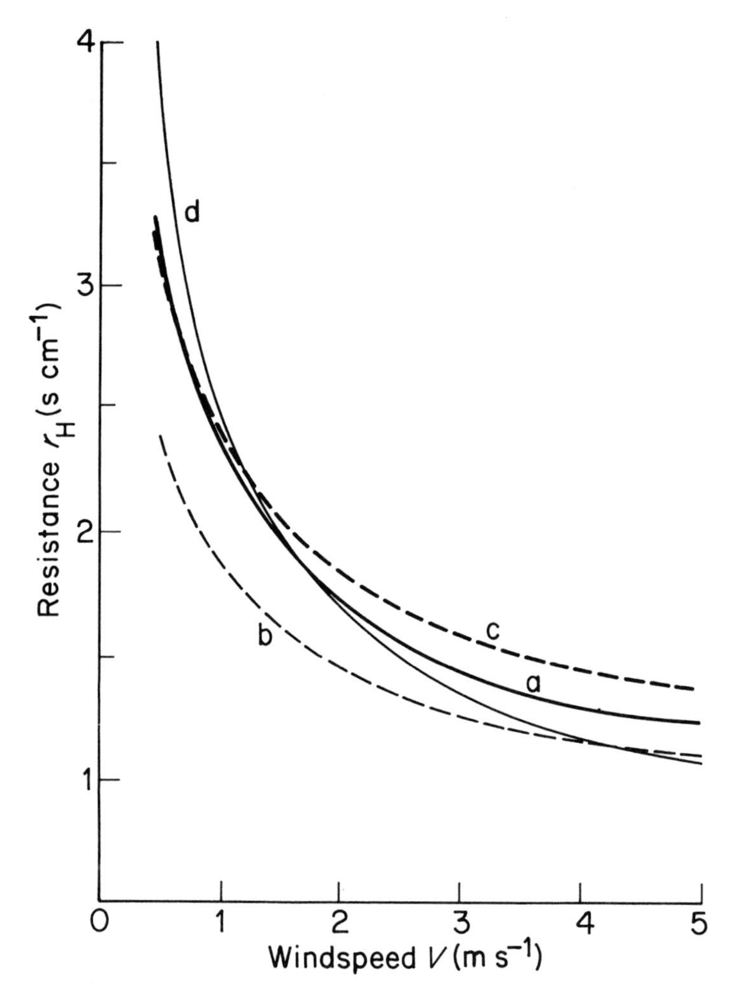

Fig. 8.5 Relations between boundary layer resistance r_H and windspeed V for cylindrical bodies. Line a, smooth isothermal cylinder (McAdams, 1954); line b, cattle (Wiersma and Nelson, 1967); line c, sheep (Monteith, 1975); line d, sheep (McArthur and Monteith, 1980a).

experiment, the heat loss from vertical subjects standing in a horizontal airstream was $8.5V^{0.5}$ W m^{-2} K^{-1} (V in m s^{-1}) and for a man with a characteristic dimension of 33 cm, this is equivalent to $A = 0.78$, $n = 0.5$ in equation (8.7). Although the corresponding values in Table A.5 are $A = 0.24$, $n = 0.6$, the two sets of coefficients give similar values of Nu for a relevant range of Re (10^4 to 10^5). Similarity with values of Nu already quoted for horizontal cylinders illustrates the usefulness of non-dimensional groups for comparing heat loss from different systems.

In a light wind, the heat loss from sheep and other animals may be governed by free convection, particularly when there is a large difference of

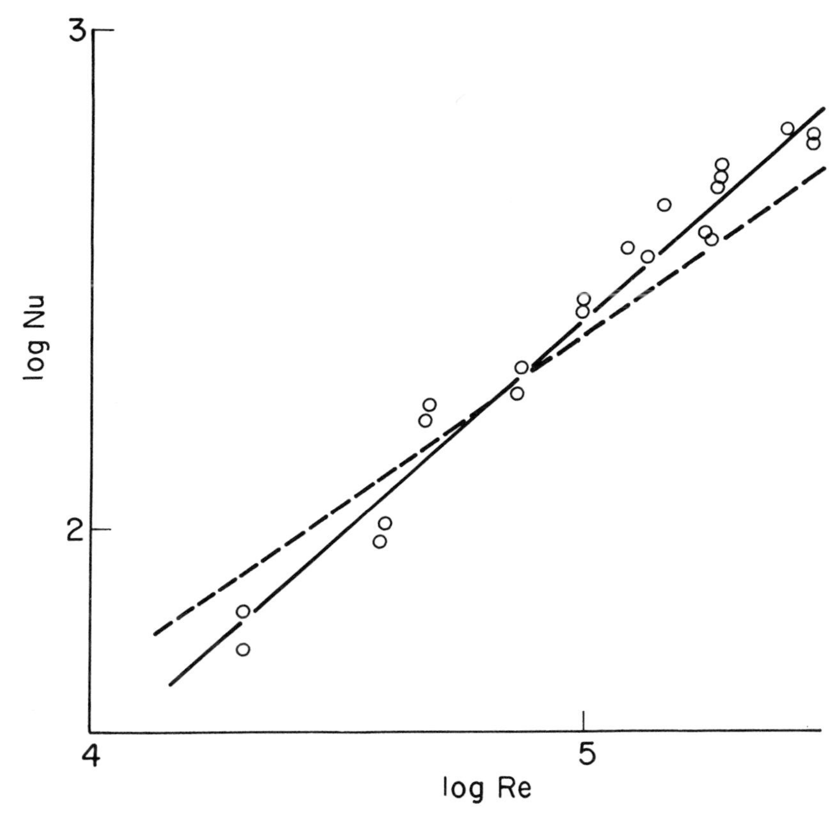

Fig. 8.6 The relation between Nu and Re for model sheep without fleece—dashed line; the relation for the same model with fleece—full line and points (from McArthur and Monteith, 1980a).

temperature between the coat and the surrounding air. When Merino sheep were exposed to strong sunshine in Australia, fleece tip temperatures reached 85°C when the air temperature was 45°C. With $d = 30$ cm, the corresponding Grashof number is about 2×10^8, so free and forced convection would be of comparable importance when $\mathrm{Re}^2 = 2 \times 10^8$, i.e. when windspeed was about 0.7 m s^{-1}. To calculate fleece temperatures in similar conditions, Priestley (1957) used a graphical method to allow for the transition from free to forced convection with increasing windspeed (see p. 124).

For erect humans, heat loss by free convection is complex because the Grashof number depends on the cube of the characteristic dimension. The nature of the convection regime therefore changes with height above the ground as shown in Fig. 8.7. The movement of air associated with free convection from the head and limbs has been demonstrated by Lewis and his colleagues using the technique of Schlieren photography (Fig. 8.8). The technique has also been used to study free convection from wheat ears and a rabbit (Fig. 8.9). The way in which the air ascending over the face of humans

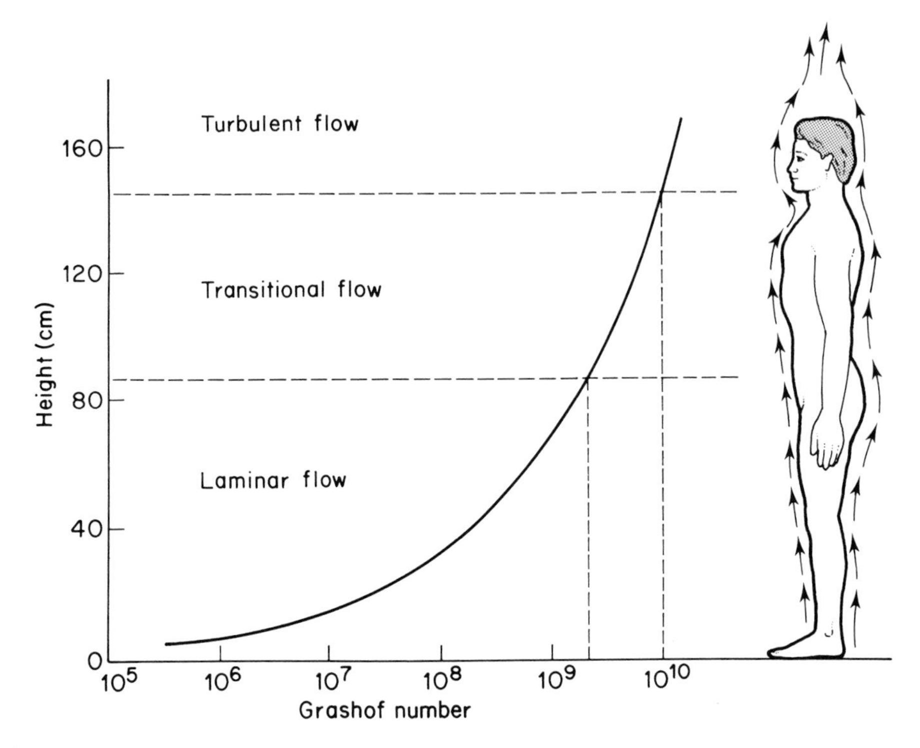

Fig. 8.7 Variation of Grashof number with vertical height over the body surface for a mean skin temperature of 33°C and an ambient temperature of 25°C (from Clark and Toy, 1975).

is deflected from the nostrils may be important in preventing the inhalation of bacteria and other pathogens. Figure 8.10 shows that the velocity and temperature profiles close to a bare leg are characteristic of the flow in free convection from vertical surfaces.

The application of conventional heat transfer analysis is much more difficult for non-cooperative subjects like piglets, particularly when the animal changes its posture and orientation with respect to windspeed in response to heat or cold stress.

Birds

Several workers have measured the heat loss from replicas of birds constructed from appropriate combinations of metal spheres and cylinders. The model of a domestic fowl used by Wathes and Clark (1981a) consisted of a copper sphere with a diameter of 265 mm to which cylinders of 15 mm diameter were soldered at points corresponding to the head and legs. The Nusselt number for forced convection was

$$\mathrm{Nu} = 2 + 0.79\ \mathrm{Re}^{0.48}$$

giving values of Nu similar to those from the standard relation for a simple

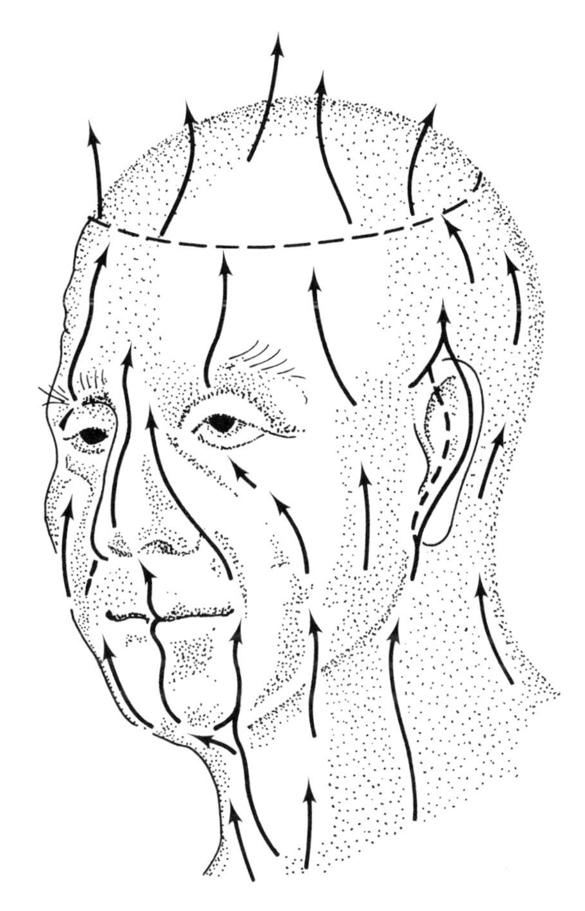

Fig. 8.8 Routes taken by the flow of naturally convected air over the human head (from Schlieren photography by Lewis *et al.*, 1969).

sphere within the experimental range of Reynolds numbers (3×10^3 to 10^4).

Insects

The convective heat loss from insect species was measured by Digby (1955) who mounted dead specimens in a wind tunnel with transparent walls. Heating was provided by radiation from an external 2 kW lamp. The difference between the temperature of the insects and the ambient air was (i) directly proportional to incident radiant energy and (ii) inversely proportional to the square root of the windspeed, as predicted for cylinders in the appropriate range of Reynolds numbers.

Evidence for the variation of excess temperature with body size was more difficult to interpret. When the maximum breadth of the thorax was taken as an index of body size d, the excess temperature was expected to change with a

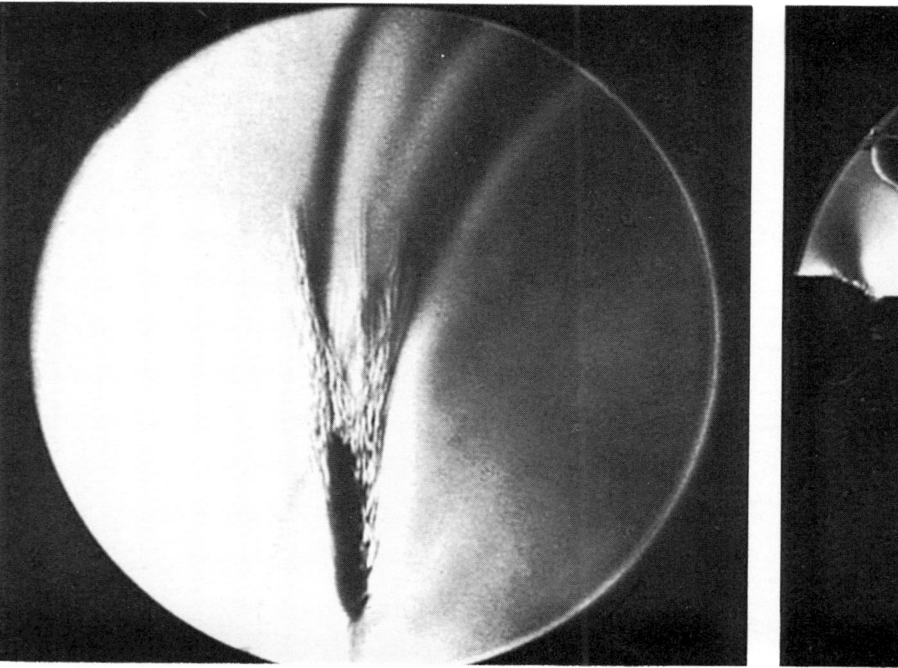

Fig. 8.9 Schlieren photographs of an ear of awned wheat and a rabbit's head showing regions of relatively warm air (light) and cooler air (dark). The wheat was irradiated with a source giving about 880 W m^{-2} net radiation. Note the disturbed air round the rabbit's nostrils, the separation of rising air over the eyebrows, and the evidence of strong heating of air round the ears (see p. 133).

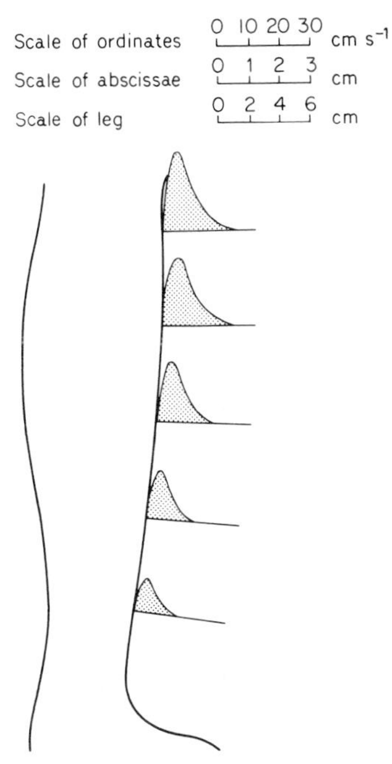

Fig. 8.10 Development of velocity profiles on the front of the leg measured with a hot wire anemometer (from Lewis *et al.*, 1969).

power of d between 0.5 (cylinders) and 0.4 (spheres). For the locusts *Schistocerca gregaria* and *Carausius morosus*, the exponent of d was 0.4 but for 20 species of Diptera and Hymenoptera, Digby's measurements support an exponent of unity, i.e. excess temperature was directly proportional to the linear dimensions of the insect. This anomalous result may be a consequence of considerable scatter in the measurements or of changes in the distribution of absorbed radiant energy with body size.

In a similar series of measurements, Church (1960) used radiation from a 20 kHz generator to heat specimens of bees and moths. He argued that the amount of energy produced by an insect in flight would be roughly proportional to body weight and therefore used radiant flux densities that were proportional to the cube of the thorax diameter. When the insects were shaved to remove hair, the excess temperature was proportional to $d^{1.4}$. As the heat loss per unit area was proportional to d, the Nusselt number was proportional to $d^{0.6}$, close to expectation for a sphere. The excess temperature of insects covered with hair was about twice the excess for denuded insects showing that the thermal resistances of the hair layer and the boundary layer were of similar size.

If the proportionality between Nu and $d^{0.6}$ is consistent with $\mathrm{Nu} = A\mathrm{Re}^n$, the index n must be 0.6 and Nu should be proportional to $V^{0.6}$. In fact, Church found that the excess temperatures of a shaved *Bombus* specimen was proportional to $V^{0.4}$ but he explained this discrepancy in terms of the difference between temperatures at the surface of and within the thorax. If all the experimental results are assumed consistent with $n = 0.6$, the measurements on the *Bombus* specimen give a Nusselt number of 0.28 $\mathrm{Re}^{0.6}$, close to the standard relation for a sphere, $\mathrm{Nu} = 0.34\ \mathrm{Re}^{0.6}$. The values of Nu from this equation are close to the values predicted from either of the relationships in Table A.5 within the appropriate range of Reynolds numbers.

Leaves

Many coniferous trees have needle-like leaves that can be treated as cylinders in order to estimate heat losses, though their cross section is seldom circular. Because the needles are closely spaced they shield each other from the wind and do not behave like isolated cylinders (see also p. 109). Engineers have developed empirical formulae for the heat loss from banks of cylinders in regular arrays and their relevance to coniferous branches has been examined by Tibbals *et al.* (1964).

To avoid the difficulty of measuring the surface temperature of real needles, branches of pine, spruce and fir were invested with dental compound and then burnt to leave a void that was filled with molten silver. The compound was then removed leaving a silver replica of the branch—a modern version of the Midas touch! For Blue spruce (*Picea pungens*), the Nusselt numbers based on the diameter of single needles were about one-third to one-half of Nu for isolated cylinders at the same Reynolds number, a measure of mutual sheltering. For White fir (*Abies concolor*), the Nusselt number in transverse flow (across four rows of needles) was about 60% greater than in longitudinal flow (across 20 or 30 rows of needles). When an average of the two modes was taken, Nu was close to the experimental values for banks of tubes and similar agreement was obtained for *Pinus ponderosa*. In still air, the Nusselt number for all three species was close to an experimental value for horizontal cylinders at the same Grashof number.

CONDUCTION

Conduction is a form of heat transfer produced by sharing of momentum between colliding molecules in a fluid, by the movement of free electrons in a metal, or by the action of inter-molecular forces in an insulator. In micrometeorology, conduction is important for heat transfer in the soil and through the coats of animals, but not in the free atmosphere where the effects of molecular diffusion are trivial in relation to mixing by turbulence.

If the temperature gradient in a solid or motionless fluid is dT/dz, the rate of conduction of heat per unit area (**G**) is proportional to the gradient and the constant of proportionality is called the thermal conductivity of the material k'. In symbols

$$\mathbf{G} = -k'(dT/dz) \tag{8.18}$$

where the negative sign is a reminder that heat moves in the direction of decreasing temperature. For a steady flow of heat between two parallel surfaces at T_2 and T_1 separated by a uniform slab of material with thickness t, integration of equation (8.18) gives

$$\mathbf{G} = -k'(T_2 - T_1)/t \tag{8.19}$$

In general, layers of motionless gas provide excellent insulation, a fact exploited by the thick coats of hair with which many mammals trap air and by man's design of clothing and double glazing. The conductivity of still air is four orders of magnitude less than the values for copper and silver.

Equation (8.18) is used to describe the vertical flow of heat in soil as discussed in Chapter 13. When it is applied to conduction in cylinders and spheres, the area through which a fixed amount of heat is conducted changes with position and the integration of equation (8.18) then yields expressions somewhat more complex than equation (8.19). For a hollow cylinder with interior and exterior radii r_1 and r_2 at temperatures T_1 and T_2 respectively, the heat flux per unit area of the outer surface is

$$\mathbf{G} = \frac{k'(T_1 - T_2)}{r_2 \ln (r_2/r_1)} \tag{8.20}$$

and for a hollow sphere the corresponding flux density is

$$\mathbf{G} = \frac{k'(T_1 - T_2)}{r_2(r_2/r_1 - 1)} \tag{8.21}$$

Equations (8.20) and (8.21) have been used to calculate heat transfer through parts of animals and birds that are approximately cylindrical or spherical.

The form of equation (8.20) has important implications for the efficiency of insulation surrounding a cylinder or any object that is approximately cylindrical such as the trunk of a sheep or a human finger. The equation predicts that, for a fixed value of the temperature difference across the insulation, the heat flow per unit length of the cylinder $2\pi r_2\mathbf{G}$ will be inversely proportional to $\ln(r_2/r_1)$. In the steady state, assuming that radiative exchange is negligible, the rate of conduction must be equal to the convective heat loss from the outer surface of the insulation. If the air is at a temperature T_3, the convective heat loss will be

$$2\pi r_2 \mathrm{Nu}(k/2r_2)(T_2 - T_3) = 2\pi r_2 \mathbf{G} = 2\pi k'(T_1 - T_2)/\ln(r_2/r_1)$$

where k is the thermal conductivity of air and Nu is the Nusselt number. By rearranging terms it can be shown that the heat loss per unit length of cylinder is

$$2\pi r_2 \mathbf{G} = \frac{2\pi k(T_1 - T_3)}{\{(k/k')\ln(r_2/r_1)\} + \{2/\mathrm{Nu}\}} \tag{8.22}$$

The Nusselt number is proportional to $(r_2)^n$ where n is 0.5 for forced and 0.75 for free convection. When r_2 increases, the heat loss $2\pi r_2\mathbf{G}$ will therefore increase or decrease depending on which of the terms in curly brackets dominates the denominator. To find the break-even point where $2\pi r_2\mathbf{G}$ is independent of r_2, equation (8.22) can be differentiated with respect to r_2 to give

$$\mathrm{Nu} = 2nk'/k$$

When the Nusselt number is greater than this critical value, the heat loss from the insulation increases as r_2 increases; when it is smaller than the critical value, the heat loss decreases with increasing r_2.

If the conductivity of air is taken as fixed and $n = 0.6$ is an approximate mean value for mixed convection, the critical value of Nu depends only on k', the conductivity of the insulating material. For animal coats and for clothing, $k \approx k'$, so the critical Nu is of the order of unity. In nature, Nusselt numbers of this size may be relevant to a furry caterpillar under a cabbage leaf on a calm night or to an animal in a burrow but, for most organisms freely exposed to the atmosphere, Nu will exceed 10. In general, therefore, the thermal insulation of animals will increase with the thickness of their hair or clothing.

The conductivity of fatty tissue on the other hand is about 12 times the conductivity of still air so the critical Nusselt number for insulation by subcutaneous fat is therefore about 14. When Nu is larger than 14, fat provides insulation in the conventional sense but in an environment where Nu is less than this critical value, a naked ape suffering from middle-aged spread might have increasing difficulty in keeping warm as his girth increased.

INSULATION OF ANIMALS

The insulation of animals has three components: a layer of tissue, fat and skin across which temperature drops from deep body temperature to mean skin temperature; a layer of relatively still air trapped within a coat of fur, fleece, feathers or clothing; and an outer boundary layer whose resistance is given by $d/(\kappa \mathrm{Nu})$ (p. 122). A comprehensive analysis of heat exchange in mammals would need to consider separately the amount of heat lost from the trunk, legs, head, etc. Because these appendages are usually less well insulated and are smaller than the trunk, they are capable of losing more heat per unit area. In practice, the loss of heat from appendages is usually small compared with the total loss from the rest of the body (although some animals subject to heat stress are believed to dissipate large amounts of heat through their ears or tails. There is some evidence that the rate of ear flapping by African elephants increases with temperature (Buss and Estes, 1971)). Average values of insulation for different species can be determined from measurements of metabolic heat production, external heat load, and the relevant mean temperature gradients.

As the insulation is a temperature difference per unit heat flux and per unit area, it is equivalent to the term $r/\rho c_p$ where ρc_p is a volumetric specific heat. To convert from units of insulation (e.g. K $m^2 W^{-1}$) to units of resistance (e.g. s cm^{-1}) and for comparison with the resistance of the boundary layer r_H, it is necessary to choose an arbitrary value of ρc_p for air, e.g. 1.22×10^3 J K^{-1} m^{-3} which is the value at 20°C. On this basis, a resistance $r = 1$ s cm^{-1} is equal to an insulation $r/\rho c_p = 0.082$ K m^2 W^{-1}.

Another unit of insulation found mainly in human studies is the clo, equal to 0.155 K m^2 W^{-1}, and therefore equivalent to 1.86 s cm^{-1} or 0.39 cm of still air (putting $r_H = t/\kappa$). The clo was originally conceived as the insulation maintaining 'a resting man, whose metabolism is 50 kcal m^{-2} h^{-1} (about 60 W m^{-2}), indefinitely comfortable in an environment of 21°C, relative humidity less than 50%, and air movement 20 ft min^{-1}'. Specialized units of this type have few merits and tend to separate a subject from other related branches of science.

Tissue

The insulation of tissue is defined as the temperature difference per unit heat flux density between deep body temperature and the skin surface. Insulation is strongly affected by the circulation of blood beneath the skin, and the constriction and dilation of blood vessels can change resistance by a factor of 2 to 3. For comparison with the thermal resistances of hair and air, values of tissue insulation found in the literature have been multiplied by the volumetric heat capacity of air at 20°C. Table 8.1 shows values ranging from a minimum of 0.3 s cm^{-1} for dilated tissue to between 1 and 2 s cm^{-1} for vaso-constricted tissue.

Table 8.1 Thermal resistances of animals. Peripheral tissue and coats

Tissue (Blaxter, 1967)	s cm^{-1}	
	Vaso-constricted	Dilated
Steer	1.7	0.5
Man	1.2	0.3
Calf	1.1	0.5
Pig (3 months)	1.0	0.6
Down sheep	0.9	0.3
Coats (Blaxter, 1967; Hammel, 1955)	s cm^{-1} per cm depth	per cent of still air
Air	4.7	100
Red fox	3.3	70
Lynx	3.1	65
Skunk	3.0	64
Husky dog	2.9	62
Merino sheep	2.8	60
Down sheep	1.9	40
Blackfaced sheep	1.5	32
Cheviot sheep	1.5	32
Ayrshire cattle: flat coat	1.2	26
erect coat	0.8	
Galloway cattle	0.9	19

Taken as an average over the whole body, the thermal conductivity of human skin during vaso-constriction appears to be about an order of magnitude greater than the conductivity of still air, i.e. about 0.2 to 0.3 W m^{-1} $°K^{-1}$, and measurements on fingers showed that the conductivity increased linearly with the rate of blood flow. The effective mean thickness of the skin during vaso-constriction is equivalent to about 2.5 cm of tissue or 0.25 cm of still air. These figures must conceal large local differences of insulation of the limbs and appendages depending on the thickness and nature of subcutaneous tissue and the degree of curvature.

Coats—mixed regimes

There are many systems both in the field and within buildings where several modes of heat transfer operate simultaneously, but one mode usually dominates and a good approximation to the rate of transfer can then be obtained by neglecting the others. In the coats of animals and in clothing, however, molecular conduction, radiation, free and forced convection all play a significant role and evidence for this is now considered.

The thermal resistance of animal coats has been measured by a number of workers and reported in units such as K m^2 W^{-1} or clo/inch. These units obscure an important physical fact; the thermal conductivity of hair, fleece and clothing is the same order of magnitude as the conductivity of still air. (A distinction made by others between 'still' air and 'dead air space' is based mainly on an arithmetical error. The two are physically identical by definition although in practice it is difficult to achieve a temperature gradient across a layer of still air without setting up a circulation of air by convection which increases in thermal conductivity.)

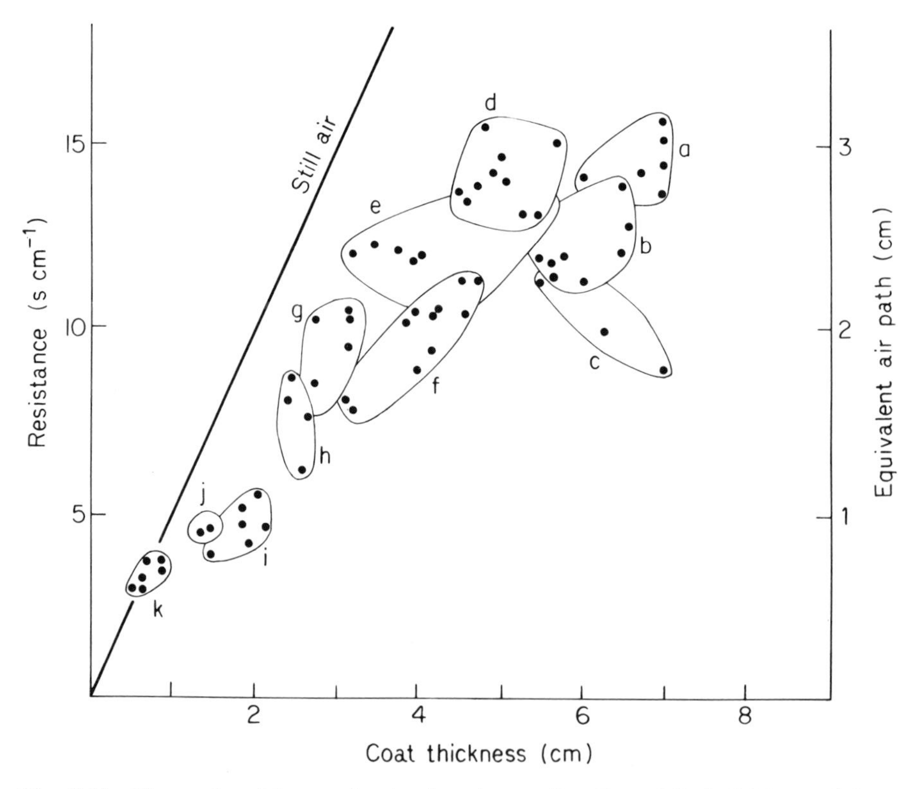

Fig. 8.11 Thermal resistance of animal coats as a function of their thickness (after Scholander *et al.*, 1950). a, dall sheep; b, wolf, grizzly bear; c, polar bear; d, white fox; e, reindeer, caribou; f, fox, dog, beaver; g, rabbit; h, marten; i, lemming; j, squirrel; k, shrew.

The thermal conductivity of air is 2.5×10^{-2} W m^{-1} K^{-1} at 20°C; or 4.8 s cm^{-1} for a layer of 1 cm; or 2.58 clo/cm; or 6.6 clo/in. Scholander *et al.* (1950) showed that the insulation per unit thickness of coat was remarkably uniform for a wide range of wild animals from shrews to bears (Fig. 8.11). The average insulation derived from his measurements is often quoted as 4 clo/in. meaning that 1 inch of fur had the same insulation as 4/6.6 or 0.6 inches of still air. In terms of still air, the efficiency of insulation is 60%. A simple calculation suggests that conduction along hair fibres was trivial but at least two other modes of heat transfer may have been responsible for the failure of the coats to behave like still air:

(1) there was significant radiative transfer from warmer to cooler layers of hair;
(2) buoyancy generated by temperature gradients was responsible for free convection.

Starting with the first possibility, Cena and Monteith (1975a) used Beer's Law to analyse the change of radiative flux with depth in samples of coat exposed to radiation. Depth within the coat was specified by an interception parameter (p), as described on p. 33. For long-wave radiation in an isothermal coat, the assumption that hairs behaved like black-bodies gave estimates of p close to those obtained by measurements of hair geometry, ranging from 4 to 5 cm for sheep to 18 cm for calf and deer with much smoother coats.

For a non-isothermal coat with a uniform temperature gradient the heat flux was found to be proportional to the gradient, and the equivalent thermal conductivity was

$$k_R = (4/3)4\sigma T^3/p$$

Putting $4\sigma T^3 = 6.3$ W m^{-2} K^{-1} ($T = 293$ K), k_R ranged from 0.004 W m^{-1} K^{-1} for calf hair to 0.02 W m^{-1} K^{-1} for fleece. These figures imply that the effective thermal conductivity of animal coats can never be as small as the value for still air and may be nearly twice that value for fleece and coats with similar structure. Corresponding values of resistance per unit depth of coat are obtained by writing $r_R = \rho c_p/k_R$ where ρc_p has some arbitrary value (e.g. 1.22 kJ m^{-3} K^{-1} for 20°C).

Even allowing for radiative transfer, the conductivity of several types of coat was substantially larger than the value predicted on the basis of transfer by molecular conduction and radiation, and the discrepancy increased with the size of the temperature gradient across the hairs. This behaviour suggested that free convection must be implicated since the effective conductivity for this mode of heat transfer is proportional to the 0.25 power of a temperature difference. The rates of heat transfer assigned to free convection were found to be consistent with air velocities within the coat of the order of 1 cm s^{-1}.

Molecular conduction, radiation and free convection therefore account for the range of values of thermal resistance in Table 8.1. Few attempts have been made to relate the insulation of an animal coat to the structure and weight of coat elements. Wathes and Clark (1981b) were able to show that the pelt resistance of the domestic fowl (i.e. feathers and skin), increased with feather mass per unit area W_f (kg m^{-2}) according to the relation

$$r = 6.5W_f + 1.5$$

where r is in s cm^{-1}. A range of W_f from 0.05 to 0.8 kg m^{-2} was obtained from pelts which had been damaged by abrasion or pecking when birds were housed in battery cages. The corresponding range of resistance from 2 to 7 s cm^{-1} implies that substantial losses of heat and therefore of productivity may be caused by poor management in this type of system. In undamaged coats, the average plumage resistance at 6 s cm^{-1} was similar to values reported for other avian species and the resistivity of 1.8 s cm^{-1} per cm coat depth was comparable with figures for sheep.

In the natural environment, forced convection must also play a significant role within coats to an extent determined by windspeed and direction with respect to an animal's body. The literature contains accounts of many experiments purporting to show that resistance of coats and clothing is reduced by a quantity proportional to the square root of V. However, careful re-analysis of the data showed that the conductance of a coat, which is the reciprocal of resistance, increased linearly with V, i.e.

$$r(V)^{-1} = r(0)^{-1} + aV \tag{8.23}$$

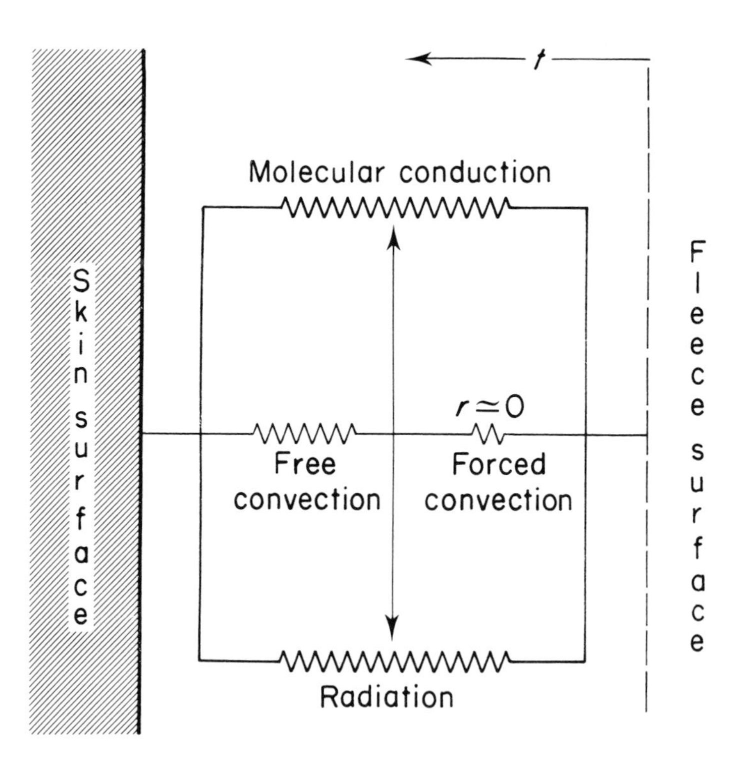

Fig. 8.12 Electrical analogue for heat transfer through animal coat with wind penetration to depth t below surface (from McArthur and Monteith, 1980b).

where $r(V)$ is resistance at windspeed V (Campbell, McArthur and Monteith, 1980). As an example, a was about 0.025 for a 3.5 cm deep fleece with $r(0) = 7 \text{ s cm}^{-1}$.

The simplest interpretation of equation (8.23) is that wind completely destroys the insulation of a depth of fleece t given by

$$t = laV/\{aV + r(0)^{-1}\}$$

where l is total fleece depth.

Figure 8.12 puts all four modes of heat transfer together and emphasizes the thermal complexity of coats and clothing.

9

MASS TRANSFER

(i) Gases and water vapour

Two modes of diffusion are responsible for the exchange of matter between organisms and the air surrounding them. Molecular diffusion operates within organisms (e.g. in the lungs of an animal or in the substomatal cavities of a leaf) and in a thin skin of air forming the boundary layer that surrounds the whole organism. In the free atmosphere, transfer processes are dominated by the effects of turbulent diffusion, although molecular diffusion continues to operate and is responsible for the ultimate degradation of turbulent energy into heat.

Turbulence is ubiquitous in the atmosphere except very close to the earth's surface on very calm nights, and the turbulent transfer of water vapour and carbon dioxide is of paramount importance for all higher forms of life. As a measure of effectiveness, the amount of carbon dioxide absorbed by a healthy green crop in one day is equivalent to all the CO_2 between the canopy and a height of 30 m. In practice, although the concentration of carbon dioxide in the atmosphere decreases between sunrise and sunset as a result of photosynthesis, this depletion rarely exceeds 15% of the mean concentration at the surface. These figures imply that turbulent transfer enables vegetation to extract CO_2 from the lowest 200 m of the atmosphere and probably from much greater heights. A small diurnal change has been reported at 500 m.

The process of mass transfer will now be described in terms of diffusion across boundary layers, through porous septa and within the free atmosphere.

NON-DIMENSIONAL GROUPS

Mass transfer to or from objects suspended in a moving airstream is analogous to heat transfer by convection and is conveniently related to a non-dimensional parameter similar to the Nusselt number of heat transfer theory. This is the **Sherwood number** Sh defined by the equation

$$\mathbf{F} = \mathrm{Sh}D(\chi_s - \chi)/d \qquad (9.1)$$

where $\mathbf{F}$ = mass flux of a gas per unit surface area (e.g. g m^{-2} s^{-1}); χ_s, χ = mean concentration of gas at the surface and in the free atmosphere (e.g.

g m^{-3}); D = molecular diffusivity of the gas in air (e.g. $m^2\ s^{-1}$). As

$$\mathrm{Sh} = \frac{\mathbf{F}}{D(\chi_s - \chi)/d} \tag{9.1}$$

the Sherwood number can be defined as the ratio of actual mass transfer **F** to the rate of transfer that would occur if the same concentration difference were established across a layer of still air of thickness d. The corresponding resistance to mass transfer is derived by comparing equation (9.1) with

$$\mathbf{F} = (\chi_s - \chi)/r$$

giving $r = d/(D\mathrm{Sh})$ (cf. $r_H = d/(\kappa \mathrm{Nu})$). Resistances and diffusion coefficients for water vapour and carbon dioxide will be distinguished by subscripts and are related by $r_V = d/(D_V \mathrm{Sh})$ and $r_C = d/(D_C \mathrm{Sh})$.

Just as the Nusselt number for forced convection is a function of Vd/ν (Reynolds number) and ν/κ (Prandtl number), the Sherwood number is the same function of Vd/ν and the ratio ν/D which is known as the **Schmidt number**, abbreviated to Sc. For example, the Sherwood number for mass exchange at the surface of a flat plate is

$$\mathrm{Sh} = 0.66\ \mathrm{Re}^{0.5}\mathrm{Sc}^{0.33} \tag{9.2}$$

cf.

$$\mathrm{Nu} = 0.66\ \mathrm{Re}^{0.5}\mathrm{Pr}^{0.33} \tag{9.3}$$

The presence of the term $0.66\ \mathrm{Re}^{0.5}$ in both expressions is a consequence of a fundamental similarity between the molecular diffusion of heat, mass and momentum in laminar boundary layers and the numbers $\mathrm{Sc}^{0.33}$ and $\mathrm{Pr}^{0.33}$ take account of differences in the effective thickness of the boundary layers for mass and heat.

For any system in which heat transfer is dominated by forced convection, the relation between Sh and Nu is given by dividing equation (9.2) by (9.3):

$$\mathrm{Sh} = \mathrm{Nu}(\mathrm{Sc}/\mathrm{Pr})^{0.33} = \mathrm{Nu}(\kappa/D)^{0.33} \tag{9.4}$$

The ratio κ/D is sometimes referred to as a **Lewis number** (Le). In air at 20°C, $(\kappa/D)^{0.33}$ is 0.96 for water vapour and 1.14 for CO_2 (see Table A.2, p. 267). The corresponding ratios of resistances (cf. p. 122) are

$$r_V/r_H = (\kappa/D_V)^{0.67} = 0.93$$

$$r_C/r_H = (\kappa/D_C)^{0.67} = 1.32$$

In free convection, the circulation of air round a hot or cold object is determined by differences of air density produced by temperature gradients, by vapour concentration gradients, or by a combination of both. If the Nusselt number is related to the Grashof and Prandtl numbers by $\mathrm{Nu} = B\mathrm{Gr}^n\mathrm{Pr}^m$, the Sherwood number will be $\mathrm{Sh} = B\mathrm{Gr}^n\mathrm{Sc}^m = \mathrm{Nu}\mathrm{Le}^m$ where m is 1/4 in the laminar regime and 1/3 in the turbulent regime. To calculate the Grashof number, it is convenient to replace the difference between the surface and air temperature $T_0 - T$ by the difference of virtual temperature (p. 12). If e_0 and e are vapour pressures at the surface and in the air and p is air pressure, the gradient of virtual temperature is

$$T_{v_0} - T_v = T_0(1 + 0.38e_0/p) - T(1 + 0.38e/p)$$
$$= (T_0 - T) + 0.38(e_0 T_0 - eT)/p \qquad (9.5)$$

where temperatures are expressed in K. The importance of the vapour pressure term when T is close to T_0 can be illustrated for the case of a man covered with sweat at 33°C and surrounded by still air at 30°C and 20% relative humidity. Then $e_0 = 5.03$ kPa and $e = 0.85$ kPa. The term $T_0 - T$ is 3 K, $0.38(e_0 T_0 - eT)/p$ is 4.9 K and the difference of virtual temperature is 7.9 K. The size of the Grashof number allowing for the difference in vapour pressure is 2.6 times the number calculated from the temperature difference alone. The corresponding error in calculating a Nusselt or Sherwood number (proportional to $\mathrm{Gr}^{0.25}$) is about -27%.

The same type of calculation is used to determine atmospheric stability when temperature and water vapour concentration are both functions of height (see p. 244).

MEASUREMENTS OF MASS TRANSFER

Plane surfaces

For laminar flow over smooth flat plates, the Sherwood number for water vapour, $0.57\ \mathrm{Re}^{0.5}$, is shown by the continuous line in Fig. 9.1. Powell (1940) obtained a similar relation $\mathrm{Sh} = 0.41\ \mathrm{Re}^{0.56}$ for circular discs with diameters between 5 and 22 cm parallel to the wind (pecked line). Thom (1968) measured evaporation from filter paper attached to the model bean leaf described on p. 106 and used bromobenzene and methylsalicylate as well as water to get a range of diffusion coefficients from 5.4×10^{-6} to 24×10^{-6} $\mathrm{m^2\ s^{-1}}$. At windspeeds exceeding 1 $\mathrm{m\ s^{-1}}$, the mass transfer of all three vapours was described by $\mathrm{Sh} = 0.7\ \mathrm{Re}^{0.5}\mathrm{Sc}^{0.33}$ i.e. within a few percent of the predicted value (eqn. 9.2). The measurements for water vapour plotted in Fig. 9.1 show that when the windspeed was less than 1 $\mathrm{m\ s^{-1}}$ ($\mathrm{Re} < 2800$), the Sherwood numbers were larger than the predicted value, possibly because the rate of mass transfer was increased by differences of density in the air surrounding the leaf.

By using radioactive lead vapour as a tracer, perfectly absorbed by the surfaces exposed to it, Chamberlain (1974) was able to measure rates of mass transfer as a function of position on model bean leaves about 11×11.5 cm. Figure 9.2 shows that when the samples were parallel to the flow in a wind tunnel, measurements of Sherwood number were consistent with the standard relation and with Thom's observations. With leaves at an angle to the flow (not shown), the local boundary layer resistance of the upwind surface increased by a factor of about four from the leading edge to the trailing edge. In contrast, the resistance over the downwind surface was greatest just behind the leading edge—a shelter effect—and then decreased with increasing distance, presumably because the surface was exposed to eddies forming in the lee of the edge. The mean resistance for the whole area of both surfaces appeared to be almost independent of exposure angle.

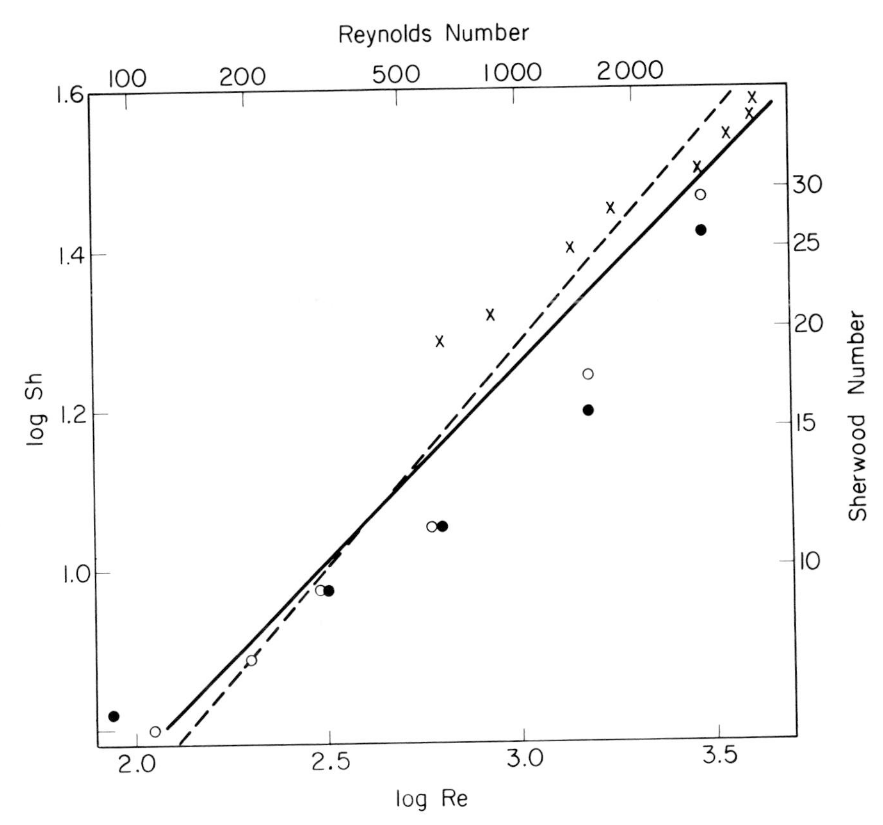

Fig. 9.1 Relation between Sherwood and Reynolds numbers for plates parallel to the airstream. Continuous line, standard relation $Sh = 0.57Re^{0.5}$; pecked line, measurements on discs by Powell (1940); X, measurements on model bean leaf by Thom (1968); O, measurements on replicas of alfalfa and Cocksfoot leaves by Impens (1965).

For real leaves in the field, wind direction is rarely constant and differences in local boundary resistance will usually be smaller (and less regular) than the previous paragraph suggests. The dependence on leaf angle of heat and mass transfer will usually be a consequence of differences in radiation absorption rather than in transfer coefficients.

When Chamberlain measured the rate of uptake of lead vapour by real bean leaves growing in a canopy, Sherwood numbers were about 25% greater than predicted from measurements on isolated models. The corresponding value of β (p. 129) is 1.25. This is consistent with measurements of evaporation from *Citrus* leaves in a canopy for which Haseba (1973) found that β increased systematically with leaf area density (area per unit volume of canopy) as Fig. 9.3 shows. For arable crops, density is often of the order of 0.1 so Fig 9.3 implies that β should be between 1.1 and 1.2 in the field, consistent with conclusions for heat. Somewhat larger values would be expected if leaves were fluttering.

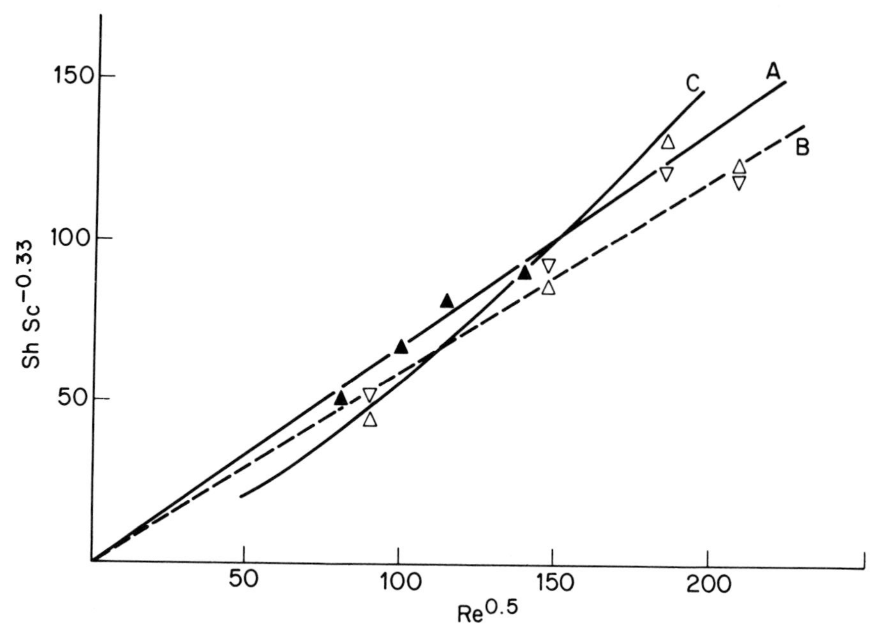

Fig. 9.2 Comparison of measurements of mass transfer between air flow and flat surfaces parallel to the flow. The non-dimensional group Sh $Sc^{-0.33}$ is plotted as a function of $Re^{0.5}$ to give the constant of equation (9.2) as a slope. ▲ small leaves; △ large leaves; ▽ large leaves (central strip only) — Chamberlain (1974); A — Pohlhausen (1921); B — Thom (1968); C — Powell (1940): (from Chamberlain, 1974).

Cylinders

For Reynolds numbers between 10^3 and 5×10^4, the Nusselt number for cylinders can be expressed as $Nu = 0.26\ Re^{0.6}Pr^{0.33}$ and the corresponding Sherwood number is $Sh = 0.26\ Re^{0.6}Sc^{0.33}$ or $Sh = 0.22\ Re^{0.6}$. Powell's measurements of the evaporation from a wet cylinder in a wind tunnel fit this relation closely and it is shown by the continuous line in Fig. 9.4. For the more restricted range of Reynolds number from 4×10^3 to 4×10^4, the relation $Nu = 0.17\ Re^{0.62}$ is often used. The corresponding Sherwood number $Sh = 0.16\ Re^{0.62}$ is shown by the pecked line in Fig. 9.4.

Nobel (1974) measured the resistance to evaporation from wet paper cylinders ($d = 2$ cm) as a function of turbulent intensity and found that turbulence of 10% was enough to decrease boundary layer resistance by 22% implying that $\beta = 1.22$ in the corresponding expression for Sherwood number. When the intensity of turbulence was increased from 10 to 70%, resistance decreased only slightly. As with similar experiments on heat transfer, observed values of β must to some extent reflect the size of turbulent eddies in relation to the size of the object as well as turbulent intensity.

Rapp (1970) established close agreement between measurements of the evaporative loss from nude men covered with sweat and values predicted

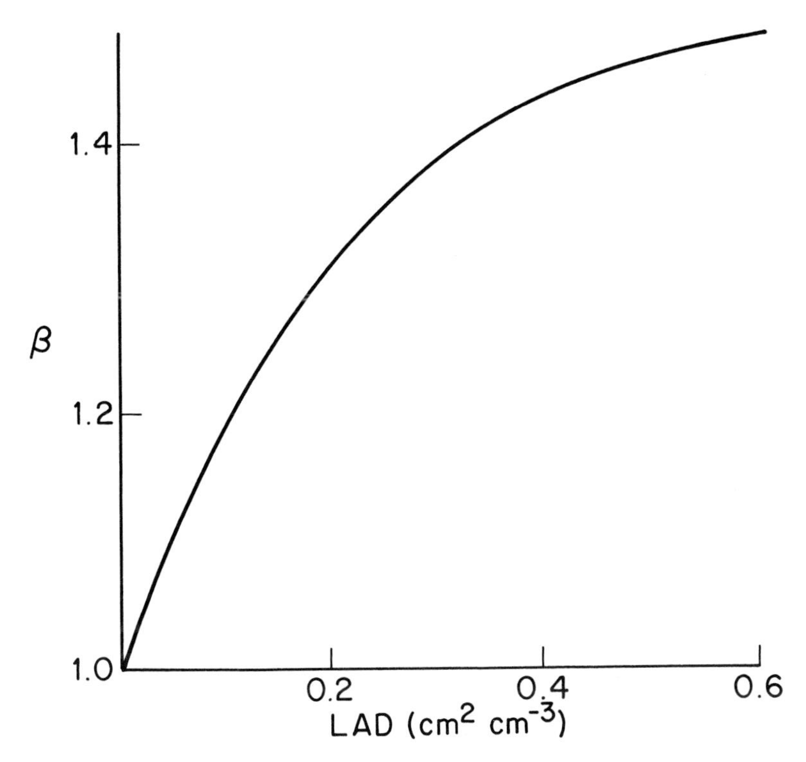

Fig. 9.3 Ratio of boundary layer resistance for an isolated *Citrus* leaf to resistance measured within a canopy of leaves of specified leaf area density (LAD); (redrawn from Haseba, 1973).

from the Sherwood number for a cylinder of appropriate diameter. The units of his calculations have been transformed to show this agreement in Fig. 9.4.

Spheres

The Nusselt number for a sphere can be expressed in the form 0.34 $Re^{0.6}$ and the corresponding Sherwood number is 0.34 $Re^{0.6}Le^{0.33}$ or 0.32 $Re^{0.6}$ for water vapour. Analysis of Powell's (1940) measurements of evaporation from wet spheres gives $Sh = 0.26\ Re^{0.59}$, about 20% less than the value predicted from heat transfer rates.

The rate of heat transfer from a 6 cm diameter balsa wood sphere was close to the predicted value when turbulence i was about 1% but was about 1.1 for $i = 30\%$ and 1.14 for $i = 70\%$ (Nobel, 1975).

VENTILATION

When mass transfer is induced by the ventilation of a system, an equation relating the mass flux to an appropriate potential gradient can be used to define a transfer resistance consistent with the values of diffusion resistance

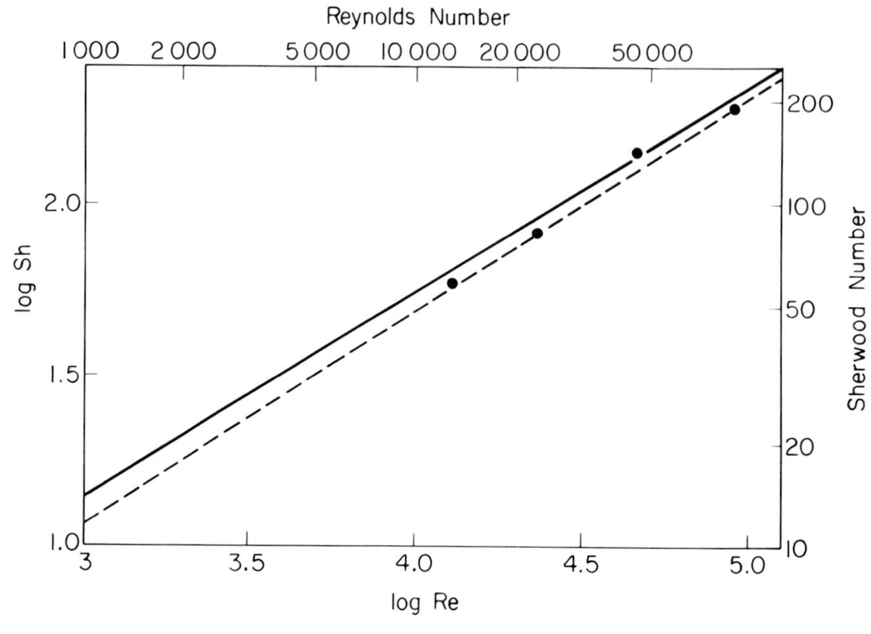

Fig. 9.4 Relations between Sherwood and Reynolds numbers for a wet cylinder at right angles to the airstream. Continuous line, $Sh = 0.22Re^{0.6}$; pecked line, $Sh = 0.16Re^{0.62}$. The points were calculated by Rapp (1970) from measurements by Kerslake on a man covered with sweat.

already discussed. Relevant examples are the exchange of carbon dioxide between the air inside and outside a glasshouse, the loss of water vapour by evaporation from the lungs of an animal, and the exchange of pollutants in open-top field chambers.

If the air in a glasshouse is well stirred so that the volume concentration of CO_2 in the internal air has a uniform value of ϕ_i (m^3 CO_2 m^{-3} air) when the external concentration is ϕ_e, the rate at which plants in the glasshouse absorb CO_2 from the external atmosphere can be written as

$$Q = \rho_c v N(\phi_e - \phi_i) \text{ g h}^{-1} \tag{9.6}$$

where v is the volume of air in the house (m^3), N is the number of air changes per hour and ρ_c is the density of CO_2 (g m^{-3}). Dividing both sides of equation (9.6) by the floor area A gives a flux of CO_2 per unit floor area

$$\mathbf{F} = Q/A = \rho_c v N(\phi_e - \phi_i)/A \tag{9.7}$$

The resistance to CO_2 diffusion can be defined by writing

$$\mathbf{F} = \rho_c(\phi_e - \phi_i)/r_c$$

and comparison of the two equations gives

$$r_c = A/vN$$
$$= (N\bar{h})^{-1}$$

where $\bar{h}$ is the mean height of the house. For example, if N is 10 air changes per hour and $\bar{h} = 3$ m, r_c is 1/30 h m^{-1} or 1.2 s cm^{-1}, comparable in size with boundary layer and stomatal resistances.

A similar approach is applicable to mass transfer from animals. If $\dot{V}$ is the volume of air respired by an animal per minute (the 'minute volume') and A is the area of skin surface, the loss of water per unit skin area is

$$\mathbf{F} = \dot{V}(\chi_s(T_b) - \chi)/60A \text{ g m}^{-2} \text{ s}^{-1} \tag{9.8}$$

where $\chi_s(T_b)$ is the water vapour concentration of air saturated at deep body temperature T_b and χ is the concentration in the environment, both expressed in g m^{-3}. Then the resistance to vapour exchange is

$$r_V = 60A/\dot{V}$$

For a man at rest $\dot{V} = 10^{-2}$ m^3 min^{-1} and if $A = 1.7$ m^2, r_V is 10^4 s m^{-1} or 100 s cm^{-1}, i.e. about two orders of magnitude larger than common values of the boundary layer resistance for a sweating nude figure. Even during very rapid respiration when V may reach 10^{-1} m^3 min^{-1}, the diffusion resistance for respiration will exceed the boundary layer resistance by an order of magnitude and, when the skin is covered with sweat, the loss of water by evaporation from the lungs will be much smaller than the cutaneous evporation rate.

To determine the response of crops to pollutant gases, cylindrical chambers with open tops (Fig. 9.5) are often used in attempts to alter the quality of air around plants growing in the field without making large changes in other

Fig. 9.5 Open-top chambers in a field of beans at Sutton Bonington. The boxes adjacent to the chambers contain fans for ventilating the chambers, and some also contain charcoal for absorbing gaseous air pollutants. The 'frustum' design at the open top reduces the rate of incursion of unfiltered air (see p. 154).

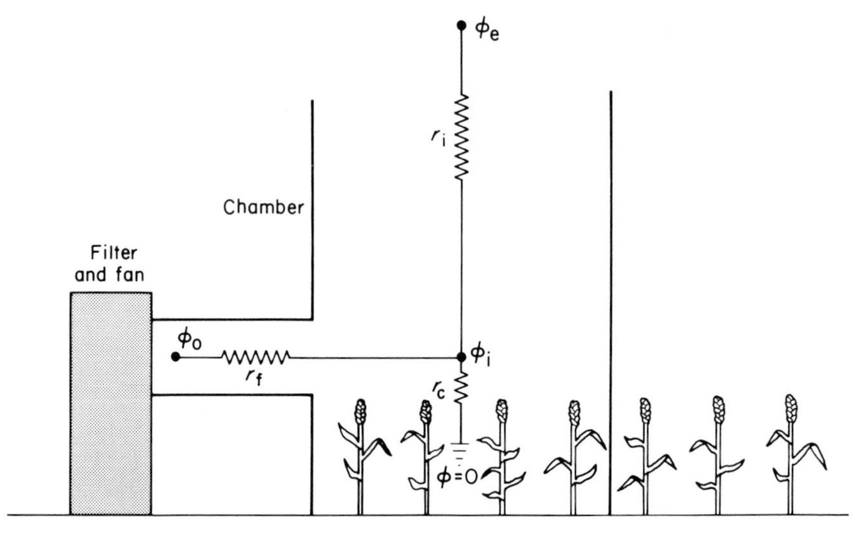

Fig. 9.6 Schematic diagram of an open-top field chamber and the equivalent resistance analogue.

aspects of the microclimate (Heagle, Body and Heck, 1973). Fans blow air into the base of chambers through filters which absorb pollutants; other chambers have fans without filters to provide 'control' treatments. However the concentration of a pollutant within a filtered open-top chamber is never zero because some unfiltered air enters by 'incursion' through the open top. Figure 9.6 shows a simple resistance analogue which can be used to estimate the concentration of the pollutant gas in the chamber. Exchange of the gas through the open top can be regarded as driven by the potential difference between the concentration ϕ_e in the outside air and the concentration ϕ_i in the chamber. The resistance restricting incursion is given by $r_i = A/vN_i$ where A is the chamber base area, v the volume, and N_i the rate of air change through the open top. Similarly air flow through the filter is driven by the potential $\phi_0 - \phi_i$ and limited by the resistance r_f describing ventilation by the fan (N_f air changes per second); ϕ_0 is the gas concentration (usually close to zero) leaving the filter. To complete the general analogue, the pollutant flux in the chamber to plants and soil where ϕ is assumed to be zero is given by $(\phi_i - 0)/r_c$, where r_c is the plant canopy resistance. It follows, from the conservation of flux, that

$$\frac{\phi_o - \phi_i}{r_f} + \frac{\phi_e - \phi_i}{r_i} = \frac{\phi_i - 0}{r_c}$$

which, for the ideal case $\phi_0 = 0$, reduces to

$$\phi_i/\phi_e = \left[1 + r_i\left(\frac{1}{r_f} + \frac{1}{r_c}\right)\right]^{-1} \tag{9.9}$$

For a set of chambers designed by Heagle, Body and Heck (1973), Unsworth, Heagle and Heck (1984a,b) estimated the resistances as follows: $r_f = 6 \text{ s m}^{-1}$

(about three air changes per minute); r_i decreased from about 50 s m^{-1} when the windspeed outside was 1 m s^{-1} to 15 s m^{-1} at a windspeed of 6 m s^{-1}; r_c for ozone uptake by a canopy of soybeans had a minimum value of about 70 s m^{-1}. Equation (9.9) then shows that the pollutant concentration in the chamber would increase from about 0.1 ϕ_e at the lowest windspeed to 0.3 ϕ_e at 6 m s^{-1}. It is unlikely that open-top chambers could be designed to be much more effective than this without substantially altering other features of microclimate such as temperature and humidity.

MASS TRANSFER THROUGH PORES

When leaves transpire, water evaporates from cell walls and escapes to the atmosphere by diffusing into sub-stomatal cavities, through stomatal pores, and finally through the leaf boundary layer into the free atmosphere. During photosynthesis, carbon dioxide molecules follow the same path but in the opposite direction. Rigorous treatments of diffusion through pores (e.g. Leuning, 1983) allow for the interaction of diffusing gases and for the difference in (dry) air pressure across pores needed to balance the difference in water vapour pressure. There is a class of problems in which these complications cannot be ignored (e.g. in precise estimates of the intercellular CO_2 concentration) but for many practical purposes the elementary treatment which follows is adequate.

It will therefore be assumed that for a leaf lamina, the resistance of stomatal pores for a particular gas depends only on their geometry, size and spacing whereas the resistance offered by the boundary layer depends on leaf dimensions and windspeed.

Some plant physiologists concerned with the simultaneous transfer of water vapour and carbon dioxide find it convenient to express fluxes in units of mol m^{-2} s^{-1} and concentration gradients in mol gas mol^{-1} air. Conductance then has the same units as flux and resistance has reciprocal units. For conversion, conductance in mol m^{-2} s^{-1} must by multiplied by m^3 mol^{-1} (0.0224 at STP) to obtain units of m s^{-1} and resistance in units of mol^{-1} m^2 s must be multiplied by mol m^{-3} (44.6 at STP) to obtain units of s m^{-1}.

Meidner and Mansfield (1968) tabulated stomatal populations and dimensions for 27 species including crop plants, deciduous trees and evergreens. The leaves of many species have between 100 and 200 stomata per mm^2 distributed on both the upper and lower epidermis (amphistomatous leaf) or on the lower surface only (hypostomatous leaf). The length of the pore is commonly between 10 and 30 μm and the area occupied by a complete stoma, including the guard cells responsible for opening and shutting the pore, ranges from 25 × 17 μm in *Medicago sativa* to 72 × 42 μm in *Phyllitis scolopendrium*.

Because stomata tend to be smaller in leaves where they are more numerous, the fraction of the leaf surface occupied by pores does not vary much between species and is about 1% on average for a pore width of 6 μm. There is much greater variation in the geometry of pores: the stomata of grasses are usually long, narrow, and aligned in rows parallel to the midrib whereas the elliptical stomata of sugar beet (*Beta vulgaris*) and broad bean

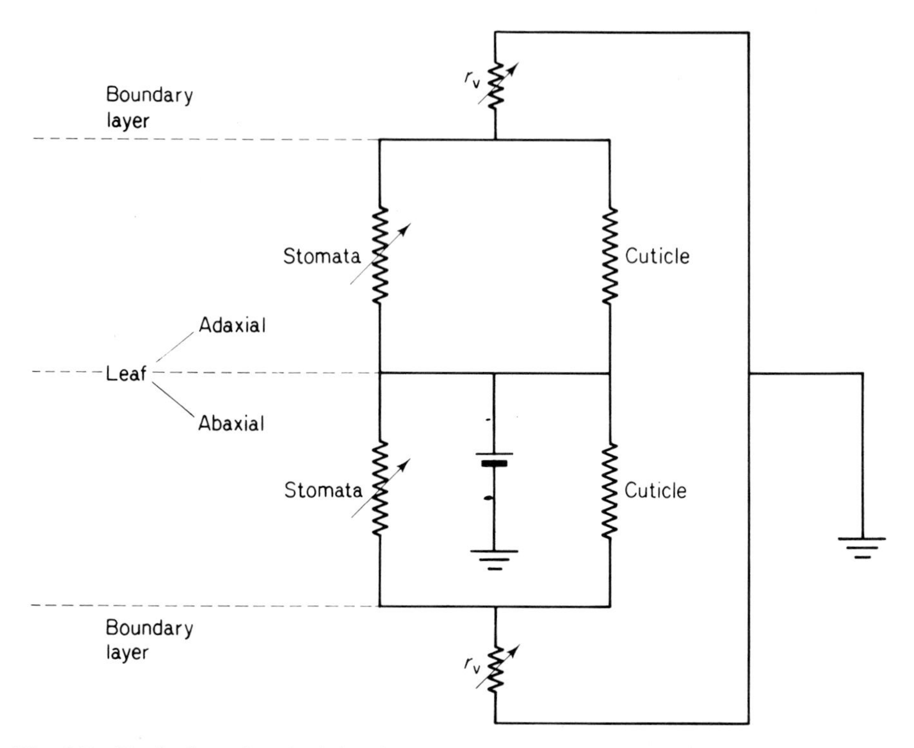

Fig. 9.7 Equivalent electrical circuit for loss of water vapour from a leaf by diffusion through the stomata and cuticle of the upper and lower epidermis.

(*Vicia faba*) are randomly oriented but uniformly dispersed over the epidermis.

The network of resistances in Fig. 9.7 is an electrical analogue for the diffusion of water vapour between the intercellular spaces and the external air. The calculation of boundary layer resistance r_V has already been discussed: values of 0.3 to 1 s cm^{-1} are expected for small leaves in a light wind. Many mesophytes have minimum stomatal resistances in the range 1 to 2 s cm^{-1} but values as small as 0.5 s cm^{-1} and as large as 4.8 s cm^{-1} have been reported for *Beta vulgaris* and *Phaseolus vulgaris* respectively. Xerophytes have larger minimum resistances up to 30 s cm^{-1}. Cuticular resistances range from 20 to 60 s cm^{-1} in mesophytes and from 40 to 400 s cm^{-1} in xerophytes. In both types of plant, the resistance of the cuticle is usually so much larger than the stomatal resistance that its role in water vapour and CO_2 transfer can generally be ignored.

Nobel (1975) found that the resistance to the loss of water vapour from the fruiting bodies of fungi (Basidiomycetes) fell between 0.3 and 0.6 s cm^{-1} but for the fruits of many tree species the range was from 30 to 7000 s cm^{-1}. The resistance generally increased with age as the cuticle became thicker, e.g. from 6 s cm^{-1} for a green orange to 1500 s cm^{-1} for mature fruit.

Entomologists use resistance networks to describe the diffusion of water vapour from within an insect either through spiracular valves (analogous to stomata) or dermal and cuticular layers (analogous to the epidermis of a leaf). Precise measurements by Beament (1958) of water loss from a cockroach nymph gave a resistance to vapour transfer of about 2×10^3 s m^{-1}, equivalent to the resistance of about 100 condensed monolayers of stearic acid (Gilby, 1980).

Pores in the eggshells of birds have a function similar to the stomata of leaves, allowing the inward diffusion of oxygen to the developing embryo and the outward diffusion of carbon dioxide and water vapour. Tullett (1984) reviewed the structure and function of eggshells. Unfortunately, avian physiologists conventionally express the porosity g of eggshells in units of water loss per egg per day per unit water vapour pressure difference across the shell. The failure to normalize for surface area obscures the fact that the diffusion resistance of eggshells is almost independent of egg size. As an illustration, the relationship between g and egg mass W (g) derived by Ar *et al.* (1974) can be written as

$$g = 37.5 \times 10^{-9} W^{0.78} \text{ g s}^{-1} \text{ kPa}^{-1}$$

For eggs with the same geometry and constant density but different size, surface area A is proportional to $W^{0.66}$ (see p. 201), and so the porosity per unit shell area g/A has only a weak dependence on egg mass ($W^{0.12}$). When calculated by the methods used to define leaf resistances, eggshell resistances to water vapour diffusion are typically between about 500 and 1000 s cm^{-1} for eggs which range in mass from about 5–500 g.

Resistance calculations

Electrical analogues provide a useful way of visualizing the process of diffusion from the intercellular spaces of a leaf through stomatal pores to the external boundary layer. Figure 9.8 shows a very simple analogue in which electrical current flows through a thin sheet of metal between two electrodes XX and YY represented by broken lines. Suppose that the resistance of the sheet is 6 ohms so that the drop in voltage between the two electrodes is 6 volts when the current flowing between them is 1 amp. The pecked lines represent points of equal voltage across the sheet and from the symmetry of the system these equipotential lines must be parallel to the electrodes. Between each pair of adjacent lines the voltage drops by 1 volt so the material between them has a resistance of 1 ohm. The path of the current is represented by the bold lines at right angles to the equipotential lines and the arrows show the direction of current flow.

In this example, the equipotential lines form a very simple pattern which could be drawn from first principles. The distribution of lines in a more complicated system can be determined by using sheets of paper which have been impregnated with graphite so that they conduct electricity although the resistance per unit length is relatively large. Low resistance electrodes of appropriate shape are drawn on the paper with silver paint and the shapes of the equipotential lines are determined with a field plotter. Figure 9.9 shows the pattern of lines on a sheet which had two parallel electrodes drawn 60 cm long and 60 cm apart. The sheet was cut along a line midway between the

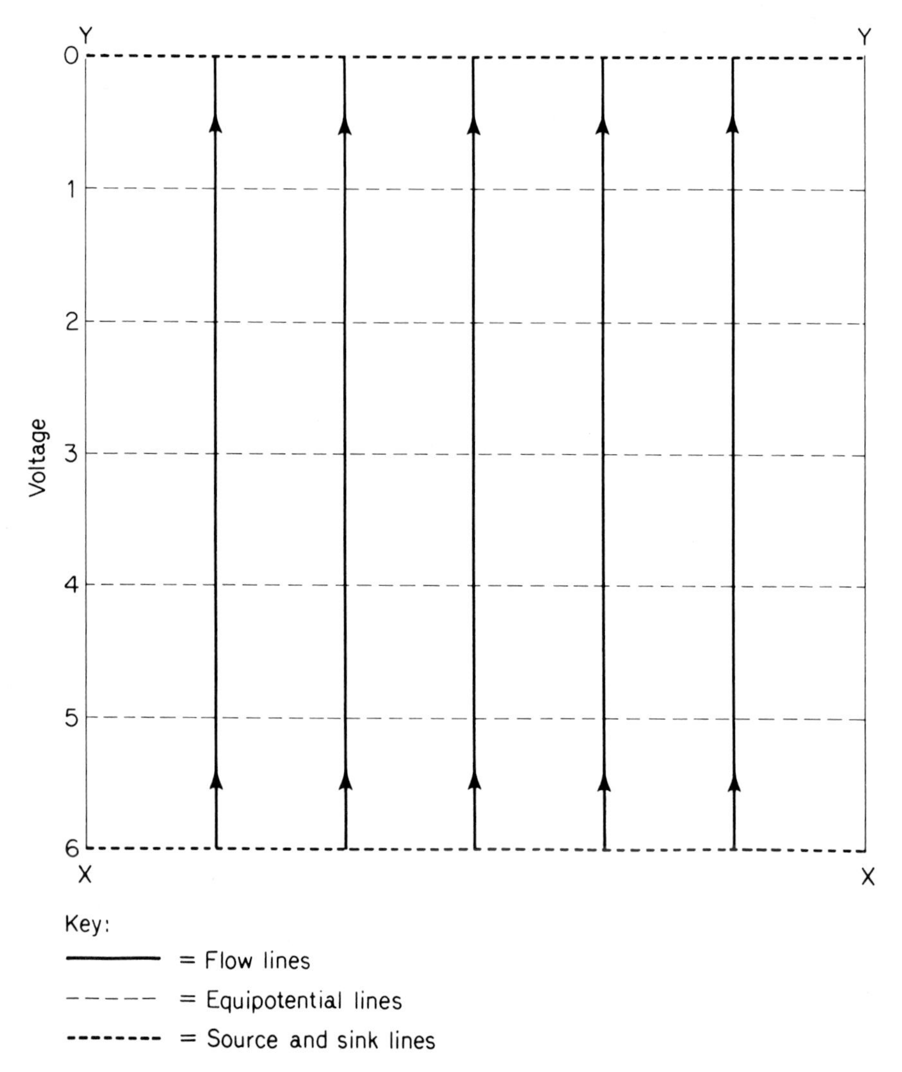

Fig. 9.8 Two-dimensional electrical analogue of diffusion between parallel planes XX and YY. The continuous lines represent flow of current and the pecked lines join points of equal potential.

electrodes so that in the centre of the sheet the flow of current was confined to a neck of conducting paper only 7 mm wide. The way in which the flow of current converged on the neck is shown by the lines of flow drawn at right angles to the equipotential lines. The additional resistance imposed by cutting the paper can be determined by counting the equipotential lines which are separated by unit resistance as in Fig. 9.8. For the complete sheet there were 6 lines (counting one electrode as a line) whereas the cut sheet was found to have 16 lines (some were omitted for clarity). The cut therefore introduced 10

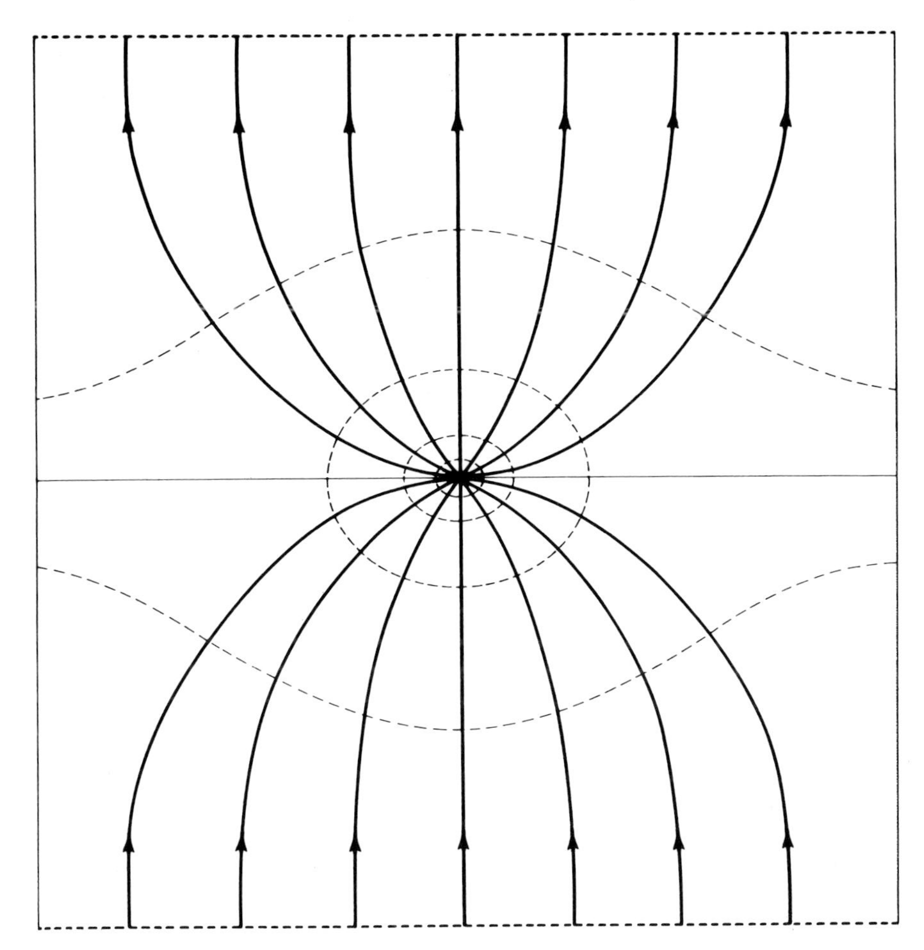

Fig. 9.9 As Fig. 9.8 when a plate with a single hole is introduced between the two planes. The plate is treated as infinitely thin.

additional units of resistance or 5 units for each side of the neck.

Figure 9.9 can be regarded as a two-dimensional analogue for the diffusion of a gas through a circular hole with a diameter d in a plate whose thickness is much less than d. According to the theory of gaseous diffusion in three dimensions, the resistance of each side of a circular hole is $r = \pi d/8D$ where D is the diffusion coefficient of the gas. This quantity corresponds to the 5 units of electrical resistance in the two-dimensional analogue. The resistance of a laminar boundary layer of thickness t is $r_a = t/D$ and this corresponds to 3 units of resistance in the analogue.

The next stage in developing a realistic stomatal analogue is to introduce a pore length l comparable with its diameter d. When an analogue was cut from conducting paper with $l = 10$ mm, $d = 7$ mm, the number of equipotential lines increased from 16 to 26. The additional 10 lines appeared within the pore at right angles to the walls and the rest of the analogue was identical to

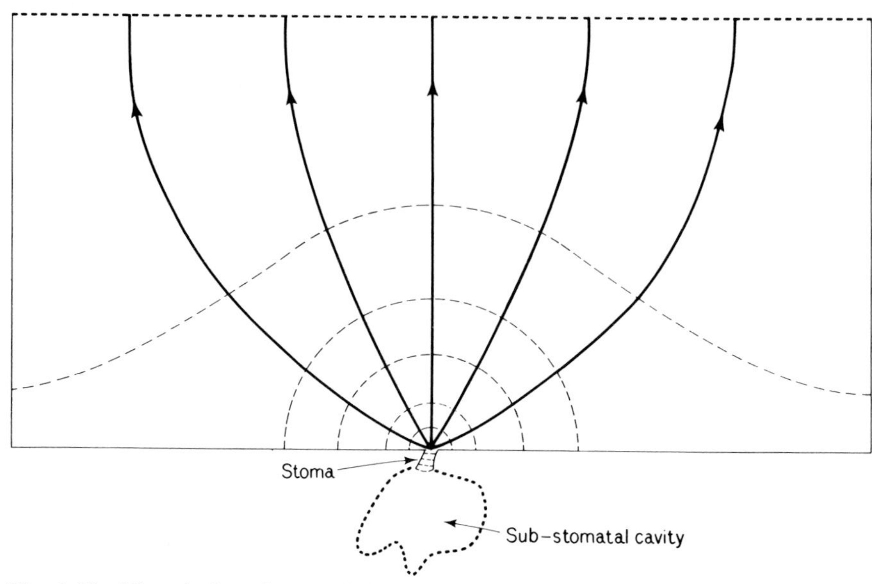

Fig. 9.10 Electrical analogue of diffusion of water vapour from a single stomatal pore. Note the absence of equipotential lines in the sub-stomatal cavity showing that the 'end-correction' can be neglected at the inner end of the pore.

Fig. 9.9. In three dimensions, the diffusion resistance equivalent to the lines within a uniform pore is $r_p = l/D$. The total resistance of a pore r_t is found by adding r_p to the resistance r_h at either end of the pore, i.e. $r_t = r_p + 2r_h$. For stomata, r_h is often smaller than r_p and is therefore referred to as an 'end-correction'.

Figure 9.10 shows the cross section of a real pore and substomatal cavity for a leaf of *Zebrina pendula*. As the cross section is not uniform the resistance of the pore cannot be calculated accurately without knowing the shape of the cross section and its area A at different distances from the end of the pore. An approximate value of r_p can be obtained from the length and mean diameter d of a pore with circular cross section or from the axes of an elliptical pore. The end-correction for a circular pore is usually assumed to be $2r_h = \pi d/4D$ where d is a representative diameter but Fig. 9.10 shows that it would be incorrect to apply a conventional end-correction to the inner end of the pore. If the substomatal cavity is assumed to be lined with cell walls from which water is evaporating (represented by the dotted electrode), the resistance of the outer end of the pore is equivalent to 6 equipotential lines, but the inner end has only 2 lines. For many leaves, $r_t = r_p + r_h = (l + \pi d/8)/\mathrm{D}$ is probably a better estimate of the resistance of a single pore than $r_t = r_p + 2r_h$.

Finally, the resistance of a multi-pore system can be estimated. As mesophytes usually have about 100 stomata per mm^2 their average separation is about 0.1 mm or 100 μm, an order of magnitude larger than the maximum diameter of the pore. At this spacing, there is little interference between the equipotential shells of individual pores (Fig. 9.11).

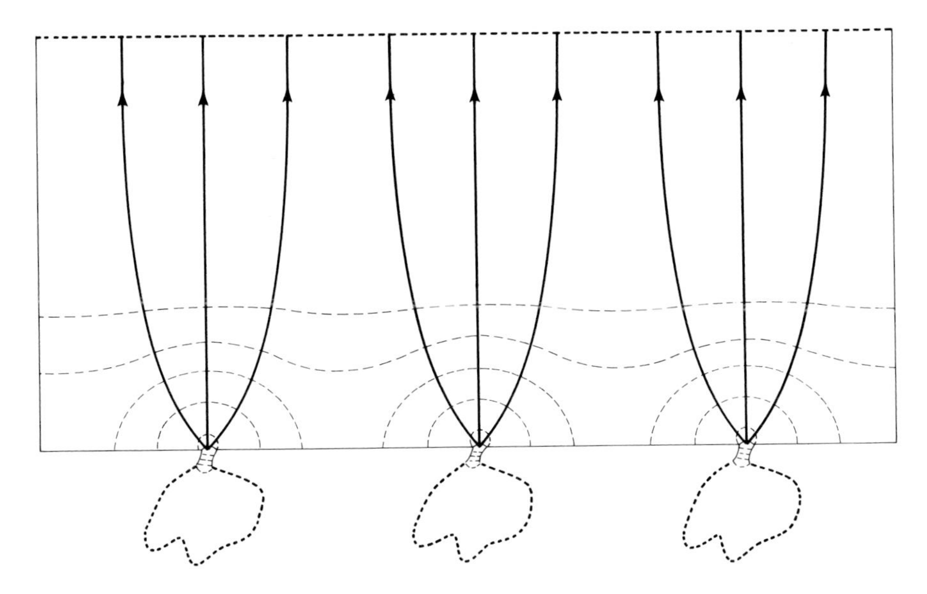

Fig. 9.11 Electrical analogue for a leaf epidermis showing three pores. Note the merging of the equipotential lines which decrease the effective end-correction for the outer end of each pore.

When there are n pores per unit leaf area, the resistance of a set of pores r_s can be readily derived from the resistance of the individual pores r_t. For example, if $\delta\chi$ is the difference of water vapour concentration maintained across a set of circular pores with a mean diameter of d, the transpiration rate can be written either as

$$\mathbf{E} = \frac{\delta\chi}{r_s} \quad \text{or as} \quad \frac{n\pi(d^2/4)\delta\chi}{r_t}$$

It follows that

$$r_s = \frac{4(l + \pi d/8)}{\pi n d^2 D}$$

Similar expressions in a variety of units were derived by Penman and Schofield (1951) and Meidner and Mansfield (1968).

Milthorpe and Penman (1967) evaluated the stomatal resistance of wheat leaves with rectangular pores. Refinements in their calculations included: (i) allowing for the stomatal slit getting shorter as the stoma closed; (ii) making the diffusion coefficient a function of the stomatal width to allow for the effect of 'slip' at the stomatal walls. This phenomenon is important when the width is comparable with the mean free path of the diffusing molecules. For example, when the width of the throat was 1 μm, the diffusion coefficient for water vapour was 88% of its value in free air; (iii) making the end-correction for the inner end of the pore 1.5 times the correction for the outer end.

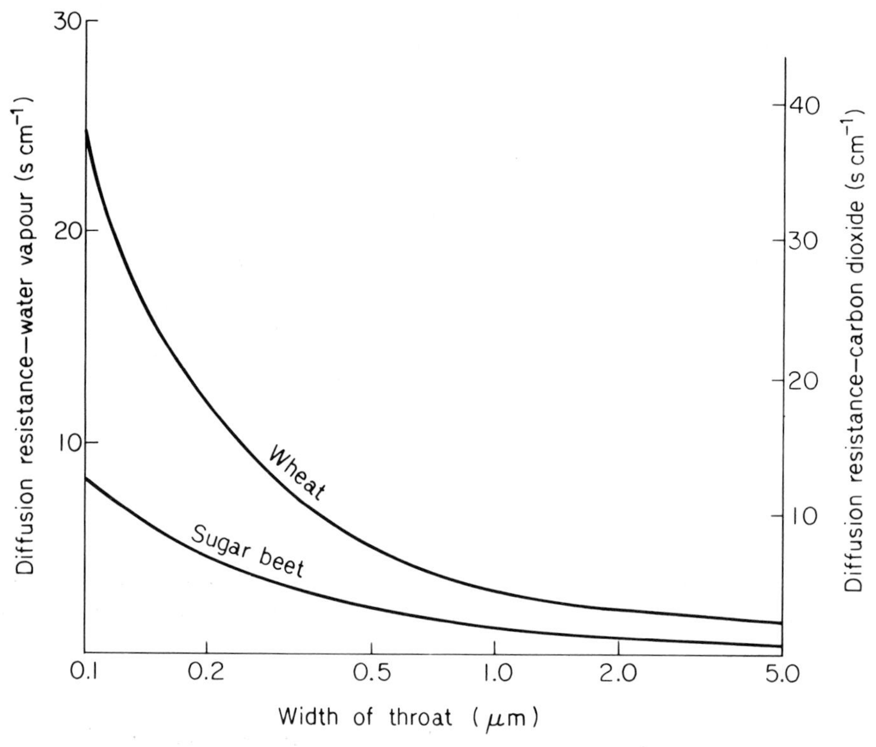

Fig. 9.12 Diffusion resistance calculated for wheat and sugar beet leaves as a function of throat width. Resistances for water vapour and carbon dioxide are given on left- and right-hand axes respectively.

(Figure 9.10 suggests that this factor should have been smaller rather than greater than unity.) Figure 9.12 shows the relation between resistance and slit width (a) from figures tabulated by Milthorpe and Penman for wheat stomata, assumed rectangular, and (b) from figures tabulated by Biscoe (1969) for sugar beet stomata, assumed elliptical.

Figure 9.12 can be used to estimate the effect of stomatal closure on the *total* resistance to diffusion of water vapour or carbon dioxide for a leaf, taking account of the distribution of stomata over the abaxial and adaxial surfaces. On the assumption that the wheat has amphistomatous leaves with the same resistance r_s on each epidermis, the total resistance for each surface is the sum of two resistances in series, i.e. $r_s + r_V$, and the resistance of the whole leaf is the sum of two resistances in parallel, i.e.

$$\left(\frac{1}{(r_V + r_s)} + \frac{1}{(r_V + r_s)}\right)^{-1} = (r_V + r_s)/2$$

When r_V is much smaller than r_s, the leaf resistance is approximately $r_s/2$. On the assumption that beet has hypostomatous leaves, the resistances of the two surfaces are $r_V + r_s$ (abaxial) and $r_V + x$ (adaxial) where x is a resistance much larger than r_s representing the resistance of the cuticle. Combining these resistances in parallel gives the total leaf resistance as

$$\left(\frac{1}{r_V+r_s}+\frac{1}{r_V+x}\right)^{-1}\approx r_V+r_s$$

which is approximately equal to r_s when r_V is much smaller than r_s. (Measurements with model bean leaves by Thom (1968) suggest that the appropriate value of r_V for a hypostomatous leaf may be about 30% smaller than the value for a flat plate of the same size because of increased exchange round the edge of the plate.)

In practice, the two surfaces of an amphistomatous leaf often have different stomatal resistances. Relevant equations have been published in the literature but are too cumbersome to reproduce here. As an additional complication, the stomata on the two surfaces may respond in different ways to levels of irradiance and of the water stress in the mesophyll tissue.

Mass transfer and pressure

Gale (1972) and others have drawn attention to the fact that the stomatal resistance of a leaf with specified stomatal geometry should be proportional to pressure because the diffusion coefficient of a gas is inversely proportional to pressure. At sea level, the normal range of pressure (95 to 102 kPa) is equivalent to a change in stomatal resistance trivial in comparison with the uncertainty of most measurements but differences of pressure are much larger between sea-level and the tops of mountains. At a height of 3000 m, for example, atmospheric pressure is only 70% of its value at sea-level implying that stomatal resistance r_s should be smaller by the same factor.

Boundary layer resistance also decreases with pressure. The aerodynamic resistance for water vapour transfer r_V is defined as t/D where the boundary layer thickness t is proportional to $\nu^{0.5}$ for a flat plate and therefore to $p^{-0.5}$. But since D is proportional to p^{-1}, r_V is proportional to $p^{0.5}$.

At first sight, the implication of the pressure dependence of r_s and r_V is that evaporation rate **E** should increase with height, all other factors being equal. This is correct if the gradient for water vapour transfer is defined in terms of a fixed difference in vapour density $\delta\chi$ since $\mathbf{E}=\delta\chi/(r_s+r_V)$. However, if the gradient is defined by a difference of specific humidity so that $\mathbf{E}=\rho\delta q/(r_s+r_V)$, the presence of density in the numerator makes **E** proportional to $p^{1/2}$ when r_V is large compared with r_s and insensitive to pressure in the more common case when r_V is small compared with r_s.

The same argument holds for rates of photosynthesis which will be effectively independent of pressure when the concentration gradient is expressed either as a volume concentration (volume CO_2 per unit volume air) or as a specific mass (g CO_2 per g air).

COATS AND CLOTHING

Few attempts have been made to determine resistances to water vapour transfer within the coats of animals and in clothing.

The physical processes responsible for the diffusion of vapour within hair were explored by Cena and Monteith (1975c) who compared rates of transfer through fibreglass and through sections of sheep's fleece, uncured and cured to remove grease. Only about 2% of the space within each sample was occupied by hair. At 16°C, the resistance of the fibreglass was about 4.3 s cm^{-1} per cm depth, close to the theoretical value for static air (i.e. the reciprocal of the molecular diffusion coefficient D). The resistivity of cured and uncured fleece was less than the value for fibreglass and the difference increased with the depth of the sample, suggesting that the transfer of vapour by diffusion was augmented by the capillary movement of water along hairs. Webster *et al.* (1985) found that the resistivity of pigeon's plumage was about twice the value for still air, presumably because the porosity of the samples was much smaller than that of fleece.

Gatenby *et al.* (1983) measured vapour concentration within the fleece of a ewe standing in a constant temperature room. The depth of fleece was about 7 cm and the concentration decreased linearly with distance from the skin at about 0.6 g m^{-3} per cm at 5°C ambient temperature, increasing to 1.0 g m^{-3} per cm at 28°C. If molecular diffusion is assumed, corresponding fluxes of latent heat range from about 3.5 to 6.0 W m^{-2}, much smaller than metabolic heat production (see Chapter 12). The real rate of latent heat transfer was probably substantially larger because of free convection associated with temperature gradients (see p. 143 and Cena and Monteith, 1975b).

Free convection also increases vapour transfer in clothing. In a study of transfer in a tropical fatigue uniform by Breckeridge cited by Campbell *et al.* (1980), the vapour conductance increased linearly with wind speed from about 0.17 cm s^{-1} in still air to about 5 cm s^{-1} at 6 m s^{-1} (equivalent to a resistance range from 6 to 1.6 s cm^{-1}). The conductance for heat transfer was larger (see p. 143) by an amount consistent with radiative transfer.

Both in man and in birds, the resistance of skin to vapour diffusion is of the order of 100 to 200 s cm^{-1} and so is much larger than the resistance of clothing or feathers. Exceptionally, measurements on premature babies lying naked in incubators revealed that skin resistance could be as small as 30 s cm^{-1} and that the associated loss of latent heat could exceed metabolic heat production if the circulating air were not humidified (Wheldon and Rutter, 1982).

10

MASS TRANSFER

(ii) Particles

Small particles are transferred in the free atmosphere by the same process of turbulent diffusion that is responsible for the mass transfer of gases. Because particles have inertia, they cannot respond to the most rapid eddy motions, but this is generally unimportant in the usual scales of atmospheric turbulence. However, inertia is important close to surfaces, when particles may be thrown against an object if the air stream which they are in changes direction rapidly. A second distinction between particles and molecules is the importance of gravitational forces, and so we will begin our discussion of mass transfer of particles with this topic.

STEADY MOTION

Sedimentation velocity

The gravitational force on a particle is the difference between the weight of the particle and the weight of the air that it displaces. This is given by the product of the volume of the particle, gravitational acceleration g, and the difference between particle density ρ and air density ρ_a. When spherical particles with radius r and volume $(4\pi r^3/3)$ fall under gravity they attain a steady sedimentation velocity V_s when the gravitational force balances the drag force, i.e. from equation (7.7)

$$(4/3)\pi r^3 g(\rho - \rho_a) = (1/2)\rho_a V_s^2 c_d \pi r^2 \tag{10.1}$$

where c_d is the drag coefficient (p. 105). For particles of natural origin, and for most pollutant particles, ρ is generally much larger than ρ_a and so equation (10.1) may be written approximately as

$$V_s^2 \simeq 8rg\rho/3\rho_a c_d \tag{10.2}$$

For particles obeying Stokes Law, i.e. those for which the particle Reynolds number Re_p is less than about 0.1 (p. 110), equation (7.8) showed that $c_d = 12\nu/V_s r$, where ν is the kinematic viscosity of air. Substituting in equation (10.2) gives

$$V_s = 2\rho g r^2/9\rho_a \nu \tag{10.3}$$

Equation (10.3) enables V_s to be calculated directly when $Re_p < 0.1$, but for particles with $Re_p > 0.1$, equation (10.2) should be used together with an estimate of c_d from Fig. 7.6 or equation (7.9) to find V_s by trial and error. For example, to find the sedimentation velocity of a rain drop, radius 100 μm, a *first approximation* can be estimated from equation (10.3), $V_s = 1.2 \text{ m s}^{-1}$. The corresponding Reynolds number is 16 (confirming that equation (10.3) is strictly not applicable), and the drag coefficient (from equation (7.9)) is about 3. The drag force on the drop (equation (10.1)) is

$$(1/2)\rho_a V_s^2 c_d \pi r^2 = 82 \times 10^{-9} \text{ N}$$

which does not balance the gravitational force

$$(4/3)\pi r^3 \rho g = 42 \times 10^{-9} \text{ N}$$

A smaller value of V_s must therefore be estimated and the drag force recalculated. The value of V_s when drag exactly balances the gravitational force may then be found by interpolation, graphically or otherwise. This method is of general applicability. Figure 10.1 shows a specific case—the variation of V_s with radius for particles of unit density $\rho = 1 \text{ g cm}^{-3}$.

Biological aerosols such as pollen and spores can be treated as spheres of unit density and so the equations given here are adequate for finding their sedimentation velocities. Fig. 10.1 shows that Stokes Law ($Re_p \lesssim 0.1$) holds to about $r = 30$ μm ($V_s = 0.1 \text{ m s}^{-1}$), a size range that includes most spores and pollen.

Particles of soil and pollutant materials are seldom spherical, and their density is seldom exactly 1 g cm^{-3} (assumed in Fig. 10.1). Such particles are often characterized by their *Stokes diameter*, which is the diameter of a sphere having the same density and sedimentation velocity. Water drops with radius exceeding about 0.4 mm are appreciably flattened as they fall, and this increases c_d. Figure 10.1 shows observed values of V_s for water drops. When $r > 2$ mm any further increase in drop weight is compensated for by an increase in deformation, and hence in c_d, and so V_s remains approximately constant. Raindrops much larger than 3 mm radius are seldom observed, as they tend to break up during their fall, but they may be maintained in strong updraughts in storm clouds.

NON-STEADY MOTION

If a particle obeying Stokes Law is projected horizontally in still air with velocity V_0 at $t = 0$, its motion is subject to the drag force $6\pi\nu\rho r V(t)$ where $V(t)$ is the velocity at time t, and to the gravitational force mg. Resolving the motion into horizontal and vertical components, and writing dx/dt and d^2x/dt^2 for the horizontal velocity and acceleration respectively, the equation of motion for horizontal displacement is

$$m\frac{d^2x}{dt^2} = -6\pi\nu\rho r\frac{dx}{dt} \qquad (10.4)$$

Similarly, for vertical motion,

$$m\frac{d^2z}{dt^2} = mg - 6\pi\nu\rho r\frac{dz}{dt} \qquad (10.5)$$

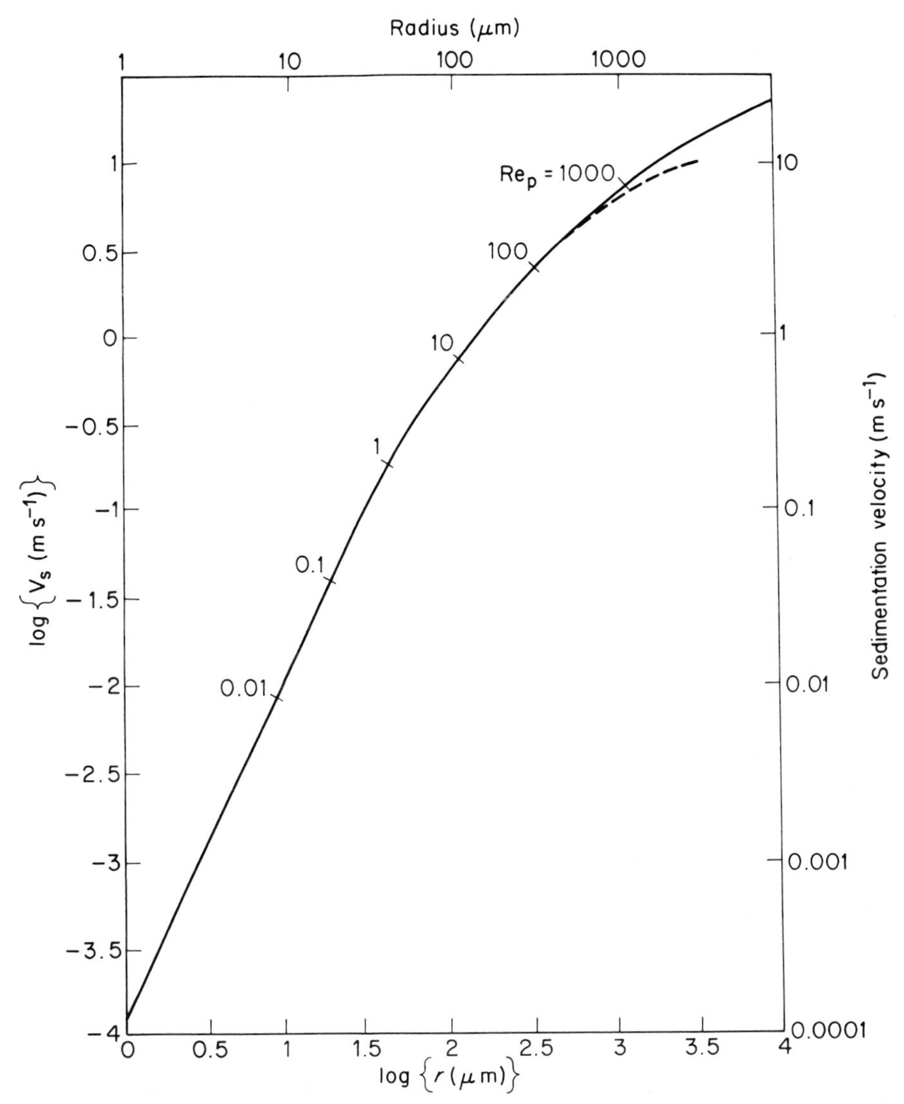

Fig. 10.1 Dependence of sedimentation velocity on particle radius for spherical particles with density 1 g cm^{-3}. Stokes' Law applies for $\text{Re}_p < 0.1$. The pecked line shows the sedimentation velocity of raindrops (from Fuchs, 1964).

The quantity $m/6\pi\nu\rho r$ has the dimension [T], and is called the **relaxation time** τ. The equations of motion may thus be written

$$\frac{d^2x}{dt^2} = \tau^{-1}\frac{dx}{dt} \tag{10.6}$$

and

$$\frac{d^2z}{dt^2} = g - \tau^{-1}\frac{dz}{dt} \tag{10.7}$$

For horizontal motion, integration of equation (10.6) gives

$$\frac{dx}{dt} = V_0 \exp\left(-\frac{t}{\tau}\right)$$

and

$$x(t) = \tau V_0[1 - \exp(-t/\tau)]$$

When the horizontal component of velocity is zero, $x = V_0\tau$ and this is called the **stopping distance** l.

Integration of the equation for vertical motion (equation (10.7)) leads to the case considered earlier for sedimentation velocity. The vertical acceleration becomes zero when $dz/dt = g\tau$ and the particle then moves at the sedimentation velocity V_s.

Thus the equations of motion lead to

$$l = V_0\tau$$

$$V_s = g\tau$$

The terms **relaxation time**, **stopping distance** and **sedimentation velocity** have a central role in particle physics.

PARTICLE DEPOSITION

In the absence of gravitational or other external forces, particles are deposited on objects by two processes: diffusion (Brownian motion, Chapter 3); and impaction. Impaction is a consequence of particles with significant inertia failing to follow precisely the streamline on which they move initially as they approach an obstacle. Figure 10.2 illustrates the trajectory of a particle near a cylinder.

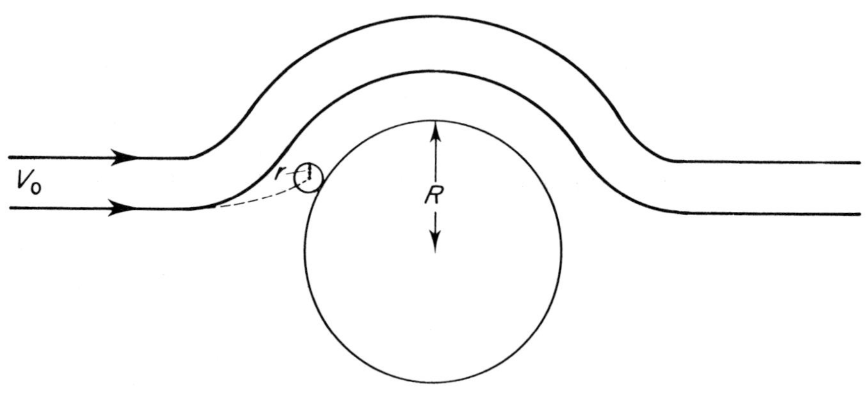

Fig. 10.2 Impaction of a particle on a cylinder.

As the streamline changes direction rapidly, the particle inertia does not allow it to turn so sharply. To hit the surface, the particle must penetrate the boundary layer within which the flow velocity decreases to zero. The probability of particle impaction therefore depends on the ratio of the particle stopping distance l to the boundary layer thickness δ. Equation 7.1 (p. 103) shows that boundary layer thickness is proportional to $(\text{dimension})^{0.5}$ and inversely proportional to $(\text{velocity})^{0.5}$; small obstacles and large flow velocities therefore favour impaction.

A good example of the importance of stopping distance and boundary layer thickness in determining impaction can be seen if soap bubbles are released upwind of an isolated tree trunk or other large cylindrical obstruction. The bubbles, with low mass and large drag, have short stopping distances, fail to penetrate the boundary layer, and move around the obstacle. In contrast, golf balls of the same size and travelling towards the obstacle at the same velocity would not slow appreciably in passing through the boundary layer, and would hit the obstacle. (They would also demonstrate dramatically another phenomenon that limits the deposition of dry particles onto surfaces—**'bounce-off'**.)

The dimensionless ratio of the stop distance of a particle to the characteristic dimension of an object (e.g. the radius of a cylinder) is called the **Stokes number**, Stk, and provides a useful way of comparing results of different impaction studies.

To describe the deposition by impaction, two further terms are commonly used. The **efficiency of impaction** c_p for particles on an object is defined as the number of impacts on the object divided by the number of particles that would have passed through the space in the same time if the object had not been there (i.e. the potential number of impacts).

The **deposition velocity** v_d is defined as number of impacts per unit area per second divided by the number of particles per unit volume in the air stream, and if the relevant area of the object is taken as the cross-sectional area exposed to the flow, then

$$c_p = v_d/V$$

where V is the flow velocity.

Figure 10.3 shows the calculated dependence of c_p on Stk for cylinders at two Reynolds numbers: at large Re the streamlines curve more sharply than at small Re, and so c_p is increased. The Figure also includes a number of observations with spores, using sticky cylinders, and a curve based on measurements of droplet impaction. Droplets impact with efficiencies close to the theoretical values although they may fragment on impact. The slightly lower values of c_p for spores may have been because some spores bounced off, or were dislodged from the cylinders even though the surfaces were sticky.

A good example in nature of the high impaction efficiency of droplets when $\text{Stk} > 10$ occurs when cloud drops, radius ~10 μm, impact on pine needles, $R \sim 0.5$ mm, when forests are enveloped in low cloud. If the drops are moving at windspeed of 5 m s^{-1}, $l \simeq 6$ mm and so $\text{Stk} \simeq 12$. The impaction efficiency of such drops on larger objects such as rain gauges is much less, and so this source of water to the forest is called **occult precipitation** (occult = hidden) because it is not recorded in normal gauges.

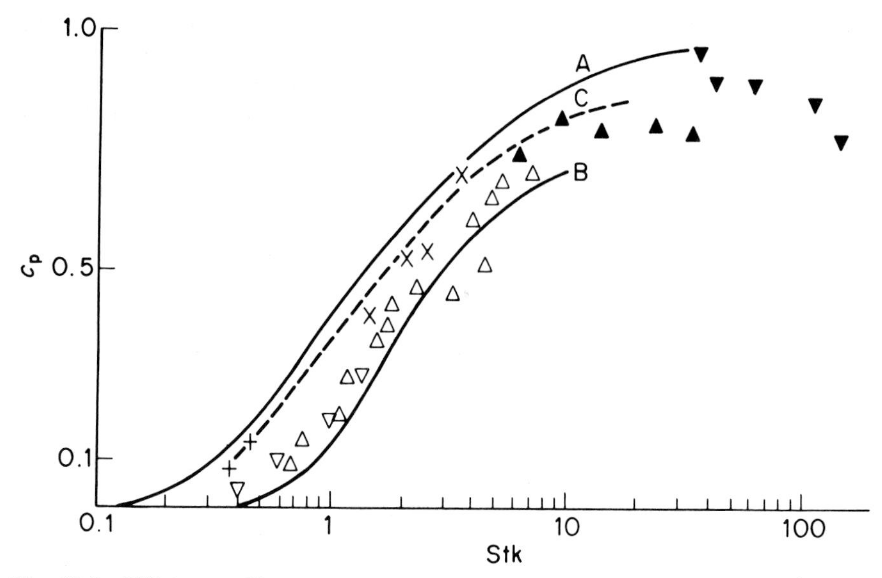

Fig. 10.3 Efficiency of impaction of particles (radius r) on cylinders (radius R) (from Chamberlain, 1975): A and B are theoretical relationships for flow Reynolds numbers Re > 100 and Re = 10 respectively. Line C is fitted to experimental measurements with droplets. The plotted points are measurements with spores impacting on sticky cylinders as follows:

Symbol	+	×	▽	△	▲	▼
r (μm)	2.3	6.4	15	15	15	15
R (mm)	0.1–0.8	0.1–0.8	10	3.3	0.4	0.09

The capture of mist (Fig. 10.4(a)), on spider's threads ($R \sim 0.1$ μm) is even more efficient than the previous example because the drops ($r = 10$ μm), much larger than the obstacle, are scarcely deflected, and any drops passing within a distance r of a thread are likely to be captured by *interception*. Interception is an important mechanism of deposition for particles with size comparable with or larger than the obstacles that they approach.

Chamberlain (1975) and Chamberlain and Little (1981) reviewed the experimental data for particle deposition on vegetation. Stickiness, or wetness of surfaces, seems to be an important factor for the retention of impacting dry particles in the size range of spores and pollen (10–30 μm radius) on leaves and stems. For example Chamberlain (1975) exposed barley straw to ragweed pollen (r about 10 μm) at windspeeds of 1.55 m s^{-1} and found that c_p increased from 0.04 to 0.31 when the straw was made sticky. The presence or absence of a soft layer to absorb particle momentum on impact, thus avoiding 'bounce-off', probably explains these results. Bounce-off is most pronounced with large particles, high velocities (large momentum), small obstacles (thin boundary layer), and large coefficients of restitution at the surface (small loss of kinetic energy on impact).

Aylor and Ferrandino (1985) used ragweed pollen (radius about 15 μm, mass 11.0 ng) and *Lycopodium* spores (radius about 10 μm, mass 3.9 ng) to study bounce-off from glass rods in a wind tunnel and from wheat stems in the tunnel and in the field. They derived a relative retention factor F defined as the ratio of the catch on non-sticky cylinders to that on sticky cylinders, and related F to the kinetic energy on impact KE_i. They showed by numerical integration of momentum equations that KE_i was related to the flow velocity u_0 in the tunnel and to Stokes number by

$$KE_i = 0.5 m_p u_0^2 [f(\text{Stk})]^2$$

where m_p is the mass of particle and the function $f(\text{Stk})$ is given by

$$f(\text{Stk}) = 0.236 \ln(\text{Stk} - 0.06) + 0.684$$

Figure 10.5(a) shows the variation of F with KE_i for *Lycopodium* and ragweed particles impacting on glass rods with diameters 3, 5 and 10 mm. For both types of particles, F was near unity for $KE_i < 10^{-12}$ J then decreased by about 2 orders of magnitude as KE_i increased to 10^{-11} J. The corresponding particle speed at the threshold for bounce-off was about 40 cm s^{-1} for *Lycopodium* and 70 cm s^{-1} for ragweed; the threshold kinetic energy for bounce-off was independent of particle diameter.

Relative retention on wheat stems in the wind tunnel (Fig. 10.5(b), filled symbols) did not decrease as steeply with KE_i as for glass rods, but the threshold for bounce-off appeared similar to that for the rods. The greater scatter is probably a consequence of variability in surface structure of the stems. Bounce-off occurs when the kinetic energy of a particle after impact exceeds the potential energy of attraction at the surface. Surface structure is likely to influence both the energy absorption on impact and the potential energy of attraction, and so variation in bounce-off is likely with biologically variable surfaces.

In the field (Fig. 10.5(b), open symbols), the average value for F for a given KE_i was smaller than in the tunnel. At low values of KE_i this was probably because turbulent variation of windspeed about the mean caused KE_i to range from sub-critical (constant F, no bounce-off) to well above super critical (F decreasing rapidly with KE_i), and time spent at super critical KE_i lowered the average value of F. At higher values of KE_i this explanation is unlikely to apply, and the lower retention observed may then have been the result of strong gusts of wind re-entraining particles that had previously deposited.

Table 10.1 Deposition of particles on segments of real leaves, and on filter paper, as a proportion of deposition on sticky artificial leaves made of PVC (from Chamberlain, 1975)

	Relative deposition					
Particle	Diameter (μm)	Grass	Plantain	Clover	Filter paper	Sticky pvc
Lycopodium spore	32	0.45	0.26	0.18	0.70	1.00
Ragweed pollen	19	0.15	0.11	0.23	0.68	1.00
Polystyrene	5	1.74	1.82	3.25	1.98	1.00
Tricresylphosphate	1	1.70	2.60	5.50	6.40	1.00
Aitken nuclei	0.08	1.06	1.70	0.86	1.54	1.00

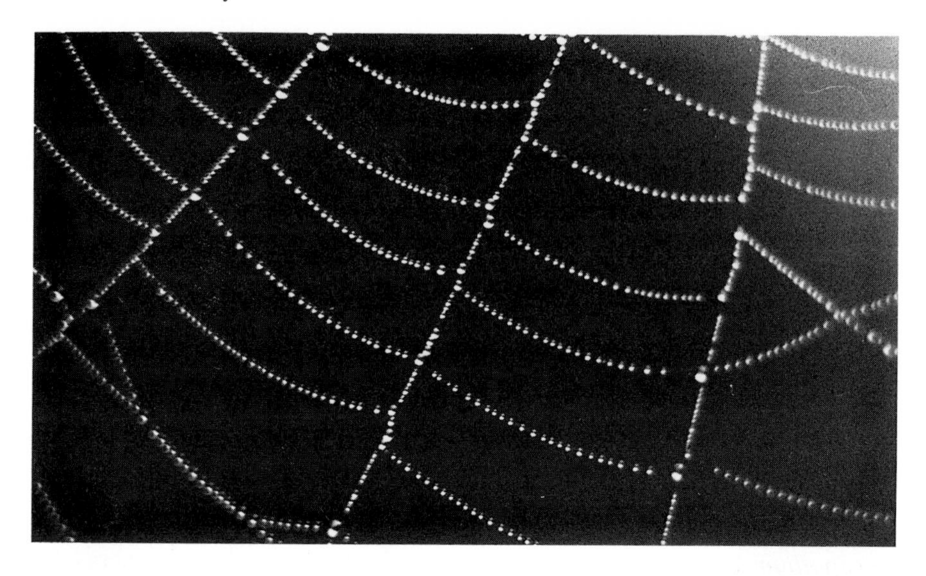

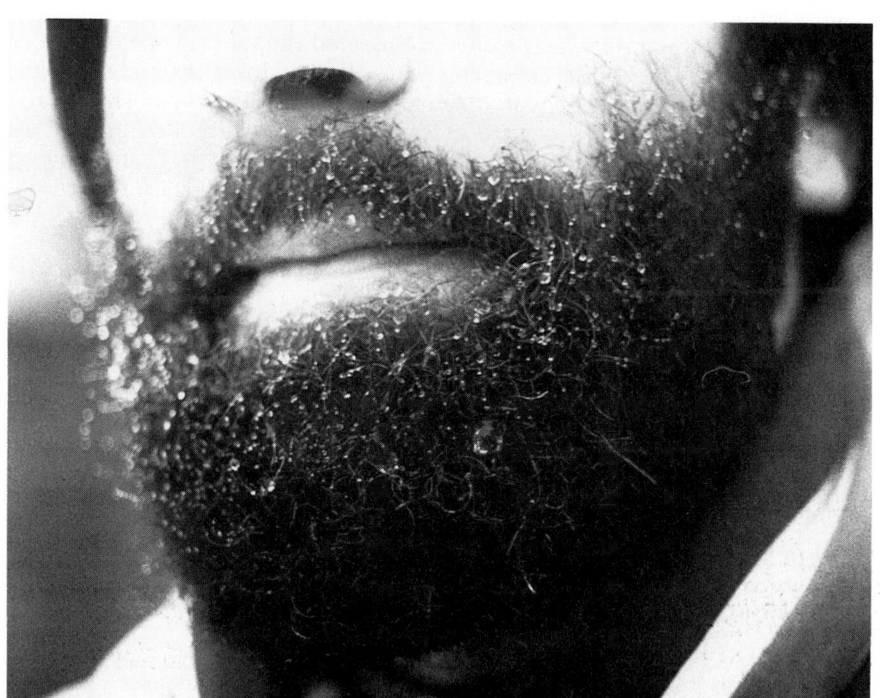

Fig. 10.4 (a) Drops of fog which have impacted on and been intercepted by threads of a spider's web. The threads of the web are at least an order of magnitude smaller than individual drops (typically 10 μm diameter) and so are efficient collectors by interception (p. 170). (b) Drops of fog collected on an author's beard after cycling. Human hair is about 50 μm diameter, so the drops visible in this picture must have coalesced from many impacted fog drops.

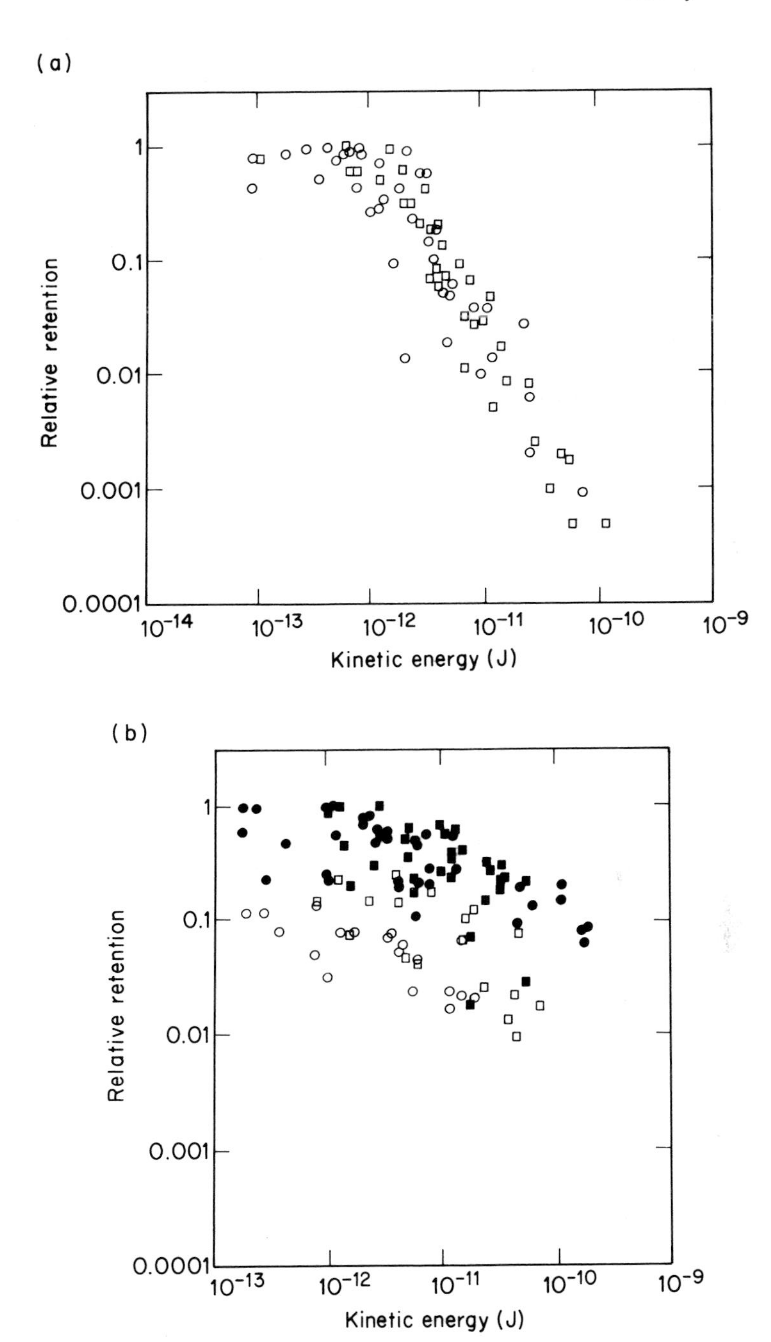

Fig. 10.5 Variation with kinetic energy at impact of the relative retention for ragweed (○) and *Lycopodium* (□) particles impacting (a) on glass rods, diameters 3, 5 and 10 mm, in a wind tunnel, and (b) on wheat stems 3–4 mm diameter in the wind tunnel (filled symbols) or in the field (open symbols).

In contrast to the factors influencing capture of particles larger than about 10 μm, the capture of small particles ($r < 10$ μm) by leaves is relatively uninfluenced by surface wetness or stickiness, but is enhanced by the presence of hairs or surface irregularities which probably act as efficient micro-impaction sites. Table 10.1 shows the relative deposition of particles of a range of sizes onto various surfaces exposed in a stand of artificial grass in a wind tunnel operating at about 4.5 m s^{-1}. The sticky surface was most efficient at capturing the spores and pollen but was not effective for the smaller particles.

The relative importance of deposition by the mechanisms of impaction, interception, sedimentation and diffusion can be assessed from Table 10.2. Displacement by Brownian motion dominates the movement of submicron size particles, whereas impaction and sedimentation increase rapidly with particle size above about 1 μm. As an illustration of the combined influence of all mechanisms on deposition, Fig. 10.6 shows how the deposition velocity of particles to an artificial sward of short grass in a wind tunnel depended on particle size. The steep decrease in v_d as r decreased below 5 μm was caused by decreasing impaction efficiency. When r was less than 0.1 μm, Brownian motion began to be effective and v_d increased. The Figure also shows

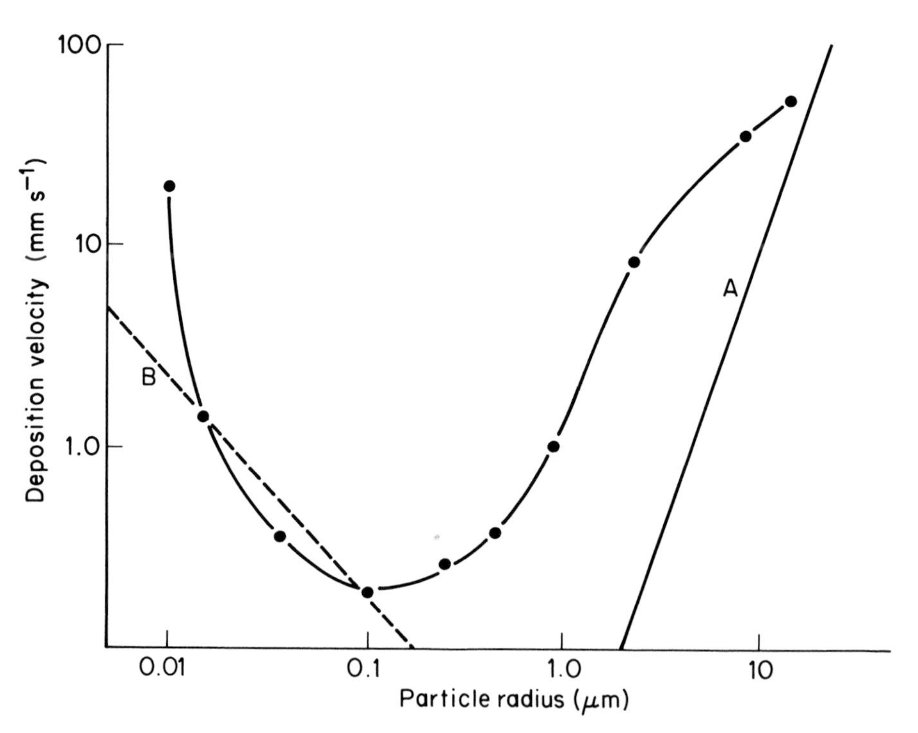

Fig. 10.6 The variation with particle radius of the deposition velocity to short grass exposed at a windspeed of about 2.5 m s^{-1} (from Little and Whiffen, 1977). Line A is the sedimentation velocity of particles with density 1 g cm^{-3}; line B was calculated from Brownian diffusion theory.

sedimentation velocity, demonstrating that sedimentation accounted for a large proportion of the deposition only when particles exceeded 10 μm radius.

Considered together, deposition processes are least efficient for particles of about 0.1–0.2 μm radius, and it is significant that man-made aerosols in this size range are found widely distributed in the Earth's atmosphere. In particular, sulphate particles in this size range, formed by the oxidation of sulphur dioxide, can be carried extremely long distances in the atmosphere, and tend to persist until they encounter conditions of high humidity in which they can grow by delinquescence and condensation into larger droplets which are more effectively deposited.

The figures in Table 10.2 are also relevant to the hazards of dust inhalation. During nasal breathing in man, particles larger than 10 μm are efficiently impacted and intercepted from fast moving air in the nose; as bronchial tubes divide, the average air velocity in each tube decreases, but particles from 1–10 μm have a strong probability of impaction when the air stream changes direction rapidly on branching, and they also sediment onto the walls of non-vertical tubes. Particles smaller than 1 μm are most likely to be deposited deep in the lungs in narrow bronchioles and alveolar sacs where the distance to a site of deposition is comparable with the distance which they can travel by Brownian diffusion during the time that they are within the lungs.

Table 10.2 Characteristic distances for particle transport

Particle Radius r (μm)	0.01	0.1	1	10
Stopping distance (μm) given initial velocity of 1 m s^{-1}	14×10^{-3}	0.23	13	1230
Distance (μm) travelled in 1 second by virtue of:				
Terminal velocity	14×10^{-2}	2.2	128	1200
Brownian diffusion	160	21	5	1.5

11

STEADY STATE HEAT BALANCE

(i) Water surfaces and vegetation

The heat budgets of plants and animals will now be examined in the light of the principles and processes considered in previous chapters. The First Law of Thermodynamics states that when a balance sheet is drawn up for the flow of heat in any physical or biological system, income and expenditure must be exactly equal. In micrometeorology, radiation and metabolism are the main sources of income; radiation, convection and evaporation are methods of expenditure.

For any component of a system, physical or biological, a balance between the income and expenditure of heat is achieved by adjustments of temperature. If, for example, the income of radiant heat received by a leaf began to decrease because the sun was obscured by cloud, leaf temperature would fall, reducing expenditure on convection and evaporation. If the leaf had no mass and therefore no heat capacity, the drop in expenditure would exactly balance the drop in income, second by second. For a real leaf with a finite heat capacity, the drop in temperature would lag behind the drop in radiation and so would the drop in expenditure, but the First Law of Thermodynamics would still be satisfied because the income from radiant heat would be supplemented by the heat given up by the leaf as it cooled.

This chapter is concerned with the heat balance of relatively simple systems in which (a) temperature is constant so that changes in heat storage are zero; and (b) metabolic heat is a negligible term in the heat budget. The heat budget of warm-blooded animals, controlled by metabolism, is considered in the next chapter and examples of diurnal changes of heat storage are considered in Chapters 13 and 15.

HEAT BALANCE EQUATION

The heat balance of any organism can be expressed by an equation with the form

$$\bar{\mathbf{R}}_n + \bar{\mathbf{M}} = \bar{\mathbf{C}} + \boldsymbol{\lambda}\bar{\mathbf{E}} + \bar{\mathbf{G}} \qquad (11.1)$$

where the bars indicate that each term is an average heat flux per unit surface area. (In the rest of this chapter, they are implied but not printed.) In this context, it is convenient to define surface area as the area from which heat is lost by convection although this is not necessarily identical to the area from which heat is gained or lost by radiation. The individual terms are

$\bar{\mathbf{R}}_n$ = net gain of heat from radiation
$\bar{\mathbf{M}}$ = net gain of heat from metabolism
$\bar{\mathbf{C}}$ = loss of sensible heat by convection
$\lambda\bar{\mathbf{E}}$ = loss of latent heat by evaporation
$\bar{\mathbf{G}}$ = loss of heat by conduction to environment

The conduction term $\bar{\mathbf{G}}$ is included for completeness but is negligible for plants and has rarely been measured for animals.

The grouping of terms in the heat balance equation is dictated by the arbitrary sign convention that fluxes directed away from a surface are positive. (When temperature decreases with distance z from a surface so that $\partial T/\partial z<0$, the outward flux of heat $\mathbf{C} \propto -\partial T/\partial z$ is a positive quantity, see p. 20.) The sensible and latent heat fluxes **C** and λ**E** are therefore taken as positive when they represent losses of heat and as negative when they represent gains. On the left-hand side of the equation, **R** and **M** are positive when they represent gains and negative when they represent losses of heat. When both sides of a heat balance equation are positive, the equation is a statement of how the total supply of heat available from sources is divided between individual sinks. When both sides are negative, the equation shows how the total demand for heat from sinks is divided between available sources.

The sections which follow deal with the size and manipulation of individual terms in the heat balance equation, with some fundamental physical implications of the equation, and with a few biological applications.

Convection and long-wave radiation

When the surface of an organism loses heat by convection, the rate of loss per unit area is determined by the scale of the system as well as by its geometry, by windspeed, and by temperature gradients. Convection is usually accompanied by an exchange of long-wave radiation between the organism and its environment at a rate which depends on geometry and on differences of radiative temperature but is independent of scale. The significance of scale can be demonstrated by comparing convective and radiative losses from an object such as a cylinder with diameter d and uniform surface temperature T_0 exposed in a wind tunnel whose internal walls are kept at the temperature T of the air flowing through the tunnel with velocity u. When Re exceeds 10^3, the resistance to heat transfer by convection increases with d according to the relation $r_H = d/(\kappa \mathrm{Nu}) \propto d^{0.4}V^{-0.6}$ (see Table A.5). The corresponding resistance to radiative transfer r_R (p. 28) is independent of d. Figure 11.1 compares r_H and r_R for cylinders of different diameters at windspeeds of 1 and 10 m s^{-1} chosen to represent outdoor conditions. Corresponding losses of heat are shown on the right-hand axis for a surface temperature excess $(T_0 - T)$ of 1K.

Because r_H is similar for planes, cylinders and spheres provided the appropriate dimension is used to calculate Nusselt numbers, a number of generalizations may be based on Fig. 11.1. For organisms on the scale of a small insect or leaf ($0.1<d<1$ cm), r_H is much smaller than r_R, implying that convection is a much more effective mechanism of heat transfer than

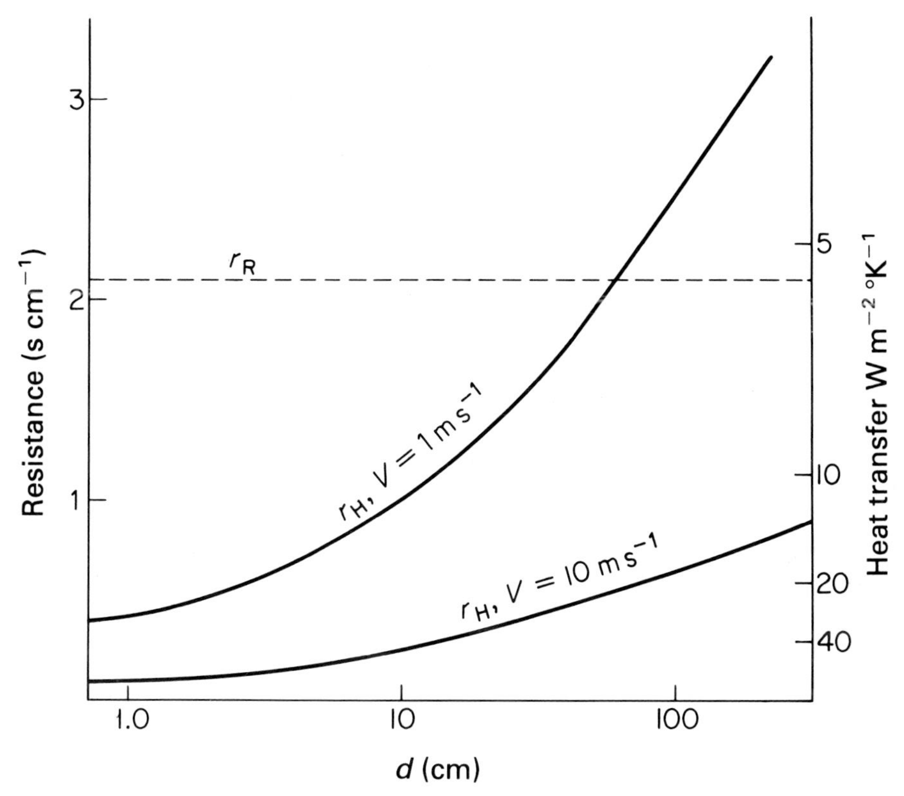

Fig. 11.1 Dependence of a resistance to convective heat transfer r_H and resistance to radiative heat exchange r_R as a function of body size represented by the characteristic dimension of a cylinder d.

long-wave radiation. The organism is tightly coupled to air temperature but not to the radiative temperature of the environment. For organisms on the scale of a farm animal or a man ($10 < d < 100$ cm), r_H and r_R are of comparable importance at low windspeeds. For very large mammals ($d > 100$ cm), r_H can exceed r_R at low windspeeds and in this state the surface temperature will be coupled more closely to the radiative temperature of the environment than to the air temperature. These predictions are consistent with measurements on locusts and on piglets, for example, and they emphasize the importance of wall temperature as distinct from air temperature in determining the thermal balance of large farm animals in buildings with little ventilation.

An organism with an emissivity of unity and a surface temperature of T_0 exchanges heat (a) by convection to air at temperature T and (b) by radiation to an environment with a mean radiative temperature equal to air temperature. The net rate at which heat is gained or lost is

$$\rho c_p\{(T - T_0)/r_H + (T - T_0)/r_R\} = \rho c_p(T - T_0)/r_{HR} \qquad (11.2)$$

where $r_{HR} = (r_H^{-1} + r_R^{-1})^{-1}$ is a combined resistance for convection (p. 22) and long-wave radiation (p. 28) formed by grouping the component resistances in parallel because the fluxes are in parallel.

HEAT BALANCE OF THERMOMETERS

Dry-bulb

As an introduction to the more relevant physics of the wet-bulb thermometer, it is worth considering the implications of the general heat balance equation for a dry-bulb thermometer. For measurements in the open, it is prudent to avoid heating a thermometer by direct exposure to sunlight so screening is employed. In one common design, a cylindrical thermometer bulb is housed in a tube through which air is drawn rapidly and for the sake of the following discussion we assume that the tube completely surrounds the bulb. If the tube itself is exposed to sunshine, its temperature, T_s, may be somewhat above the temperature of the air (T) and of the thermometer(T_t).

The net long-wave radiation received by the thermometer from the housing (assuming $\varepsilon = 1$) is

$$\mathbf{R}_n = \sigma(T_s^4 - T_t^4) = \rho c_p(T_s - T_t)/r_R \tag{11.3}$$

and the loss of heat by convection from the bulb to the air is

$$\mathbf{C} = \rho c_p(T_t - T)/r_H \tag{11.4}$$

In equilibrium, $\mathbf{R}_n = \mathbf{C}$ and rearrangement of terms gives

$$T_t = \frac{r_H T_s + r_R T}{r_R + r_H} \tag{11.5}$$

implying that the temperature recorded by the thermometer is a weighted mean between the temperature of the air which is needed and the temperature of its housing. There are two main ways in which the difference between apparent and true air temperature can be minimized:

(i) by making r_H much smaller than r_R, either by adequate ventilation or by choosing a thermometer with very small diameter (see Fig. 11.1). The numerator in equation (11.5) then tends to $r_R T$ and the denominator to r_R.

(ii) by making T_s very close to T, e.g. by painting the screen white, by introducing insulation between outer and inner surfaces or by increasing ventilation on both sides of the screen.

In the standard Assmann psychrometer, the screen is a double-walled cylinder, nickel-plated on the outer surface and aspirated at about 3 m s^{-1}. Since the diameter of the mercury-in-glass thermometer it contains is about 3 mm, it is effectively decoupled from its radiative environment (Fig. 11.1).

Wet-bulb

The concept of 'wet-bulb temperature' is central to the environmental physics of systems in which latent heat is a major heat-balance component and it has two distinct connotations: the thermodynamic wet-bulb temperature, which is a theoretical abstraction; and the temperature of a thermometer covered with a wet sleeve, which, at best, is a close approximation to the thermodynamic wet-bulb temperature.

A value for the thermodynamic wet-bulb temperature can be derived by considering the behaviour of a sample of air enclosed with a quantity of pure water in a container with perfectly insulating walls. This is an adiabatic system within which the sum of sensible and latent heat remains constant. The initial state of the air can be specified by its temperature T, vapour pressure e and total pressure p. Provided e is smaller than $e_s(T)$, the saturated vapour pressure at T, water will evaporate and both e and p will increase. The increase of latent heat in the system represented by the increase in water vapour concentration must be balanced by a decrease in the amount of sensible heat derived by cooling the air. The process of humidifying and cooling continues until the air becomes saturated at a temperature T' which, by definition, is the thermodynamic wet bulb temperature. The corresponding saturated vapour pressure is $e_s(T')$.

To relate T' and $e_s\,(T')$ to the initial state of the air, the initial water vapour concentration is approximately $\rho\varepsilon e/p$ when p is much larger than e (see p. 12). When the vapour pressure rises from e to $e_s(T')$, the total change in latent heat content per unit volume is $\lambda\rho\varepsilon[e_s(T')-e]/p$. The corresponding amount of heat supplied by cooling unit volume of air from T to T' is $\rho c_p(T-T')$. (A small change in the heat content of the water vapour is included in more rigorous treatments but is usually unimportant in micrometeorological problems.) Equating latent and sensible heat

$$\lambda\rho\varepsilon\{e_s(T')-e\}/p = \rho c_p(T-T') \tag{11.6}$$

and rearranging terms gives

$$e = e_s(T') - (c_p p/\lambda\varepsilon)(T-T') \tag{11.7}$$

The group of terms $(c_p p/\lambda\varepsilon)$ is often called the **'psychrometer constant'** for reasons which appear shortly, but it is not constant (because p is atmospheric pressure and λ changes somewhat with temperature) nor is it exact (because of the approximations made). The psychrometer constant is often assigned the symbol γ and, at a standard pressure of 101.3 kPa, has a value of about 66 Pa K^{-1} at 0°C increasing to 67 Pa K^{-1} at 20°C.

Another useful quantity which has the same dimensions as γ is the change of saturation vapour pressure with temperature or $\partial e_s(T)/\partial T$, usually given the symbol Δ (or s) (see p. 10). This quantity can be used to obtain a simple (but approximate) relation between the saturation vapour pressure deficit $D = e_s(T) - e$ and the wet-bulb depression $B = T - T'$.

When the saturation vapour pressure at wet-bulb temperature is written as

$$e_s(T') \approx e_s(T) - \Delta(T-T') \tag{11.8}$$

where Δ is evaluated at a mean temperature of $(T+T')/2$, the psychrometer equation (11.7) becomes

$$e \approx e_s(T) - (\Delta+\gamma)(T-T') \tag{11.9}$$

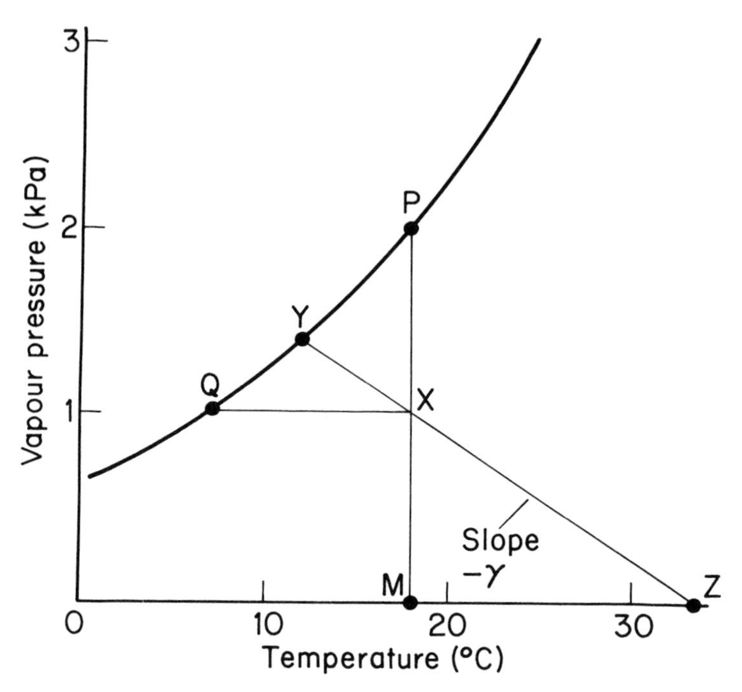

Fig. 11.2 The relation between dry bulb temperature, wet bulb temperature, equivalent temperature, vapour pressure and dew point. The point X represents air at 18°C and 1 kPa vapour pressure. The line YXZ with a slope of $-\gamma$ gives the wet bulb temperature from Y (12°C) and the equivalent temperature from Z (33.3°C). The line QX gives the dew point temperature from Q (7.1°C). The line XP gives the saturation vapour pressure from P (2.1 kPa).

or

$$D \approx (\Delta + \gamma)B \tag{11.10}$$

Equation (11.7) can be represented graphically by plotting $e_s(T)$ against T (Fig. 11.2). The curve QYP represents the relation between saturation vapour pressure and temperature and the point X represents the state of any sample of air in terms of e and T. Suppose the wet bulb temperature of the air is T' and that the point Y represents the state of air saturated at this temperature. The equation of the straight line XY joining the points (T,e), $(T',e_s(T'))$ is

$$e - e_s(T') = \text{slope} \times (T - T') \tag{11.11}$$

Comparison of equations (11.7) and (11.11) shows that the slope of XY is $-\gamma$. The wet bulb temperature of any sample of air can therefore be obtained by drawing a line with slope $-\gamma$ through the appropriate coordinates T and e to intercept the saturation curve at a point whose abscissa is T'.

If a sample of air in the state given by X were moved towards Y, the path XY shows how temperature and vapour pressure would change in adiabatic evaporation, i.e. with the total heat content of the system constant. Similarly, starting from Y and moving to X, the path YX shows how T and e would change if water vapour were condensed adiabatically from air that was initially saturated. As condensation proceeded, the temperature of the air would rise until all the vapour had condensed. This state is represented by the point Z at which $e = 0$. The corresponding temperature T_e is called the **'equivalent temperature'** of the air. As Z has coordinates $(T_e, 0)$, the equation of the line ZX can be written in the form

$$T_e = T + e/\gamma \tag{11.12}$$

Alternatively, the equation of YZ can be written

$$T_e = T' + e_s(T')/\gamma \tag{11.13}$$

showing that the equivalent and wet bulb temperature are uniquely related. Both T' and T_e remain unchanged when water is evaporated or condensed adiabatically within a sample of air.

Moving now from principles to a real wet-bulb thermometer, it is necessary to account for the finite rate at which heat is lost by evaporation and gained by convection and radiation.

Suppose that a bulb covered with a wet sleeve has a temperature T_w when it is exposed to air at temperature T and surrounded by a screen at air temperature. The rate at which heat is gained by convection and radiation is

$$\mathbf{C} + \mathbf{R}_n = \rho c_p((T - T_w)/r_{HR} \tag{11.14}$$

and the rate at which latent heat is lost is

$$\begin{aligned} \boldsymbol{\lambda}\mathbf{E} &= (\lambda\rho\varepsilon/p)\{e_s(T_w) - e\}/r_V \\ &= \rho c_p\{e_s(T_w) - e\}/\gamma r_V \end{aligned} \qquad 11.15$$

In equilibrium, $\boldsymbol{\lambda}\mathbf{E} = \mathbf{R}_n + \mathbf{C}$ from which

$$e = e_s(T_w) - \gamma(r_V/r_{HR})(T - T_w) \tag{11.16}$$

It is often convenient to regard $(\gamma r_V/r_{HR})$ (or simply $(\gamma r_V/r_H)$ in some circumstances) as a modified psychrometer constant, written γ^*.

Comparing equations (11.7) and (11.16), it is clear that the measured wet-bulb temperature will not be identical to the thermodynamic wet-bulb temperature unless $r_V = r_{HR}$. Because $r_V = (\kappa/D)^{0.66} r_H = 0.93 r_H$ (p. 147), this condition implies that

$$0.93 r_H = (r_H^{-1} + r_R^{-1})^{-1} \tag{11.17}$$

from which $r_H = 0.075 r_R$. At 20°C, $r_R = 2.1$ s cm^{-1} and $r_H = 0.17$ s cm^{-1}. A wet-bulb thermometer will therefore record a temperature above or below the thermodynamic wet-bulb temperature depending on whether r_H is greater or less than this value.

Because both r_V and r_H are functions of windspeed and r_R is not, γ^* decreases with increasing windspeed and when r_V is much less than r_R tends to a constant value independent of windspeed, viz.

$$\gamma^* = \gamma(r_V/r_H) = 0.93\gamma \tag{11.18}$$

In the Assmann psychrometer, regarded as a standard for measuring vapour pressure in the field, the resistances corresponding to the specification already given are $r_V = 0.149$ s cm^{-1}, $r_H = 0.156$ s cm^{-1} giving $\gamma^* = 63$ Pa K^{-1}. A much more detailed discussion of psychrometry leading to a standard value of 62 Pa K^{-1} is given by Wylie (1979). With this and similar instruments, the error involved in using γ instead of γ^* is often negligible in micrometeorological work.

A further source of psychrometer error not considered here is the conduction of heat along the stem of the thermometer which can be minimized by using a long sleeve and/or a thermometer with very small diameter.

HEAT BALANCE OF SURFACES

Wet surface

The physics of the wet-bulb thermometer is the key to solving a wide range of problems concerned with the exchange of sensible and latent heat between wet surfaces and their environment. In this context, 'wet' can mean covered with pure water or with a salt solution.

We consider first a surface of pure water over which air is moving. In the free atmosphere, temperature and vapour pressure are T and e and corresponding potentials at the water surface are T_0 and $e_s(T_0)$. Resistances for heat and vapour transfer between the surface and the point where T and e are measured are r_H and r_V respectively.

Using the standard convention that fluxes away from surfaces are inherently positive, the surface will *gain* heat by convection at a rate given by

$$\mathbf{C} = \rho c_p(T - T_0)/r_H \qquad (11.19)$$

and will *lose* latent heat at a rate given by equation (11.15) replacing T_w by T_0 and γr_V by $\gamma^* r_H$ to give

$$\boldsymbol{\lambda}\mathbf{E} = \rho c_p\{e_s(T_0) - e\}/\gamma^* r_H \qquad (11.20)$$

To start with the simplest heat balance, the system will be treated as adiabatic so that

$$\boldsymbol{\lambda}\mathbf{E} + \mathbf{C} = 0 \qquad (11.21)$$

If $e_s(T_0)$ were simply a linear function of T, it would now be possible to eliminate T_0 from equations (11.19) to (11.21) and to evaluate $\boldsymbol{\lambda}\mathbf{E}$ and $\mathbf{C}$ as functions of the temperature and vapour pressure of the airstream and of the two resistances. To achieve this objective, it is legitimate to *assume* a linear relation over a narrow range of, say, 10 K. The saturation vapour pressure at T_0 may then be related to the corresponding pressure at air temperature by using equation (11.8) in the form

$$e_s(T_0) \approx e_s(T) - \Delta(T - T_0) \qquad (11.22)$$

where Δ must be evaluated at T because T_0 is unknown at this stage of the analysis.

Substituting this (approximate) value of $e_s(T_0)$ in equation (11.20) and eliminating T_0 using equations (11.19) and (11.21) gives

$$\boldsymbol{\lambda}\mathbf{E} = -\mathbf{C} = \frac{\rho c_p\{e_s(T) - e\}r_H^{-1}}{\Delta + \gamma^*} \tag{11.23}$$

Provided the process is adiabatic, **C** is the rate at which heat released when air is cooled to the wet-bulb temperature T_0 (Fig. 11.1). More precisely, it is the heat released by unit volume of air per unit area of surface and per unit time. (Unit volume per unit area and time has dimensions of m s^{-1} which is the reciprocal of the resistance unit used to specify the rate of exchange in these equations.)

Suppose now that the process of heat exchange is not adiabatic. If the surface receives additional heat by radiation at a rate $\mathbf{R}_n$, the heat balance equation becomes

$$\boldsymbol{\lambda}\mathbf{E} + \mathbf{C} = \mathbf{R}_n \tag{11.24}$$

The solution of equations (11.19), (11.20) and (11.24) is, using eq. 11.22,

$$\boldsymbol{\lambda}\mathbf{E} = \frac{\rho c_p\{e_s(T) - e\}r_H^{-1}}{\Delta + \gamma^*} + \frac{\Delta \mathbf{R}_n}{\Delta + \gamma^*} \tag{11.25}$$

By comparing equations (11.23) and (11.25), it is seen that (11.25) still contains the adiabatic term unchanged, but cooling a parcel of air to wet-bulb temperature is now only part of the story. The surface receives additional energy from radiation and may therefore be warmer than the wet-bulb temperature. When a parcel of air stays saturated as it is warmed by an amount δT, the sensible heat content of the air increases in proportion to δT, and equations (11.20) and (11.22) imply that the latent heat content increases in proportion to $(\Delta/\gamma^*)\delta T$. It follows that the fraction of $\mathbf{R}_n$ allocated to sensible heat will be $\mathbf{R}_n/(1 + \Delta/\gamma^*)$ and the complementary fraction allocated to latent heat will be $(\Delta/\gamma^*)\mathbf{R}_n/(1 + \Delta/\gamma^*)$. The second term in equation (11.25) can therefore be identified as the diabatic component of latent heat loss associated with the additional supply of heat from radiation (Fig. 11.3).

To recap, in order to estimate the rate at which a wet surface exchanges sensible and latent heat with the air above it, it is necessary to know the temperature of the surface. When this is not given, it can be eliminated from the equations describing the system by assuming that the saturation vapour pressure is a linear function of temperature—a valid approximation when the temperature range is small. (Alternatively, a solution could be found by iteration within a computer program.) H. L. Penman (1948) was the first to demonstrate this procedure and his formula is often written by combining the terms in equation (11.25) to give

$$\boldsymbol{\lambda}\mathbf{E} = \frac{\Delta \mathbf{R}_n + \rho c_p\{e_s(T) - e\}r_H^{-1}}{\Delta + \gamma^*} \tag{11.26}$$

(See also (11.26a) for a related form of this equation, p. 187.) The complementary expression for sensible heat loss found by putting $\mathbf{C} = \mathbf{R}_n - \boldsymbol{\lambda}\mathbf{E}$ is

$$\mathbf{C} = \frac{\gamma^* \mathbf{R}_n - \rho c_p\{e_s(T) - e\}r_H^{-1}}{\Delta + \gamma^*} \tag{11.27}$$

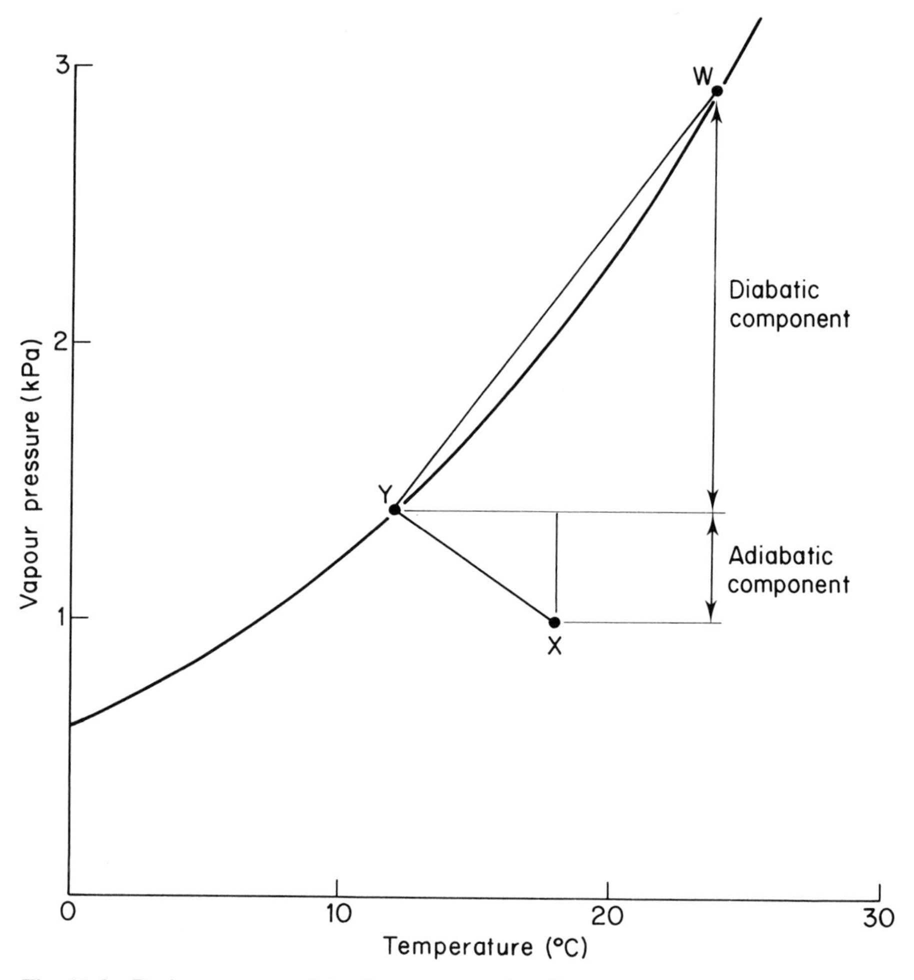

Fig. 11.3 Basic geometry of the Penman equation (from Monteith, 1981a). A parcel of air at X is cooled adiabatically to Y (cf. Fig. 11.2) and heated diabatically to W. Corresponding quantities of latent heat are given in equation (11.25).

To find an expression for surface temperature, equations (11.19) and (11.27) are combined to give

$$T = T_0 + \underset{\text{(diabatic)}}{\frac{(\gamma^* r_H/\rho c_p)\mathbf{R}_n}{\Delta + \gamma^*}} - \underset{\text{(adiabatic)}}{\frac{\{e_s(T) - e\}}{\Delta + \gamma^*}} \qquad (11.28)$$

This equation shows that the temperature of a wet surface will be warmer or cooler than that of the air passing over it depending on whether the diabatic term proportional to $\mathbf{R}_n$ is greater or less than the adiabatic term proportional to $\{e_s(T) - e\}$.

A minor defect in the original Penman equation was that $\mathbf{R}_n$ was a specified quantity although its exact value is always a (weak) function of surface temperature. This difficulty can be overcome by the expedient of replacing $\mathbf{R}_n$ by $\mathbf{R}_{ni}$, the radiation that the surface would receive if it were at air temperature (p. 57). The fact that the surface is not usually at air temperature is taken into account by using the resistance for convective and radiative heat loss in parallel, r_{HR}, in place of r_H (p. 179). Then

$$\boldsymbol{\lambda}\mathbf{E} = \frac{\Delta \mathbf{R}_{ni} + \rho c_p\{e_s(T) - e\}r_{HR}^{-1}}{\Delta + \gamma^*} \tag{11.26a}$$

where $\gamma^* = \gamma(r_H + r_V)/r_{HR}$.

To calculate the evaporation from an open water surface for periods of a week or more, Penman estimated net radiation, saturation deficit, temperature and windspeed from climatological data, used an empirical function of windspeed to estimate r_H, and assumed $r_V = r_H$. He also made the assumption, implicit in the derivation given here, that heat storage in the water was negligible compared with the value of $\mathbf{R}_n$. For relatively shallow tanks of water of the type used by Penman this assumption is valid when the averaging period is several days. The greater the depth of water, the greater the averaging period must be for heat storage to be safely neglected, and for very deep lakes the period would have to be at least a year. As depth and heat storage increase, the month of maximum evaporation moves later and later in the year until it is out of phase with the annual radiation cycle.

To conclude this section, the dependence of evaporation rate on weather is briefly considered. It is clear from inspection of equation (11.26) that the rate of evaporation from water increases linearly with the absorption of net radiation and with the value of the saturation deficit $\{e_s(T) - e\}$. It also increases with windspeed because r_H decreases with increasing wind. Because $\boldsymbol{\lambda}\mathbf{E}$ is a function of Δ which increases with temperature, evaporation rate depends on temperature. Differentiation of equation (11.26) with respect to T (with saturation deficit kept constant) shows that the fractional change of $\boldsymbol{\lambda}\mathbf{E}$ is related to the fractional change of T by

$$\frac{1}{\boldsymbol{\lambda}\mathbf{E}}\frac{\partial(\boldsymbol{\lambda}\mathbf{E})}{\partial T} = \left(\frac{\Delta}{\Delta + \gamma^*}\right)\left\{\frac{\mathbf{R}_n}{\boldsymbol{\lambda}\mathbf{E}} - 1\right\}\frac{1}{\Delta}\frac{\partial \Delta}{\partial T} \tag{11.29}$$

It follows that $\boldsymbol{\lambda}\mathbf{E}$ will increase or decrease with temperature depending on whether its initial value is less than or greater than $\mathbf{R}_n$ (i.e. on whether the surface is cooler or warmer than the air). However, because $\Delta/(\Delta + \gamma^*)$ is less than 1, because $\mathbf{R}_n/\boldsymbol{\lambda}\mathbf{E}$ is often close to unity, and because $(\partial\Delta/\partial T)/\Delta$ is only about 0.3% at 20°C, the temperature dependence of $\boldsymbol{\lambda}\mathbf{E}$ is negligible.

Finally, equation (11.28) implies that increasing windspeed (decreasing r_H) will always decrease surface temperature in a system where γ^* is effectively independent of windspeed.

Leaf

The equations for the sensible and latent heat exchange of a leaf are formally identical to those presented for a wet surface provided the resistances for heat

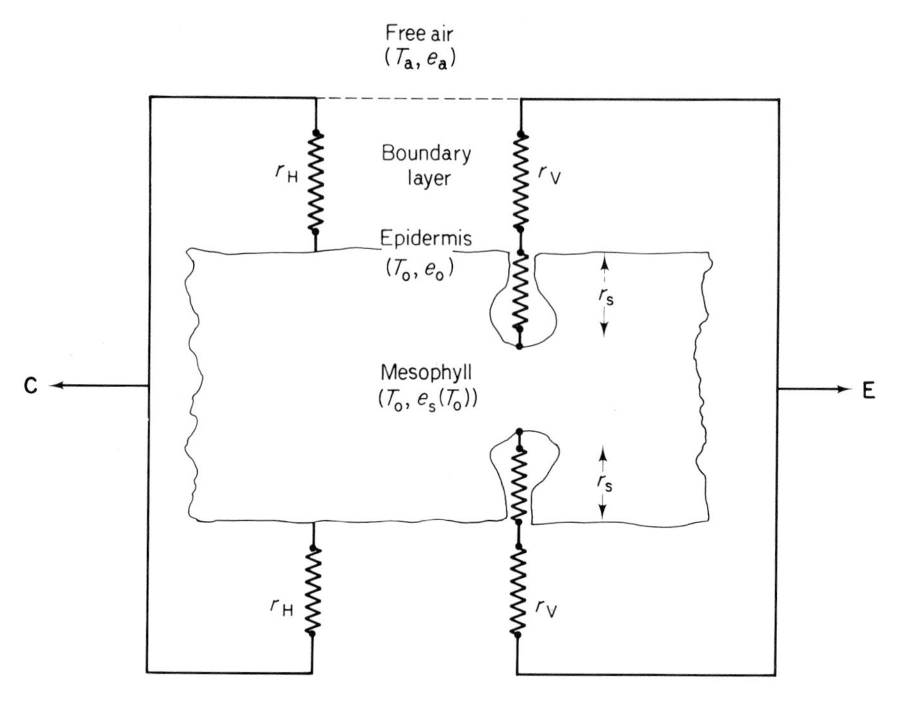

Fig. 11.4 Electrical analogue for transpiration from a leaf and leaf heat balance (see also Fig. 9.6).

and water vapour transfer are specified in an appropriate way. Figure 11.4 shows an equivalent circuit in which the resistance for heat transfer is r_H for each side of the leaf and $r_H/2$ for the two sides in parallel. The resistance to vapour transfer for each side of the leaf is the sum of a boundary layer resistance r_V and a stomatal resistance r_s. The ratio γ^*/γ therefore assumes values of $(r+r_s) \div (r_H/2)$ for a hypostomatous leaf and $(r_V+r_s)/2 \div r_H/2$ for an amphistomatous leaf with the same stomatal resistance on both surfaces. In general

$$\begin{aligned} \gamma^* &= n\gamma(r_V + r_s)/r_H \\ &\approx n\gamma(1 + r_s/r_H) \end{aligned} \tag{11.30}$$

where $n = 1$ (amphistomatous leaf) or $n = 2$ (hypostomatous leaf).

To apply equation (11.26) to a leaf, it is necessary to assume that the mean temperature of the epidermis, which is the site for sensible heat exchange, is the same as the mean temperature of wet cell walls which are the site of latent heat exchange. Most leaves are so thin that this assumption is fully justified. It is also safe to assume that the metabolic term in equation (11.1) is negligible compared with $\mathbf{R}_n$. During the day when a net amount of energy is stored by photosynthesis, $\mathbf{M}/\mathbf{R}_n$ is never more than a few percent and it is even less at night when heat is generated by respiration.

Despite the fact that equation (11.26) can be applied to a leaf as well as to a wet surface, the two systems differ significantly in the following ways.

(i) Dependence of transpiration rate on radiation and saturation deficit. For a water surface, evaporation rate is finite even when *either* net radiation *or* saturation deficit is zero and increases linearly with either of these variables when the other is held constant. For leaves in their natural environment, r_s depends strongly on solar radiation and, in the absence of light, stomata are usually closed so that transpiration is effectively zero. (Evaporation may occur very slowly through a waxy cuticle.) Moreover, substantial evidence has accumulated both from the field and from work in controlled environments that many plants close their stomata as saturation deficit increases, presumably as a mechanism for conserving water. When this response is evinced, transpiration rate will not increase in proportion to saturation deficit and may even reach a maximum value beyond which it decreases as the air gets drier still. The physical basis for this response must lie in the behaviour of the guard

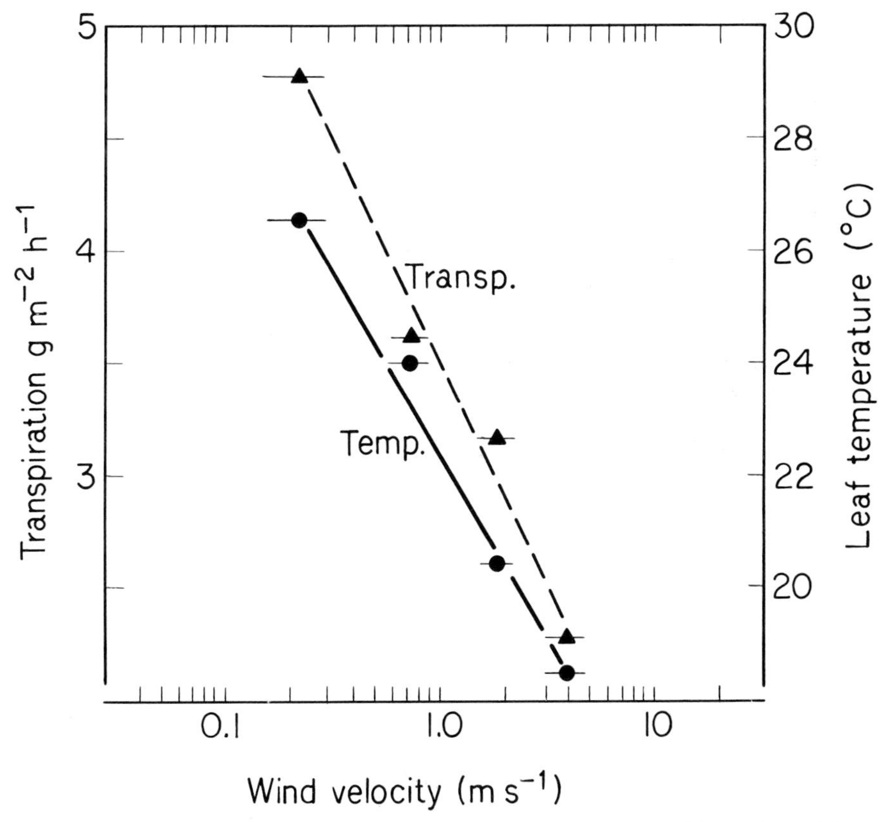

Fig. 11.5 The change of transpiration rate and leaf temperature with windspeed for a *Xanthium* leaf exposed to radiation of 700 W m^{-2} at an air temperature of 15°C and 95% relative humidity (from Mellor *et al.*, 1964).

cells which open and close the stomatal pore but the mechanism is still a matter for debate.

(ii) Dependence of transpiration and temperature on windspeed.

With increasing windspeed, the rate of evaporation from a wet surface always increases and surface temperature decreases. For a leaf, it can be shown by differentiating equation (11.26a) with respect to r_{HR} that $\mathbf{\lambda E}$ is independent of r_{HR} and hence of windspeed when $\mathbf{\lambda E/C} = \Delta/(n\gamma)$. (Note that equation (11.26a) is used in preference to (11.26) for this analysis to avoid the complication that $\mathbf{R}_n$ is a function of surface temperature and therefore of r_H). When $\mathbf{\lambda E/C}$ exceeds this critical value, an increase of windspeed increases the latent heat loss at the expense of the sensible loss in such a way that $\mathbf{\lambda E + C}$ stays constant. This behaviour is expected intuitively because the evaporation from a free water surface always increases with windspeed. When $\mathbf{\lambda E/C}$ is smaller than the critical value,

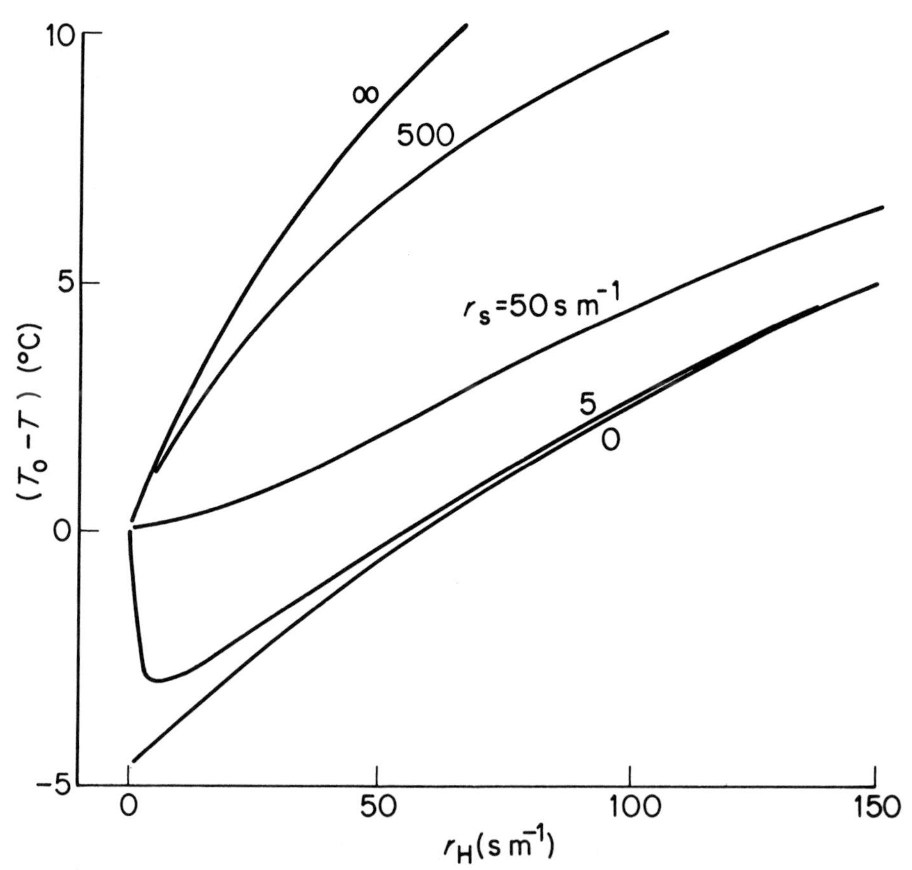

Fig. 11.6 Predicted difference between surface and air temperature for an amphistomatous leaf in sunlight with specified boundary layer and stomatal resistances (for both laminae in parallel). Assumed microclimate: $\mathbf{R}_{ni} = 300$ W m^{-2}, $T = 20$°C, saturation deficit 1 kPa (from Monteith, 1981b).

however, an increase of windspeed increases **C** at the expense of **λE**: evaporation decreases as windspeed increases. This behaviour has been demonstrated in the laboratory (Fig. 11.5) and inferred from measurements in the field. One of the practical implications is that providing shelter does not necessarily benefit plants in saving water.

The dependence of leaf temperature on r_H and by implication on windspeed is shown in Fig. 11.6 for an arbitrary set of weather variables corresponding to bright sunshine in a temperate climate. When stomata are shut (resistance assumed infinite), the temperature excess of the leaf surface increases somewhat more slowly than r_H (i.e. the response is concave to the resistance axis) because allowance was made for the decrease of $\mathbf{R}_n$ with increasing surface temperature. When r_s is very large (500 s m^{-1} is a representative figure for partly closed stomata), curvature becomes much greater because, provided $\boldsymbol{\lambda}\mathbf{E}/\mathbf{C} < \Delta/\gamma$, sensible heat loss decreases (and latent heat loss increases) as windspeed declines. When stomata are fully open ($r_s = 50$ s m^{-1}) the curvature becomes convex to the

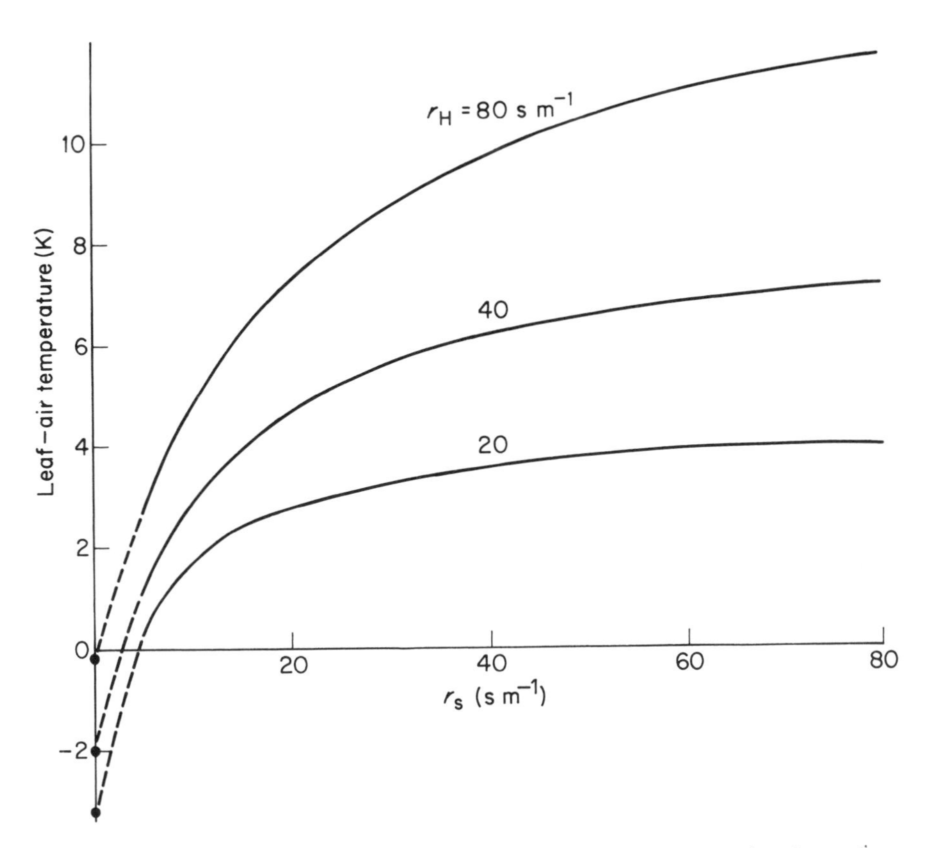

Fig. 11.7 Excess of leaf over air temperature as function of stomatal and aerodynamic resistances when $\mathbf{R}_{ni} = 300$ W m^{-2}, $D = 1$ kPa, $T = 20$°C (see also Fig. 11.6).

resistance axis (for small resistances at least) because in this regime $\mathbf{\lambda E/C} > \Delta/\gamma$ and sensible heat loss increases with decreasing windspeed.

During rain, drops are intercepted by foliage and 1–2 mm of water may be retained within the canopy. Subsequently, the rate of evaporation of this water is faster than the transpiration rate because it is not limited by the stomatal resistance. In regions where rain is frequent, the total loss of water by evaporation from forests can be substantially increased by the direct evaporation of intercepted water (Calder, 1977; Shuttleworth, 1988). The line marked $r_s = 0$ corresponds to a leaf with water covering its surface and therefore behaving like a wet-bulb exposed to radiation. In very strong wind, the limiting value of $T - T_0$ is the wet-bulb depression of the air, $T - T_w$. When r_s is given the unrealistically small value of 5 s m^{-1}, the dependence of surface temperature on r_H is almost the same as for water when r_H exceeds 50 s m^{-1} but $T_0 - T$ has a minimum at a very small value of r_H and approaches zero as r_H approaches zero. By writing equation (11.28) in terms of $\mathbf{R}_{ni}$ and r_{HR} and evaluating $\partial T/\partial r_{HR}$, it can be shown that a minimum temperature is reached when

$$\frac{r_{HR}^2}{r_s}\left\{\left(1+\frac{r_s}{r_{HR}}\right)+\frac{\Delta}{\gamma}\right\}=\frac{\rho c_p D}{\gamma \mathbf{R}_n} \tag{11.31}$$

The weather assumed for Fig. 11.6 gives a value of about 30 s m^{-1} for the right-hand side of equation (11.31). Because stomatal resistances are very rarely less, it must be very unusual for a leaf to get warmer as windspeed increases but a wet-bulb thermometer could behave like this if it were over-ventilated so that part of the wick became slightly dry and presented a small additional resistance to vapour transfer.

(iii) Relation between leaf and air temperature.
Equation (11.28) can be used to explain why the excess of leaf temperature over air temperature in bright sunshine is often reported to be large in cold climates and small (or even negative) in hot climates. Two features of the equation are implicated. First, because Δ increases with temperature, $T_0 - T$, if positive, will tend to increase with decreasing temperature. Strong radiative heating at low temperatures may be important for the survival of arctic species growing in a very short summer. Second, saturation deficit is commonly much larger in hot than in cold climates so the negative term in equation (11.28) is larger. It follows that a leaf of a tropical plant adapted to keep stomata open at air temperatures in the range 30 to 40°C should be able to maintain a tissue temperature close to air temperature when water is available to its roots.

(iv) Transpiration rate and stomatal resistance. Increasing the stomatal resistance of a leaf decreases the rate of evaporation and increases the sensible heat loss when $\mathbf{R}_n$ is constant. Stomatal closure therefore increases the temperature of leaf tissue as shown in Fig. 11.7. When air temperature is in the range 20 to 30°C, many species have a minimum stomatal resistance of 100–200 s m^{-1} so in bright sunshine and in a breeze, small leaves are expected to be only 1–2°C hotter than the surrounding air. Greater excess temperatures are observed on very large leaves in a light wind because r_H is large. From the same set of calculations, it can be shown that the relative humidity of air in contact with the epidermis will usually be similar to the relative humidity of the ambient air, and this feature of leaf microclimate may have important implications for the activity of fungi which need a very high relative humidity to reproduce and grow.

Dew

When $\mathbf{R}_n$ is negative at night, condensation will occur on a leaf when the numerator of equation (11.26) is negative, i.e. when $-\Delta\mathbf{R}_n$ exceeds $\rho c_p\{e_s(T)-e\}r_H$. The rate of dew formation can be calculated from the formula putting $\gamma^* = \gamma(r_V/r_H)$. When the air is saturated the predicted maximum rate of dew formation on clear nights is about 0.06 to 0.07 mm per hour but may be much less in unsaturated air (Fig. 11.8). These estimates are consistent with the maximum quantities of dew observed on leaves and on artificial surfaces—about 0.2 to 0.4 mm per night depending on site and circumstances. As these quantities are an order of magnitude smaller than potential evaporation rates, dew rarely makes a significant contribution to the water balance of vegetation even in arid climates.

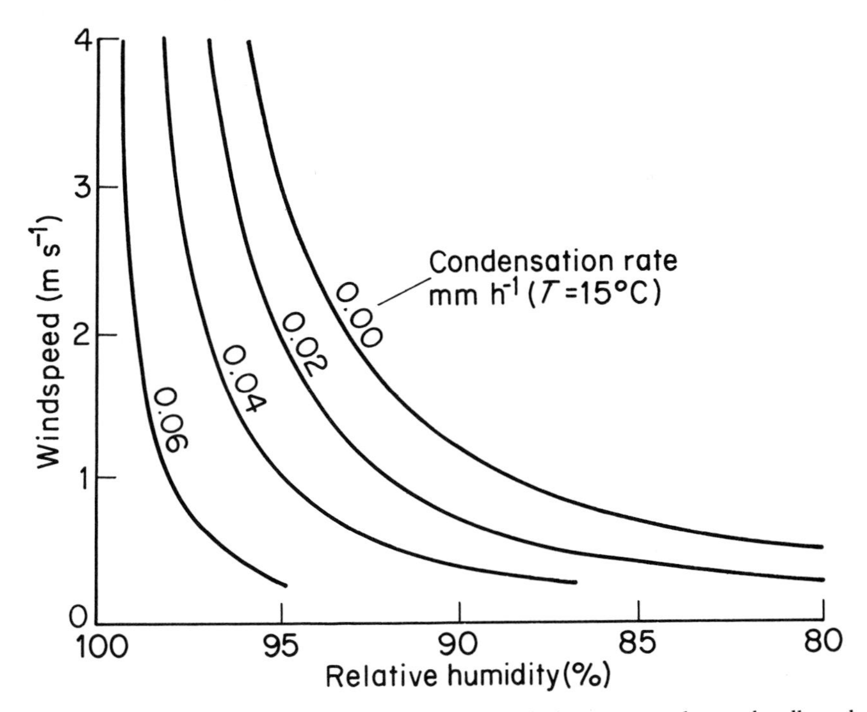

Fig. 11.8 The rate of condensation on a horizontal plane exposed to a cloudless sky at night when the air temperature is 15°C, as a function of windspeed and relative humidity (from Monteith, 1981b).

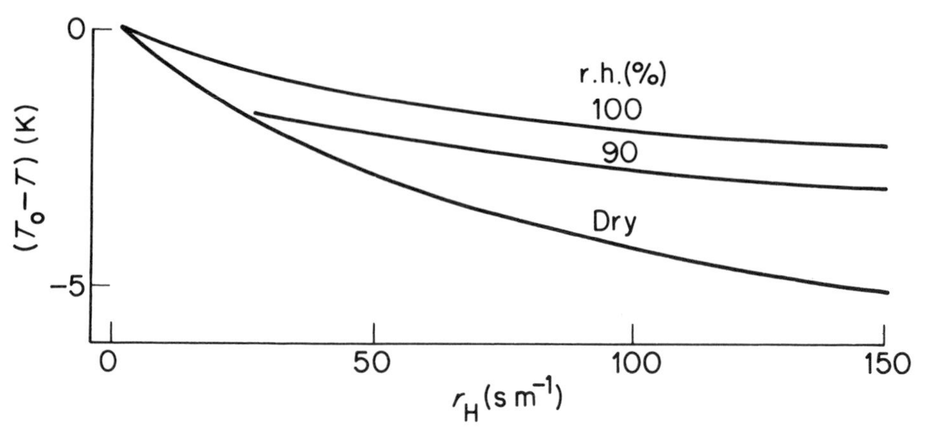

Fig. 11.9 Predicted difference between surface and air temperature for a leaf with specified boundary layer resistance in the dark ($r_s = \infty$). Assumed microclimate: $\mathbf{R}_n = -100$ W m^{-2}; $T = 10$°C. Dew formation occurs when r.h. is 100 or 90%. The 'dry' curve is appropriate when the r.h. is too low to allow condensation (from Monteith, 1981b).

Fig. 11.10 Hoar frost on leaves of *Helleborus corsicus* (p. 195). Note the preferential formation of ice on the spikes. The exchange of heat and water vapour is faster round the edge of a leaf than in the centre of the lamina because the boundary layer is thinner at the edge (p. 102). A faster exchange of heat implies that the spikes should be somewhat warmer than the rest of the leaf, i.e. closer to air temperature at night. A faster rate of mass exchange implies that the spikes should collect hoar frost faster when their temperature is below the frost-point temperature of the air.

Because $\mathbf{R}_n$ is negative at night, the temperatures of leaves is always less than air temperature. Figure 11.9 shows the dependence of $T_0 - T$ on r_H on the same scale as Fig. 11.6. For a dry leaf, the temperature difference is about 5 K in light winds in the absence of dew but is less than 2 K when dew forms from air with a relative humidity more than 90%. When minimum temperature is critical, dew may therefore be important for the thermal regime of cold- or frost-sensitive plants as well as for the water supply of plants in very dry areas. When leaf temperatures falls below the frost-point of the air, hoar frost, rather than dew, is deposited (Fig. 11.10).

DEVELOPMENTS FROM THE PENMAN EQUATION

All the equations derived here for an isolated leaf can be applied to a uniform stand of vegetation provided the resistances of the system are described in an appropriate way. The complications which arise from this procedure are described in Chapter 15. To complete the present chapter, we shall consider ways in which the Penman equation can be extended to apply to surfaces whose wetness is specified either in terms of a fixed relative humidity or a fixed wet-bulb depression. The concept of equilibrium evaporation arising from the latter type of specification is also discussed.

Specified surface humidity

When a chemical compound is dissolved in water, the free energy of the water molecules is reduced and there is a corresponding reduction in the vapour pressure of air in equilibrium with the water surface. Equation (2.33) gives the unique relation between the free energy of a solution (which depends on the molar concentration of the compound and on the degree of dissociation of its ions) and the relative humidity h of air in equilibrium with the solution. In deriving the heat balance equation for the surface of a solution, the saturated vapour pressure term $e_s(T)$ has to be replaced wherever it occurs by $he_s(T)$ so the Penman equation becomes

$$\lambda \mathbf{E} = \frac{\Delta' \mathbf{R}_n + \rho c_p \{he_s(T) - e\}/r_H}{\Delta' + \gamma^*} \tag{11.32}$$

where $\Delta' = h\Delta$. This form of the equation was used by Calder and Neal (1984) to estimate annual evaporation and surface temperature for the Dead Sea, assuming $h = 0.75$.

Equation (11.32) is valid for any surface at which the free energy of water can be treated as a constant. In principle, it could be applied to bare soil, making h a function of the water content at the surface, but this is not a practical method of estimating evaporation from soil because water content changes very rapidly with depth below the surface. Another possible application is the drying of hay or straw where h may depend on the porosity of the material as well as on the free energy of cell contents.

Specified surface wet-bulb depression

Slatyer and McIlroy (1961) derived a form of the Penman equation in which the adiabatic term was obtained by assuming that a parcel of air at an initial wet-bulb depression of B made contact with a surface and was cooled (adiabatically) until it reached equilibrium with the surface in terms of its temperature and vapour pressure. It then had a smaller wet-bulb depression B_0. When the rate of heat and vapour exchange was specified by the resistance $r_H = r_V$, the equation became

$$\lambda \mathbf{E} = \frac{\Delta \mathbf{R}_n}{\Delta + \gamma} + \rho c_p (B - B_0)/r_H \tag{11.33}$$

As the saturation vapour pressure deficit is given by $D \approx B(\Delta + \gamma)$ (p. 182), equation (11.33) is equivalent to

$$\lambda \mathbf{E} = \frac{\Delta \mathbf{R}_n + \rho c_p (D - D_0)/r_H}{\Delta + \gamma} \tag{11.34}$$

where D_0 is the saturation deficit of air in equilibrium with the surface.

Equilibrium evaporation

The Slatyer-McIlroy equation has not been widely adopted because the term B_0 (or D_0) depends not only on the surface resistance to vapour transfer but on prevailing weather too. However, equation (11.34) draws attention to the fact that the adiabatic (second) term in Penman's equation (11.26) represents a lack of equilibrium between the state of the atmosphere at a reference height as given by D and the corresponding state of air in equilibrium with the surface as given by D_0. Priestley and Taylor (1972) suggested that air moving over an extensive area of uniform surface wetness (but not necessarily water) should come into equilibrium with the surface when $D = D_0$ giving the equilibrium rate of evaporation as

$$\lambda \mathbf{E}_q = \frac{\Delta \mathbf{R}_n}{\Delta + \gamma} \tag{11.35}$$

However, the measurements which they reviewed convinced them that, on average, the latent heat of evaporation from water or from well-watered vegetation exceeded $\lambda \mathbf{E}_q$ by a factor of 1.26. Some workers have tried to demonstrate experimentally that the factor is exactly 1.26 while others have shown that it can be extremely variable!

In an attempt to examine the properties of the atmosphere responsible for the fact that the loss of latent heat from well-watered vegetation invariably *exceeds* the equilibrium rate of equation (11.35), McNaughton and Spriggs (1986) explored the behaviour of the Planetary Boundary Layer, about 1 km deep, within which temperature and humidity change diurnally in response to the input of sensible and latent heat at the surface. The size of the 'constant' in the Priestley-Taylor equation appears to depend on the way in which the Boundary Layer receives dry air from above by entrainment and increases in depth during the day as a consequence of the input of heat both from above (when an inversion is present) and from below (when the ground is warm relative to the air above it).

McNaughton (1986) has also pointed out that the saturation deficit D is the atmospheric potential which controls the non-equilibrium flux of heat and water vapour. The gradient of D can be written as

$$\begin{aligned} \partial D/\partial z &= \partial\{e_s(T) - e\}/\partial z \\ &\approx \Delta \partial T/\partial z - \partial e/\partial z \\ &\propto \Delta \mathbf{C} - \gamma \boldsymbol{\lambda} \mathbf{E} \end{aligned} \tag{11.36}$$

The process of equilibrium between a relatively dry air mass and a moist surface can therefore be described in terms of a decrease in the *atmospheric* value of saturation deficit D complemented by an increase in the *surface* value D_0 until the two quantities are the same. When the gradient of D vanishes, $\Delta \mathbf{C} = \gamma \boldsymbol{\lambda} \mathbf{E}$ which is equivalent to equation (11.35) when $\mathbf{C} = \gamma \mathbf{R}_n/(\Delta + \gamma)$. When moist air moves over dry ground, equilibration proceeds in the opposite direction.

Coupling

McNaughton and Jarvis (1983) extended the concept of an equilibrium evaporation rate by re-writing the Penman-Monteith equation in the form

$$\mathbf{E} = \Omega \mathbf{E}_q + (1 - \Omega)\mathbf{E}_i \tag{11.37}$$

where $\mathbf{E}_q$ is defined by equation (11.35) and

$$\Omega = (\Delta + \gamma)/(\Delta + \gamma^*) \tag{11.38}$$

It follows that $\Omega \mathbf{E}_q$ is the rate of evaporation that would obtain if the heat budget of a surface were dominated by the diabatic (radiative) term. This condition tends to be satisfied when short well-watered vegetation is exposed to bright sunshine, humid air and a light wind. The evaporation rate is then effectively independent of the saturation deficit of the ambient air and the surface may be described as 'decoupled' from the prevailing weather. The term 'coupling' must not be taken too literally, however, because $\mathbf{E}_q$ depends on the absorption of radiation and is slightly dependent on temperature (through Δ).

The complementary quantity $(1 - \Omega)\mathbf{E}_i$ is the rate of evaporation 'imposed' by the environment when the surface is 'fully coupled' to the prevailing weather, i.e. when the diabatic term $\rho c_p D/r_H$ is much larger than the adiabatic term, a condition satisfied when r_H is very small (rough surface, strong wind) and $\mathbf{R}_n$ is also small. In these conditions, the saturation deficit within the canopy is effectively equal to the value (D) at a reference height in the free atmosphere so that

$$\boldsymbol{\lambda} \mathbf{E}_i \approx \rho c_p D/(\gamma r_c) \tag{11.39}$$

For vegetation freely supplied with water, the value of the decoupling coefficient Ω depends mainly on windspeed and surface roughness and ranges from about 0.1 to 0.2 for forests (strong coupling) to 0.8 to 0.9 for short crops (decoupled) (Jarvis and McNaughton, 1986). When vegetation starts to run short of water, Ω decreases because γ^* increases, implying better coupling between vegetation and air. In this case, however, the rate of water loss is more and more determined not by the state of the atmosphere but by the rate at which roots can abstract water. During any period when the rate of extraction is approximately constant, plants in a stand have to adjust the canopy resistance r_c so that it is approximately proportional to the imposed saturation deficit. 'Coupling' then implies that it is stomatal resistance rather than transpiration rate that responds to the state of the atmosphere.

12

STEADY STATE HEAT BALANCE

(ii) Animals

The heat balance of a leaf or stand of vegetation is determined mainly by its environment and over periods of a few hours the temperature of tissue cannot be controlled except to a limited extent by the movement of leaves with respect to the sun's rays. In contrast, warm-blooded animals exposed to a changing environment are able to control deep body temperature within narrow limits by adjusting the rate at which heat is produced by metabolism or dissipated by evaporation. They are therefore classed as **homeotherms** (from the Greek *homoios*—similar, and *therme*—heat). Cold-blooded animals or **poikilotherms** (*poikilos*—various) form a class intermediate between vegetation and homeotherms. They have relatively slow metabolic rates and no thermostat but tend to avoid extremes of heat and cold by seeking shade or shelter. Lizards, tortoises and bees, for example, appear to prefer environments where exposure to solar radiation raises body temperature to between 30 and 35°C, close to the range in which homeotherms operate.

The heat balance equation of any animal can be written in the general form

$$\mathbf{M} + \mathbf{R}_n = \mathbf{C} + \lambda\mathbf{E} + \mathbf{G} \tag{12.1}$$

where each term refers to the gain or loss of heat per unit of body surface area, **M** is the rate of heat production by metabolism and **G** is conduction to the substrate on which the animal is standing or lying. The latent heat loss $\lambda\mathbf{E}$ is the sum of components representing losses from the respiratory system $\lambda\mathbf{E}_r$ and from the skin when sweat evaporates, $\lambda\mathbf{E}_s$. The terms $\mathbf{R}_n$ and **C** have the same significance as previously.

Methods of calculating the radiative exchange of animals were considered in Chapter 5. The size and significance of the remaining terms will now be considered separately as an introduction to the 'thermo-neutral diagram' which shows how they are related for homeotherms.

HEAT BALANCE COMPONENTS

Metabolism (M)

Standard measurements of basal metabolic rate are made when a subject has been deprived of food and is resting in an environment where metabolic rate

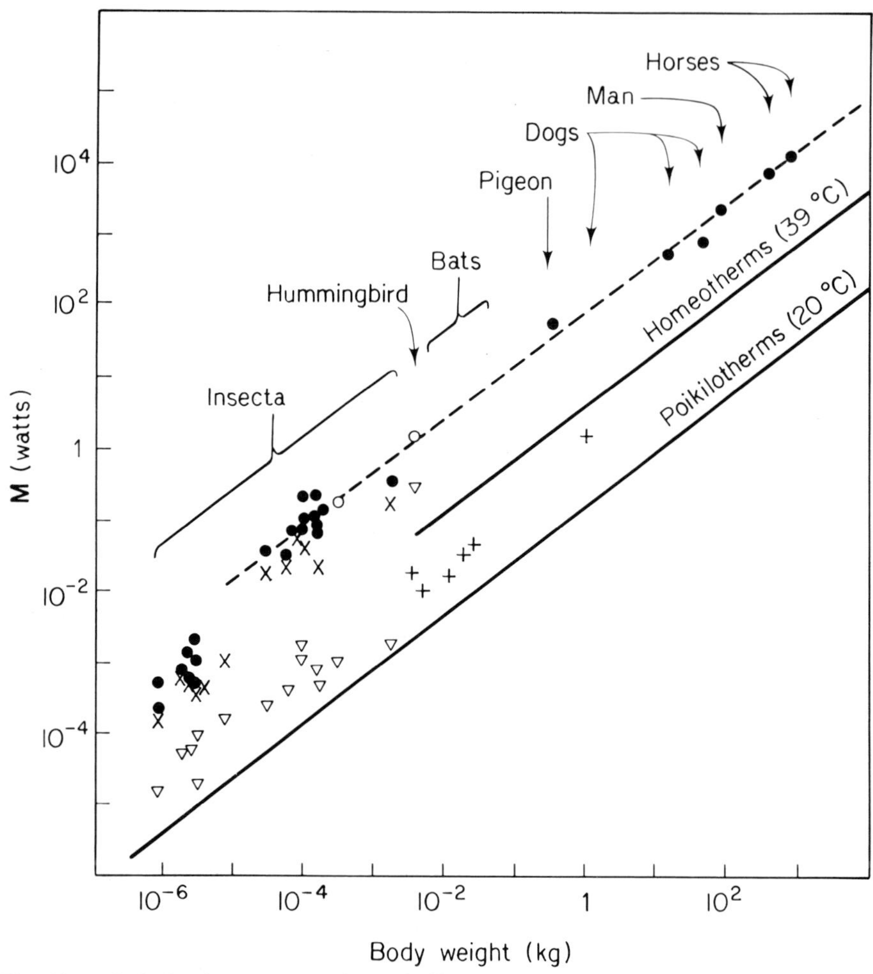

Fig. 12.1 Relation between basal metabolic rate of homeotherms (upper continuous line), maximum metabolic rate for sustained work by homeotherms (upper pecked line) and basal rate for poikilotherms at 20°C (lower continuous line) (from Hemmingsen, 1960).

is independent of external temperature. The basal metabolism of an animal $\mathbf{M}_b$ (expressed here in watts and not watts per unit surface area as elsewhere) can be related to body mass W by the relation

$$\mathbf{M}_b = BW^n \tag{12.2}$$

where B is a constant, implying that $\ln \mathbf{M}_b$ is a linear function of ln W (Fig. 12.1).

According to many sets of measurements by animal physiologists, $n = 0.75$ and Kleiber (1965) suggested that, for *intra*-specific work, a value of 70 kcal day^{-1} per $\text{kg}^{0.75}$ (3.4 W per $\text{kg}^{0.75}$) should be adopted for B. Hemmingsen (1960) found that B = 1.8 W per $\text{kg}^{0.75}$ was a better value for *inter*-specific comparisons over a range of mass from 0.01 to 10 kg. His review also showed that the metabolism of a poikilotherm kept at a body temperature of 20°C is about 5% of the value for a homeotherm of the same mass with a deep body temperature of 39°C. Hibernating mammals metabolize energy at about the same rate as poikilotherms of the same mass.

Referring metabolic rates to a power of body mass is awkward in studies of heat transfer where fluxes are referred to unit area. Conveniently, Prothero (1984) found that n was closer to 2/3 than to 3/4 when bias produced by circadian rhythms was removed from metabolic measurements. For any set of objects with identical geometry but different in size, surface area is proportional to the 2/3 power of mass, assuming constant mass per unit volume. It follows that if basal metabolic rates are proportional to the 2/3 power of mass, they should be proportional to surface area, at least when animals of like geometry are compared. This proportionality has been demonstrated for mammals ranging in size from a mouse (0.02 kg) to an elephant (1400 kg). The average basal metabolic rate for a mammal per unit of body surface area is about 50 W m^{-2}. This is much smaller than the flux of short-wave radiation absorbed by a dark-coated animal in bright sunshine (say 300 W m^{-2}) but is comparable with the energy that might be absorbed under shade. On a cloudless night, the basal metabolic rate is comparable to the net loss of heat by long-wave radiation from a surface close to air temperature.

For walking or climbing, the efficiency with which man and domestic animals use additional metabolic energy is about 30%. To work against gravity at a rate of 20 W m^{-2}, for example, metabolism must increase by about 60 W m^{-2}. The power expended in walking increases with velocity and, in man, reaches 700 W at 2 m s^{-1}.

For rapid forms of locomotion such as running or flying, the work done against wind resistance is proportional to the wind force times the distance travelled. When the drag coefficient of a moving body is independent of velocity, the drag force should be proportional to V^2 (p. 105), and as the distance travelled is proportional to V, the rate of energy dissipation should increase with V^3. However, studies on birds by Tucker (1969) revealed a range of velocities (e.g. 7 to 12 m s^{-1} for gulls) in which the rate of energy production was almost independent of V, implying a remarkable decrease of drag coefficient with increasing velocity.

Latent heat (λE)

In the absence of sweating or panting, the loss of latent heat from an animal is usually a small fraction of metabolic heat production and takes place both from the lungs as a result of respiration and from the skin as a result of the diffusion of water vapour, sometimes politely referred to as 'insensible perspiration'. For a man inhaling completely dry air, the vapour pressure difference between inhaled and exhaled air is about 5.2 kPa and the heat used for respiratory evaporation $\mathbf{\lambda E_r}$ is the latent heat equivalent of about 0.8 mg of water vapour per ml of absorbed oxygen (Burton and Edholm, 1955). With round figures of 2.4 J mg^{-1} for the latent heat of vaporization of water and 21 J ml^{-1} oxygen for the heat of oxidation, $\mathbf{\lambda E_r/M}$ is about 10%. When air with a vapour pressure of 1.2 kPa is breathed, the difference of vapour pressure decreases to 4 kPa and $\mathbf{\lambda E_r/M}$ is about 8%.

In the absence of sweating, the latent heat loss from human skin ($\boldsymbol{\lambda}\mathbf{E}_s$) is roughly twice the respiratory loss, implying a total evaporative loss of the order of 25 to 30% of **M** depending on vapour pressure. When sweating, however, a man can produce about 1.5 kg of fluid per hour, equivalent to 600 W m^{-2} when the environment allows sweat to evaporate as fast as it is produced. More commonly, the rate of evaporation is restricted to a value determined by resistances and vapour pressure differences. Excess sweat then drips off the body or soaks into hair and clothing.

Sheep lack glands of the type that allow man to sweat profusely but species such as cattle can lose substantial amounts of water by sweating. Sheep and dogs compensate for their inability to sweat by panting in hot environments. For cattle exposed to heat stress, the respiratory system can account for 30% of total evaporative heat loss, the remaining 70% coming from the evaporation of sweat from the skin surface and from wetted hair. The maximum evaporative heat loss from ruminants is only a little larger than metabolic heat production whereas a sweating man can lose far more heat by evaporation than he produces metabolically if he is exposed to strong sunlight.

Interspecific differences in **λE/M** and in mechanisms of evaporation may play an important part in adaptation to dry environments. Relatively small values of **λE/M** have been reported for a number of desert rodents which appear to conserve water evaporated from the lungs by condensation in the nasal passages where the temperature is about 25°C (Schmidt-Nielsen, 1965). The respiratory system operates as a form of counter-current heat exchanger. In contrast, measurements of total evaporation from a number of reptiles yield figures ranging from 4 to 9 mg of water per ml of oxygen. These figures imply that nearly all the heat generated by metabolism was dissipated by the evaporation of water, lost mainly through the cuticle.

For different animal species, the amount of body water available for evaporation depends on body volume whereas the maximum rate of cutaneous evaporation depends on surface area. At one extreme, insects have such a large surface : volume ratio that evaporative cooling is an impossible luxury. Man and larger animals can use water for limited periods to dissipate heat during stress and, in general, the larger the animal the longer it can survive without an external water supply.

Convection (C)

The left side of Figure 12.2. demonstrates how resistance to convective heat transfer between an organism and its environment decreases as body size decreases, a significant relation in animal ecology (see also Fig. 11.1). Schmidt-Nielsen (1965) regarded both convective and radiative exchanges as proportional to surface area and concluded that, in the desert, 'the small animal with its relatively larger relative surface area is in a much less favourable position for maintaining a tolerably low body temperature'. It is true that small animals are unable to keep cool by evaporating body water but when convection is the dominant mechanism of heat loss they can lose heat more rapidly (per unit surface area) than large animals exposed to the same windspeed. In this context, the main disadvantage of smallness is microclima-

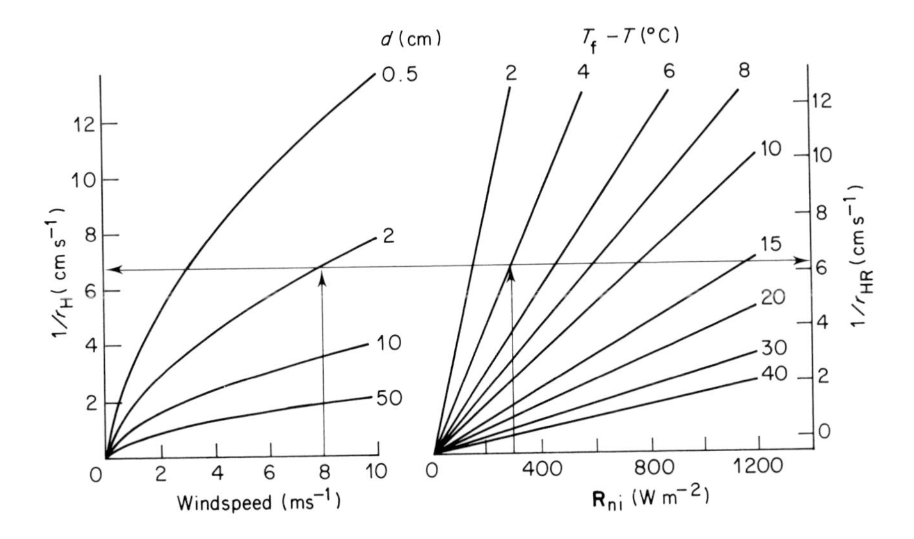

Fig. 12.2 Diagram for estimating thermal radiation increment when windspeed, body size, and net radiation are known. For example, at 8 m s^{-1} an animal with $d = 2$ cm has $r_H{}^{-1} = 6.7$ cm s^{-1} and $r_{HR}{}^{-1} = 6.2$ cm s^{-1}. When $\mathbf{R}_{ni} = 300$ W m^{-2}, $T_f - T$ is 4°C.

tic: because windspeed increases with height above the ground, small animals moving close to the surface are exposed to slower windspeeds than larger mammals. As r_H is approximately proportional to $(d/V)^{0.5}$, an animal with $d = 5$ cm exposed to a wind of 0.1 m s^{-1} will be coupled to air temperature in the same way as a much larger animal with $d = 50$ cm exposed to wind at 1 m s^{-1}. Insects and birds in flight and tree-climbing animals escape this limitation.

Conduction (G)

Few attempts have been made to measure the conduction of heat from an animal to the surface on which it is lying. Mount (1968) measured the heat lost by young pigs to different types of floor material and found that the rate of conduction was strongly affected by posture as well as by the temperature differences between the body core and the substrate. When the temperature of the floor (and air) was low, the animals assumed a tense posture and supported their trunks off the floor, but as the temperature was raised, they relaxed and stretched to increase contact with the floor. Figure 12.3 shows the heat loss per unit area, recalculated from Mount's data for new born pigs in a relaxed posture. As the heat flow was approximately proportional to the temperature difference, the thermal resistance for each type of floor can be calculated from the slope of the lines. The resistances are about 8, 17 and 58 s cm^{-1} for concrete, wood and polystyrene respectively. As the corresponding resistances for convective and radiative transfer are usually around 1 to 2 s cm^{-1} (Fig 11.1), it follows that heat losses by conduction to the floor of an animal house are likely to be significant only when it is made of a relatively

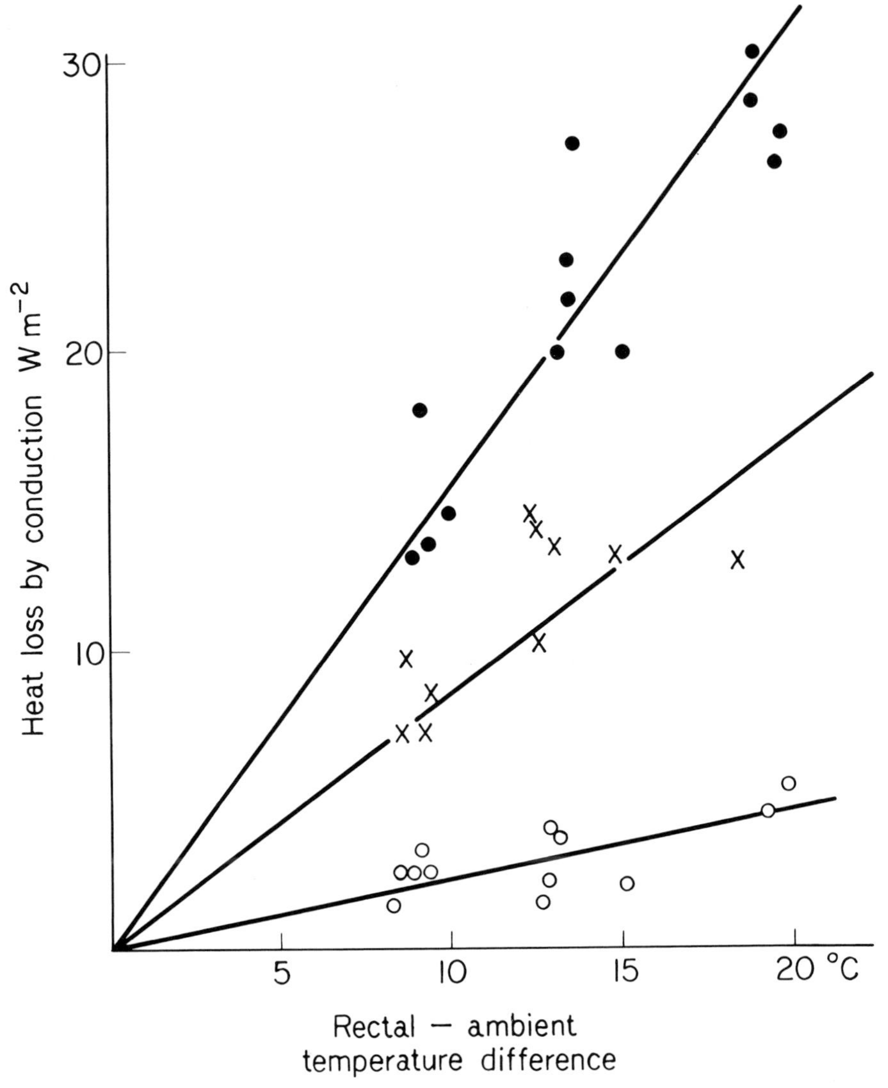

Fig. 12.3 Measurements of heat lost by conduction from a pig to different types of floor covering expressed as watts per square metre of total body area (after Mount, 1967). ●—concrete; X—wood; ○—polystyrene.

good heat conductor such as concrete. Conduction will be negligible when the floor is wood or concrete covered with a thick layer of straw.

Gatenby (1977) measured the conduction of heat beneath a sheep with a fleece length of about 2 cm lying on grass in the open. When the deep soil temperature was 10°C, downward fluxes of about 160 W per m^2 of contact

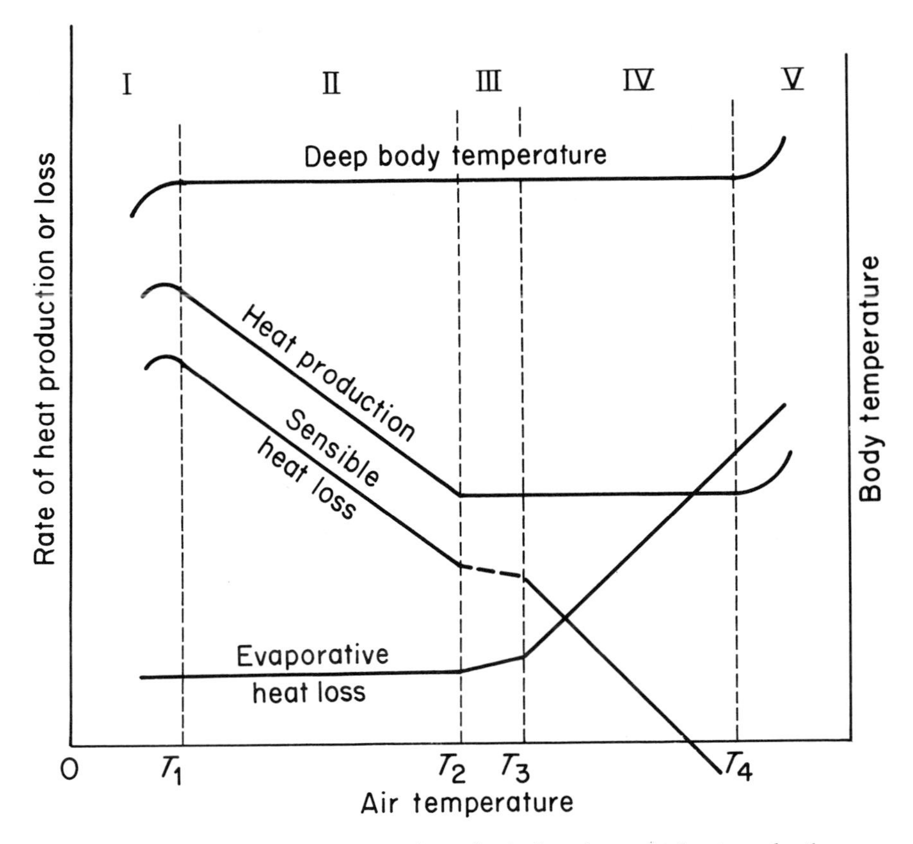

Fig. 12.4 Diagrammatic representation of relations between heat production, evaporative and non-evaporative heat loss and deep-body temperature in a homeothermic animal (from Mount, 1979).

surface were measured when the sheep lay down, equivalent to about 40 W per m^2 of body surface and therefore comparable with the loss by convection. Energy requirements for free-ranging livestock therefore depend to some extent on the time spent lying, on soil temperature and on the thermal properties of the ground.

THE THERMO-NEUTRAL DIAGRAM

The fundamental relation between the metabolic heat production of a homeotherm and the temperature of its environment is often represented by a thermo-neutral diagram (Fig. 12.4), for which measurements of metabolism can be obtained in one of two ways. In 'direct' calorimetry, the subject is placed in a calorimeter, usually with wall temperature equal to air temperature to simplify estimates of radiative transfer, and the flow of heat through the wall is measured with transducers. For 'indirect' calorimetry, measurements of oxygen consumption are widely used to determine heat loss from the subject. Ways of defining the effective temperature of more complex environments are considered later.

To simplify interpretation of the thermo-neutral diagram without concealing major details, conduction of heat to the ground will be neglected and the animal (or man) will be assumed covered by a uniform layer of hair (or clothing). It is convenient to work with isothermal net radiation $\mathbf{R}_{ni}$ (p. 187) because this flux is independent of surface temperature and to introduce a combined resistance for convection and long-wave radiative exchange r_{HR} (p. 179).

The heat balance equation for the surface of a coat whose mean temperature is T_0 then becomes

$$\mathbf{M} + \mathbf{R}_{ni} - \boldsymbol{\lambda}\mathbf{E}_r - \boldsymbol{\lambda}\mathbf{E}_s = \rho c_p(T_0 - T)/r_{HR} \tag{12.3}$$

where T is air temperature.

If the resistance of the coat is r_c, if it is impermeable to radiation of all wavelengths, and if the evaporation of sweat is confined to the surface of the skin with mean temperature T_s, the flux of sensible heat through the coat is

$$\mathbf{M} - (\boldsymbol{\lambda}\mathbf{E}_r + \boldsymbol{\lambda}\mathbf{E}_s) = \rho c_p(T_s - T_0)/r_c \tag{12.4}$$

The flux of heat through the skin surface is

$$\mathbf{M} - \boldsymbol{\lambda}\mathbf{E}_r = \rho c_p(T_b - T_s)/r_d \tag{12.5}$$

where r_d is the resistance of the body tissue and T_b is the body temperature.

Eliminating T_0 and T_s from equations (12.3) to (12.5) gives the relation between heat balance components as

$$\mathbf{M} + (r_{HR}/r_t)\mathbf{R}_{ni} = \rho c_p(T_b - T)/r_t + \boldsymbol{\lambda}\mathbf{E}_r + \boldsymbol{\lambda}\mathbf{E}_s(r_{HR} + r_c)/r_t \tag{12.6}$$

where $r_t = r_{HR} + r_c + r_b$. This equation can be used to explore the physiologically significant air temperatures shown in Fig. 12.4 if it is written in the form

$$T = T_b - \{r_t\mathbf{M} + r_{HR}\mathbf{R}_{ni} - r_t\boldsymbol{\lambda}\mathbf{E}_r - (r_{HR} + r_c)\boldsymbol{\lambda}\mathbf{E}_s\}/\rho c_p \tag{12.7}$$

Equations (12.6) and (12.7) will now be used to examine five discrete regimes incorporated in the thermo-neutral diagram.

I $T < T_1$

At an air temperature of T_1, sometimes referred to as the 'cold limit', the body produces heat at a maximum rate $\mathbf{M}_{max}$. Assuming that the sum of the two latent heat terms in equation (12.7) is $0.2\mathbf{M}_{max}$ (see p. 201), the value of T_1 is given by

$$T_1 \approx T_b - \{0.8\, r_t\mathbf{M}_{max} + r_{HR}\mathbf{R}_{ni}\}/\rho c_p \tag{12.8}$$

Table 12.1 contains values of measurements of T_1 summarized by Mount (1979) for newborn and mature individuals of three species. Comparisons emphasize the value of hair in establishing the size of r_t and therefore of the product $r_t\mathbf{M}_{max}$. Estimates of this quantity given in the Table were derived by assuming that $T_b = 37°C$ for all species, that $\mathbf{R}_{ni}$ was zero in the experimental

Table 12.1 Cold limit and lower critical temperature for three species (from Mount, 1979) with corresponding values of $r_t\mathbf{M}$ (see text)

	Cold limit T_1 (°C)	Critical temperature T_2 (°C)	$r_t\mathbf{M}_{max}$ (kJ m^{-3})	$r_t\mathbf{M}_{min}$ (kJ m^{-3})
Man				
newborn	27	33	15	6
mature	14	28	35	14
Pig				
newborn	0	34	56	5
mature	−50	10	131	41
Sheep				
newborn	−100	30	206	11
mature	−200	20	355	86

conditions and that $\rho c_p = 1200$ J m^{-3} K^{-1}.

When air temperature falls below the cold limit, heat is dissipated faster than it is made available by metabolism and radiation so that the steady-state heat balance represented by equation (12.7) cannot be achieved and body temperature must then decrease with time (see p. 177). This depresses the metabolic rate so that temperature falls still further, leading to death by hypothermia if the process is not reversed.

II $T_1 < T < T_2$

In this regime, sensible heat loss decreases linearly with increasing air temperature, inducing an identical decrease in metabolic rate provided the latent heat component of the heat balance remains constant. Differentiation of equation (12.6) then gives

$$\frac{\partial \mathbf{M}}{\partial T} = -\frac{\rho c_p}{r_t} \qquad (12.9)$$

If $r_t = 600$ s m^{-1}, for example, $\partial \mathbf{M}/\partial T^{-1}$ is -2 W m^{-2} K^{-1}.

Equation (12.9) is valid between the cold limit T_1 and a temperature T_2 at which the metabolic rate reaches a minimum value $\mathbf{M}_{min}$, usually somewhat larger than the basal metabolic rate (p. 200) on account of physical exertion and/or the digestion of food. Repeating the assumptions used to find an approximation for T_1 gives

$$T_2 \approx T_b - \{0.8\, r_t \mathbf{M}_{min} + r_{HR}\mathbf{R}_{ni}\}/\rho c_p \qquad (12.10)$$

Observed values of T_2 and estimates of $r_t\mathbf{M}_{min}$ are in Table 12.1.

Livestock exposed to a temperature below T_2 need more feed to achieve the same liveweight gain or production of milk, eggs, etc. than a comparable animal kept in the thermo-neutral zone. The temperature T_2 is therefore referred to as the (lower) critical temperature.

III $T_2 < T < T_3$

This is the zone of 'least thermoregulatory effort' for all animals and it is also identified as the comfort zone for man. As temperature rises above T_2, metabolic rate remains constant so the heat balance equation can be satisfied only if (a) the total resistance to sensible heat loss r_t decreases and/or (b) the latent heat loss $\lambda\mathbf{E}$ increases. In many animals, including man, a decrease of r_t is a consequence of blood vessels near the skin surface dilating so that blood circulates faster, effectively decreasing the tissue resistance r_d (see p. 141). This process is sometimes accompanied by a small increase of sweat rate.

IV $T_3 < T < T_4$

In this regime, **M** and r_t are constant and the maintenance of thermal equilibrium as temperature rises requires that the decrease of sensible heat loss should be balanced by an identical increase of latent heat loss. This increase is made possible by more rapid production and evaporation of sweat and/or by faster breathing and other mechanisms which increase respiratory evaporation (p. 202). The temperature range T_2 to T_4, in which heat production is at its minimum value and is independent of air temperature, is usually called the thermo-neutral zone.

V $T > T_4$

As temperature rises further, loss of latent heat cannot increase indefinitely. The rate of evaporation of sweat is limited by the aerodynamic resistance of the body and therefore by windspeed and the rate of evaporation from the respiratory system is likewise limited by the rate of panting and the volume of air inhaled and exhaled. Both rates of evaporation depend on the vapour pressure of the atmosphere. When $\lambda\mathbf{E}$ reaches a maximum (or even at a somewhat lower temperature as Fig. 12.4 suggests), body temperature starts to rise, a primary symptom of hyperthermia. Differentiating equation (12.7) with respect to T when $\lambda\mathbf{E}$ is constant gives

$$1 = \frac{\partial T_b}{\partial T} - \frac{\partial \mathbf{M}}{\partial T_b}\frac{\partial T_b}{\partial T}\frac{r_t}{\rho c_p}$$

or

$$\frac{\partial T_b}{\partial T} = \{1 - r_t(\partial \mathbf{M}/\partial T_b)/\rho c_p\}^{-1} \tag{12.11}$$

The requirement that $\partial T_b/\partial T$ should be positive implies that

$$\frac{\partial \mathbf{M}}{\partial T_b} < \rho c_p / r_t \tag{12.12}$$

In man, the increase of metabolic rate with temperature is about 7% per K so if $\mathbf{M} = 150\ \mathrm{W\ m^{-2}}$, r_t should not exceed $170\ \mathrm{s\ m^{-1}}$ ($\rho c_p = 1200\ \mathrm{J\ m^{-3}\ K^{-1}}$) implying good ventilation and a minimum of clothing, familiar requirements for avoiding extreme discomfort in tropical climates. When equation (12.12) is not satisfied, body temperature and metabolic rate tend to rise uncontrollably and fatally.

SPECIFICATION OF THE ENVIRONMENT

The thermo-neutral diagram has been used mainly to summarize observations of metabolic rate in livestock or in man exposed to the relatively simple environment of a calorimeter. In the real world, metabolic rate depends on several microclimatic factors in addition to air temperature, notably radiation, windspeed and vapour pressure. Attempts have therefore been made to replace the simple measure of actual air temperature used in the previous discussion with an effective temperature which incorporates the major elements of microclimate. Two examples are now given, the first dealing with the radiative component of the heat balance equation.

Radiation increment

When the radiant flux is expressed in terms of net isothermal radiation, the heat balance equation may be written

$$\mathbf{R}_{ni} + (\mathbf{M} - \boldsymbol{\lambda}\mathbf{E}) = \rho c_p(T_0 - T)/r_{HR} \qquad (12.13)$$

where $\boldsymbol{\lambda}\mathbf{E}$ is here assumed to be a relatively small term, independent of T. Suppose that the thermal effect of radiation on an organism can be handled by substituting for air temperature an effective temperature T_f such that

$$(\mathbf{M} - \boldsymbol{\lambda}\mathbf{E}) = \rho c_p(T_0 - T_f)/r_{HR} \qquad (12.14)$$

Then eliminating $\mathbf{M} - \boldsymbol{\lambda}\mathbf{E}$ from equations (12.11) and (12.12) gives

$$T_f = T + \{\mathbf{R}_{ni} r_{HR}/\rho c_p\} \qquad (12.15)$$

where the term in curly brackets is similar to the radiation increment used by Burton and Edholm (1955), Mahoney and King (1977) and many others.

Figure 12.2 provides a graphical method of estimating the temperature increment $T_f - T$ when windspeed and net radiation are known. Given the characteristic dimension of an organism d and the velocity of the surrounding air V, the corresponding value of r_H can be read from the left-hand vertical axis and r_{HR} derived from equation (11.2) can be read from the right-hand axis. From the right-hand section of the Figure, the coordiates of r_{HR} and $\mathbf{R}_{ni}$ define a unique value of $T_f - T$. An example is shown for $V = 8$ m s^{-1}, $d = 2$ cm, which gives $1/r_H = 6.8$ cm s^{-1} from the left-hand axis and $1/r_{HR} = 6.3$ cm s^{-1} from the right-hand axis. At $\mathbf{R}_{ni} = 300$ W m^{-2}, $T_f - T = 4$ K.

In a system where the resistances are fixed, the relation between temperature gradients and heat fluxes can be displayed by plotting temperature against flux as in Fig. 12.5. By definition, a resistance is proportional to a temperature difference divided by a flux, and is therefore represented by the slope of a line in the Figure. From a start at the bottom left-hand corner, T_f is determined by drawing a line (1) with slope $r_{HR}/\rho c_p$ to intercept the line $x = \mathbf{R}_{ni}$ at $y = T_f$. The equation of this line is

$$T_f - T = r_{HR}\mathbf{R}_{ni}/\rho c_p \qquad (12.16)$$

The temperature of the surface T_0 is now found by drawing a second line

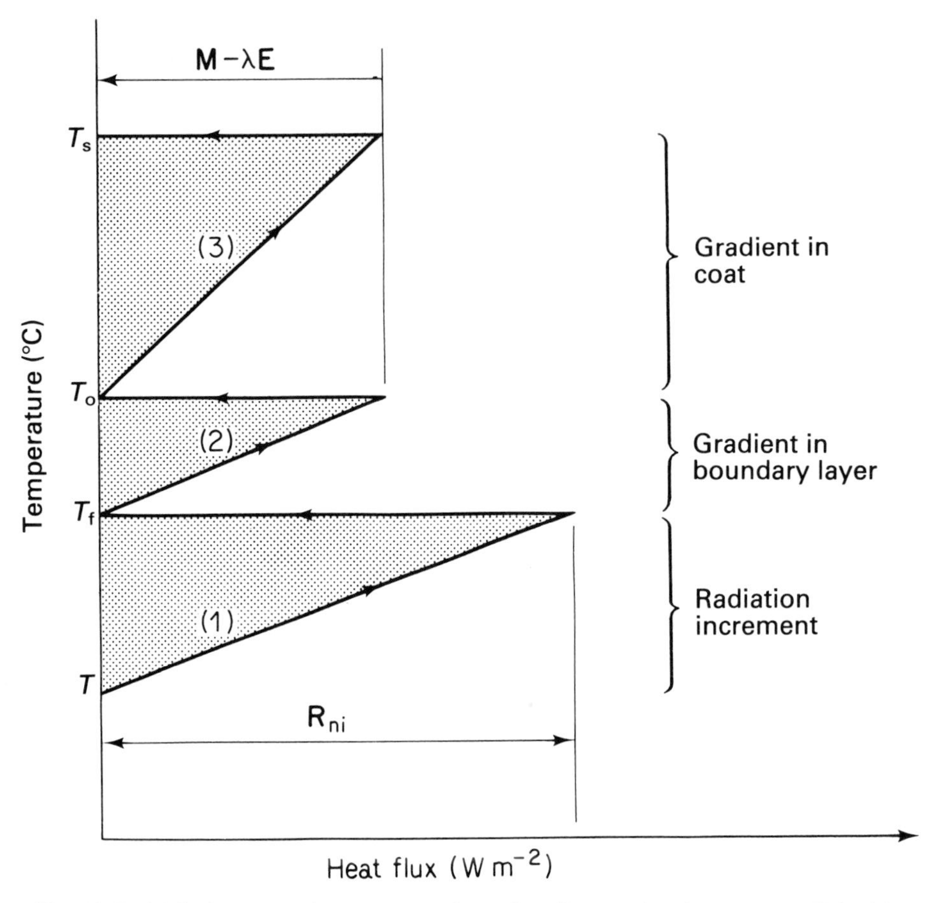

Fig. 12.5 Main features of temperature/heat-flux diagram for dry systems. T_s is skin temperature, T_0 coat surface temperature, T_f effective environment temperature, and T air temperature.

(2) with the same slope as (1) to intersect $x = \mathbf{M} - \boldsymbol{\lambda}\mathbf{E}$ at $y = T_0$. The equation of this line is

$$T_0 - T_f = r_{HR}(\mathbf{M} - \boldsymbol{\lambda}\mathbf{E})/\rho c_p \tag{12.17}$$

Finally, for an animal covered with a layer of hair, a mean skin temperature T_s can be determined if the mean coat resistance r_c is known. Provided that evaporation is confined to the surface of the skin and the respiratory system, the increase of temperature through the coat is represented by the line (3) whose equation is

$$T_s - T_0 = r_c\,(\mathbf{M} - \boldsymbol{\lambda}\mathbf{E})/\rho c_p \tag{12.18}$$

This form of analysis can be used to solve two types of problem. When the environment of an animal is prescribed in terms of windspeed, temperature and net radiation, it is possible to use a thermo-neutral diagram to explore the physiological states in which the animal can survive. Conversely, when physiological conditions are specified, a corresponding range of environmental conditions can be established, forming an ecological 'niche' or 'climate space' (Gates, 1980). Examples now given for a locust, a sheep and a man are based on case studies reported in the literature.

CASE STUDIES

Locust

In a monograph describing the behaviour of the Red Locust (*Nomadacris septemfasciata*), Rainey, Waloff and Burnett (1957) derived radiation budgets and heat balances for an insect of average size basking on the ground or flying. Both states are represented in Fig. 12.6. At the bottom of the diagram, the basking locust is exposed to an air temperature of 20°C but the effective temperature of the environment is about 25°C ($\mathbf{R}_{ni}$ = 150 W m^{-2}, r_{HR} = 0.38 s cm^{-1}). Because $\mathbf{M} - \boldsymbol{\lambda}\mathbf{E}$ is trivial, the surface temperature is only a fraction of a degree above T_f. The flying locust is exposed to a larger radiant flux density (230 W m^{-2}) but with a relative windspeed of 5 m s^{-1} the value of r_{HR} is 0.33 s cm^{-1}, somewhat smaller than for the basking locust. The increment $T_f - T$ is therefore only slightly larger than for basking. The difference between body surface and effective temperature is relatively large for the flying locust because the rate of metabolism is much larger than the basal rate and from measurements of oxygen consumption in the laboratory is estimated to be 270 W m^{-2}. The graph predicts that a flying locust should be about 12°C hotter than the surrounding air, consistent with maximum temperature excesses measured by sticking hypodermic thermocouple needles into locusts captured in the field.

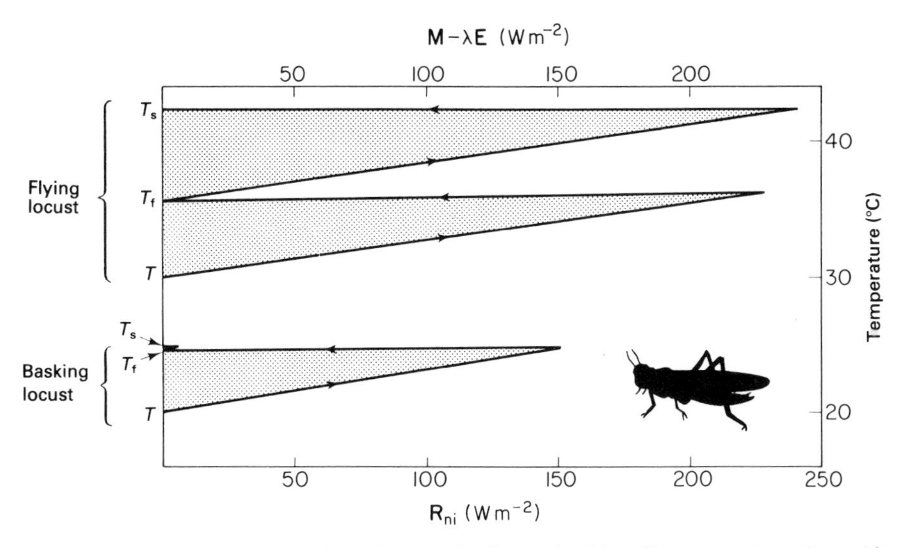

Fig. 12.6 Temperature/heat-flux diagram for locust basking (lower section of graph) and flying (upper section).

Sheep

Figure 12.7 shows the heat balance of sheep with fleeces 1, 4 and 8 cm long exposed to

(i) An air temperature of -10°C and a net radiative *loss* of $\mathbf{R}_{ni} = -50$ W m^{-2}.
(ii) An air temperature of 40°C and a radiative *gain* of 160 W m^{-2}.

In both cases the animal is supposed to behave like a cylinder with a diameter of 50 cm exposed to a wind of 2 m s^{-1}, a special case of the conditions analysed by Priestley (1957).

For the cold state, T_f is found by drawing a line with slope $r_{HR}/\rho c_p$ to give $T - T_f = 3$°C, at $\mathbf{R}_{ni} = -50$ W m^{-2}. From $T_f = -13$°C, three lines were drawn corresponding to the total resistance $(r_{HR} + r_c)$ for the three values of fleece length taking r_c as 1.5 s cm^{-1} per cm length from Table 8.1. For the heat balance equation in the form

$$T_b - T_f = (\mathbf{M} - \boldsymbol{\lambda}\mathbf{E})(r_{HR} + r_c)/\rho c_p \tag{12.19}$$

and if $T_b = 37$°C, the lines intercept $T = 37$°C at three values of $\mathbf{M} - \boldsymbol{\lambda}\mathbf{E}$; 42, 74 and 182 W m^{-2}. As the average daily metabolism of a healthy, well-fed sheep is expected to be about 60 or 70 W m^{-2}, the graph implies that a fleece at least 4 cm long is needed to withstand effective temperatures between -10 and -15°C.

For the hot state, the top section of the graph shows that T_f is 50°C in the

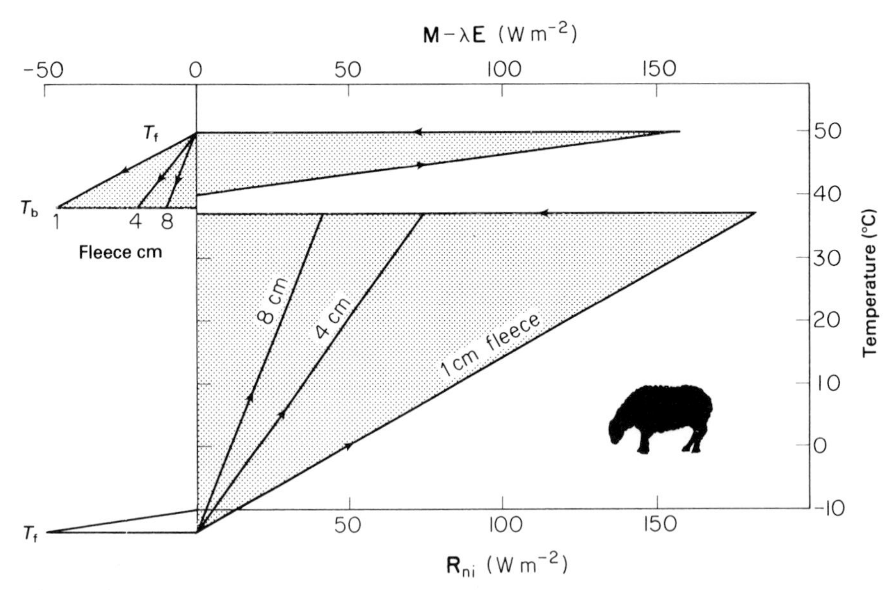

Fig. 12.7 Temperature/heat-flux diagram for sheep with fleece lengths of 1, 4 and 8 cm exposed to air temperatures of -10°C and net radiation of -50 W m^{-2} (lower section); and 40°C with net radiation 160 W m^{-2} (upper section).

conditions chosen. Thermal equilibrium cannot be achieved when T_f exceeds T_b unless $\mathbf{M}-\lambda\mathbf{E}$ is negative, i.e. unless more heat is lost by evaporation than is generated by metabolism. Assuming $T_b = 38°C$ in this case, $\mathbf{M}-\lambda\mathbf{E}$ would need to be -46 W m^{-2} if the fleece length is 1 cm, decreasing to -10 W m^{-2} for an 8 cm fleece, i.e. there would be a flow of sensible heat into the sheep from its environment which would decrease as insulation increased. Studies on sheep in controlled environments show that $\lambda\mathbf{E}$ can reach 90 W m^{-2} under extreme heat stress but, even during a period of minimum activity, $\mathbf{M}$ is unlikely to be less than 60 W m^{-2}. A figure of -30 W m^{-2} can therefore be taken as a lower limit for $\mathbf{M}-\lambda\mathbf{E}$. The diagram implies that a minimum fleece length of about 2 cm would be needed to withstand the conditions that were chosen to represent heat stress.

Man

The radiation and heat balance of men working in Antarctica were studied by Chrenko and Pugh (1961), and Fig. 12.8 is based on their analysis for a man wearing a black sweater standing facing the sun. The air temperature was only $-7.5°C$ but, because the sun was 22° above the horizon, the radiative load on vertical surfaces facing the sun was exceptionally large. As the wind was light, r_{HR} was relatively large, about 1 s cm^{-1}. The top left-hand side of the diagram, referring to the sunlit chest, was constructed from measured

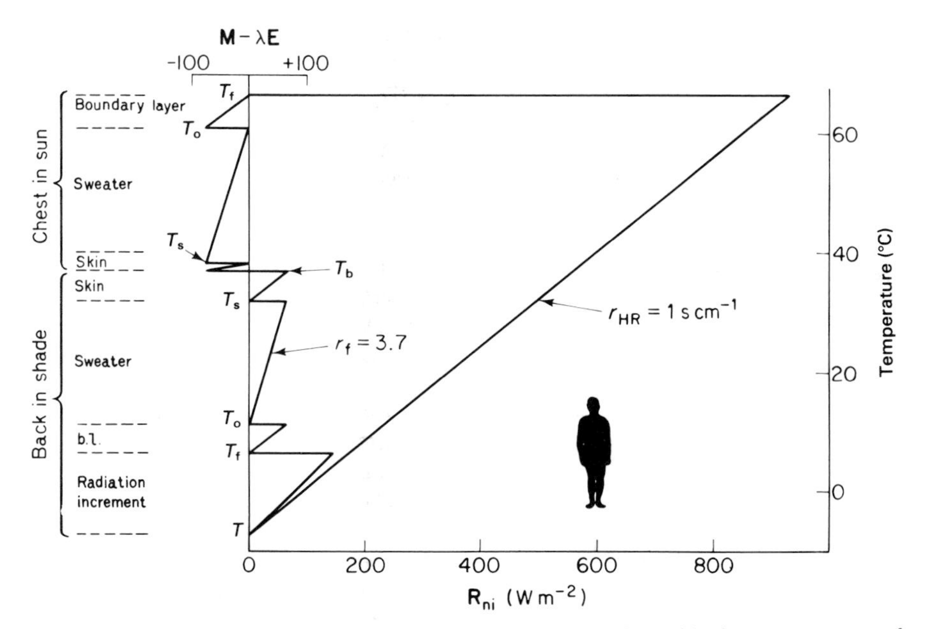

Fig. 12.8 Temperature/heat-flux diagram for a man wearing a black sweater exposed to arctic sunlight and air temperature of $-7°C$. The lower part of the diagram shows the equivalent temperature and sweater surface temperature on his back (in shade) and the upper part shows the same temperature on his chest (in full sun).

temperatures and from the heat flow through the clothing measured with a set of heat flow transducers. The radiation increment was 74°C, the surface of the sweater was at 61°C and the skin was at 38°C. The conduction of heat into the body was −75 W m^{-2} and, assuming the resistance of the skin over the chest was 0.2 s cm^{-1}, the deep body temperature would be 37°C. The same deep body temperature can be reached starting from the bottom left-hand side of the diagram representing the temperature gradients on the man's back. As the net radiation on the back was only 145 W m^{-2}, T_f was only 6.5°C (cf. 74°C on the chest); the outer surface of the sweater was at 12°C and the skin was at 32°C. There was an outward flow of heat through the skin of 60 W m^{-2} and, assuming $T_b = 37$°C, the resistance of the skin is about 1 s cm^{-1}, consistent with the value for vaso-constricted tissue in Table 8.1.

These measurements demonstrate a flow of solar energy through the trunk from chest to back and it would be necessary to integrate this heat flow over the whole body to determine $\overline{\mathbf{M}} - \overline{\boldsymbol{\lambda}\mathbf{E}}$. Such an exercise is obviously impractical. The heat balance of an animal in a natural environment can be established when $\overline{\mathbf{M}} - \overline{\boldsymbol{\lambda}\mathbf{E}}$ is measured directly or is estimated from relevant laboratory studies but the converse operation of determining $\overline{\mathbf{M}} - \overline{\boldsymbol{\lambda}\mathbf{E}}$ when other terms in the heat balance are known is confined to men or animals in a calorimeter or controlled environment chamber.

Apparent equivalent temperature

A second type of environmental index is needed when the loss of latent heat from an organism is predominantly by the evaporation of sweat from the skin. When the loss of sensible heat occurs from the same surface, temperature and vapour pressure can be combined in a single variable which may be called the **apparent equivalent temperature**. To derive this quantity, the heat balance equation is written,

$$\mathbf{R}_{ni} + \mathbf{M} = \frac{\rho c_p(T_0 - T)}{r_{HR}} + \frac{\rho c_p(e_0 - e)}{\gamma r_V} \qquad (12.20)$$

where T_0 and e_0 are mean values for the skin surface and T and e refer to the air.

If γ is replaced by $\gamma^* = \gamma(r_V/r_{HR})$ equation (12.20) can be written

$$\mathbf{R}_{ni} + \mathbf{M} = \rho c_p(T_{eo}{}^* - T_e{}^*)/r_{HR} \qquad (12.21)$$

where $T_e{}^*$ is the apparent equivalent temperature of ambient air given by $T + e/\gamma^*$ and therefore equal to the equivalent temperature derived on p. 183 when $r_V = r_{HR}$. By analogy with the radiation increment, e/γ^* may be regarded as a humidity increment. The mean value of the apparent equivalent temperature at the surface is $T_{eo}{}^*$.

In principle, the apparent equivalent temperature should be used in a thermo-neutral diagram in place of conventional temperature when the metabolic rate of an animal is measured in an environment where vapour pressure changes as well as temperature. Extending this process a stage further, the apparent equivalent temperature of the environment can be modified to take account of the radiation increment by writing

$$T_{eR}^* = T_e^* + \mathbf{R}_{ni} r_{HR}/\rho c_p \qquad (12.22)$$

The quantity T_{eR}^* is an index of the thermal environment which allows the heat balance equation to be reduced to the form

$$(1-x)\mathbf{M} = \rho c_p(T_{eo}^* - T_{eR}^*)/r_{HR} \qquad (12.23)$$

where x is the fraction of metabolic heat dissipated by respiration. A thermo-neutral diagram using T_{eR}^* as a thermal index rather than T or T_e^* would be valid for changes in radiant heat load as well as in vapour pressure and temperature.

This type of formulation provides a relatively straightforward way of investigating the heat balance of a naked animal whose skin is either dry or entirely wetted by sweat. The intermediate case of partial wetting is more difficult to handle because the value of T_e^* depends on fractional wetness which is not known *a priori*. A complete solution of the relevant equations for an animal leads to a complex expression within which the structure of the original Penman equation can be identified (McArthur, 1987).

Sweating man

To illustrate how apparent equivalent temperatures can be used, the heat balance of a sweating man will be analysed graphically. The net heat **H** to be dissipated will be taken as the sum of the metabolic heat load **M** and the isothermal net radiation $\mathbf{R}_{ni}$ which can be calculated on the assumption that a body intercepts radiation like a cylinder (Chapter 5). The resistance to vapour transfer is $d/(D\mathrm{Sh})$ where the characteristic dimension d is often taken as 34 cm for a standard man, and the resistances to heat transfer are $d/(\kappa\mathrm{Nu})$ and r_R in parallel. For forced convection in a windspeed of about 2 m s^{-1}, r_H is 1 s cm^{-1} and with $r_R = 2.1$ s cm^{-1}, $r_{HR} = 0.68$ s cm^{-1}. The corresponding value of r_V is 0.9 s cm^{-1} so $\gamma^* = \gamma r_V/r_{HR} = 87$ Pa K^{-1}. These values will be taken as standard in the following discussion.

The rate at which sweat evaporates from a man in a given environment cannot be determined without knowing how fast it is produced. The maximum rate of sweating that a normal man can sustain for several hours is about 1 kg h^{-1} and if his surface area is 1.8 m^2 the equivalent rate of evaporative heat loss is 375 W m^{-2}. The rate of sweating is determined partly by skin temperature and partly by metabolic rate. When a subject reports that a particular combination of clothing and environment is 'comfortable' and there is no visible evidence of sweat, his mean skin temperature is usually about 32 to 33°C.

Laboratory experiments show that many subjects regard the environment as very uncomfortable when skin temperature rises above about 35°C (95°F) and as intolerable when the skin temperature reaches about 37°C (98°F). When γ^* is 87 Pa K^{-1} the corresponding equivalent temperatures for a wet surface are 100°C and 109°C. Figure 12.9 shows these limits on a diagram of the type already used to present 'dry' heat balances, but the equivalent temperature T_e^* now replaces T as the ordinate. The left-hand side of the diagram shows how T_e^* is related to the temperature and vapour pressure of

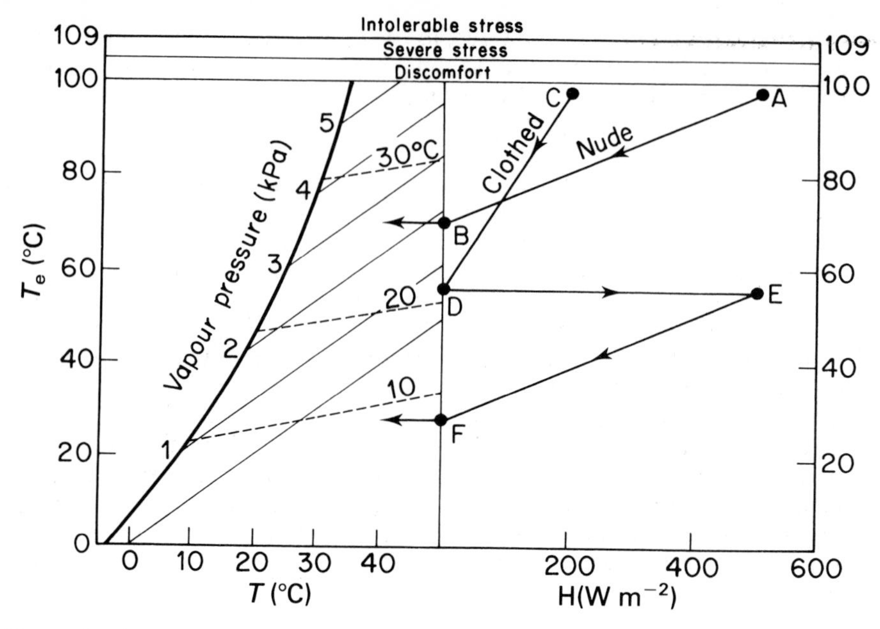

Fig. 12.9 Apparent equivalent temperature and heat-flux diagram for a clothed and a nude man (right-hand section) and the relation between apparent equivalent temperature, vapour pressure and air temperature (left-hand section). The nude man with a total heat of 500 W m^{-2} can avoid discomfort if the apparent equivalent temperature of the environment is less than the value at B (70°C) but the clothed man with a heat load of only 200 W m^{-2} must stay in an environment with an apparent equivalent temperature less than 28°C (point F). The pecked lines are isotherms of wet bulb temperature.

the ambient air. Three lines of constant wet bulb temperature (10, 20, 30°C) are plotted as a reminder that T' is closely related to T_e* when $\gamma \approx \gamma^*$.

The right-hand side of the diagram shows gradients of equivalent temperature for a nude and for a clothed subject, both assumed to be sweating freely when **M** is 200 W m^{-2} (light work) and $\mathbf{R}_{ni}$ is 300 W m^{-2} (bright sunshine). The mean value of T_{e0}* for the skin of the nude man is assumed to be 98°C. The equivalent temperature of the air T_e* must therefore satisfy equation (12.21) where $\rho c_p/r_{HR} = 17.7\ \mathrm{W\,m^{-2}\,K^{-1}}$. Thus when $\mathbf{H} = \mathbf{R}_{ni} + \mathbf{M} = 500\ \mathrm{W\,m^{-2}}$, $T_{eo}{}^* - T_e{}^*$ is 28.5 K, T_e* is 69.5°C and equation (12.21) is represented by the line AB. Reference to the left side of the figure shows the range of air temperature and humidity for which T_e* is 70°C, e.g. 35°C, 3 kPa. The increase of $\mathbf{R}_{ni} + \mathbf{M}$ which would make the same environment seem either 'severe' or 'intolerable' can be derived from the figure by extending the line BA upwards. Conversely if $\mathbf{R}_{ni} + \mathbf{M}$ is constant, the effect of increasing T_e* by raising T or e is found by displacing AB upwards without changing its slope.

The effect of clothing can be demonstrated on the same diagram if γ^* is assumed to have the same value between the skin and the surface of the clothing as it has in the free atmosphere and if all the radiation from the environment is assumed to be intercepted at the surface of the clothing. Then the flux of heat from the wet skin (at an equivalent temperature $T_{ec}{}^*$) to the surface of the clothing (at $T_{e0}{}^*$) is given by

$$\mathbf{M} = \rho c_p(T_{e0}{}^* - T_{ec}{}^*)/r_c \qquad (12.24)$$

The diffusion resistance of the clothing r_c is taken as 2.5 s cm^{-1} (equivalent to about 1 clo, p. 140) so the factor $(\rho c_p/r_c)$ is 4.8 W m^{-2} K^{-1}. The line CD drawn with this slope shows that when $T_{e0}{}^*$ is 98°C, $T_{ec}{}^*$ is 56°C. To represent the gradient of $T_e{}^*$ from the surface of the clothing to the ambient air, the line EF is drawn parallel to AB. The presence of clothing decreases the equilibrium value of $T_e{}^*$ from 70°C to 27°C and, because CD is much steeper than AB, a severe or intolerable heat stress would be imposed by a relatively small increase of metabolic rate.

The atmospheric conditions for thermal equilibrium can be represented to a good approximation by a wet bulb temperature of 25°C for the nude subject and of 10°C for the clothed subject. The wet bulb temperature has frequently been used as an index of environmental temperature in human studies. Figure 12.9 confirms that it is a good index for a fully wetted skin provided $r_{HR} \approx r_V$ but it is inappropriate when the skin temperature is below the limit for rapid sweating.

13
TRANSIENT HEAT BALANCE

In the last two chapters, it was possible to treat temperature as constant in time because, at every point in the systems considered, the input of heat energy was assumed to balance the output exactly. This state of equilibrium is rare in natural environments except in the depths of caves and in deep water. In the habitats of most plants and animals, air temperature has a marked diurnal variation, more or less in phase with radiation, with superimposed short-term fluctuations associated with changes in cloudiness and with turbulence. In this chapter, we consider the equations which describe how a system responds thermally to changes of external temperature, initially making the simplifying assumption that the system itself is isothermal and contains no source of heat. More complex cases are treated in detail by Gates (1980).

Examples are described for the three types of temperature change illustrated in Fig. 13.1: an instantaneous or 'step' change; a 'ramp' change at steady rate; and an harmonic oscillation.

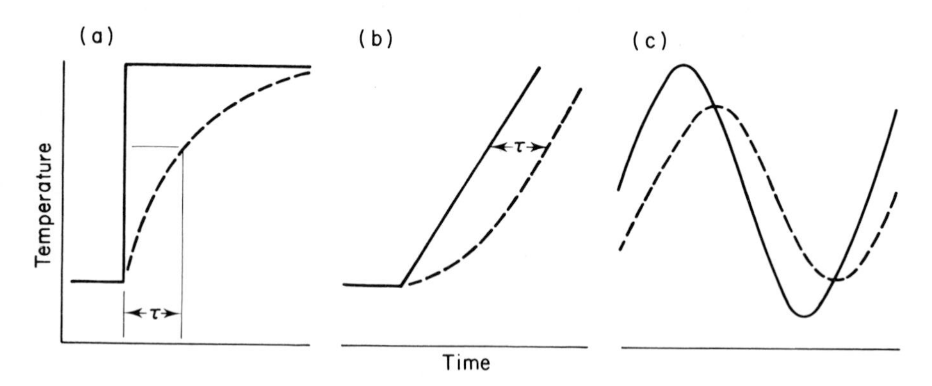

Fig. 13.1 Change of surface temperature (dashed line) in response to change of environmental temperature (full line). (a) Step change (equation (13.10)); τ is the time for a fractional change of $1 - e^{-1}$ or 0.63. (b) Ramp change (equation (13.12)); τ is the constant time lag established after the term $\exp(-t/\tau)$ becomes negligible. (c) Harmonic oscillation for the case $\phi = \pi/4$ (equation (13.16)) (from Monteith, 1981b).

TIME CONSTANT

If we take first the simplest case of a system in which the latent heat exchanges are negligible, the steady state heat budget is simply

$$\mathbf{R}_{ni} = \mathbf{C} = \rho c_p (T_0 - T)/r_{HR} \qquad (13.1)$$

where $\mathbf{R}_{ni}$ and $\mathbf{C}$ are fluxes per unit area, T_0 is mean surface temperature, T is air temperature and r_{HR} is the corresponding resistance for the loss of sensible heat and long-wave radiation. Introducing an effective temperature defined by equation (12.15), i.e.

$$T_f = T + \{\mathbf{R}_{ni} r_{HR}/\rho c_p\} \qquad (13.2)$$

the heat budget equation can be reduced to

$$T_0 = T_f \qquad (13.3)$$

A similar equation can be written for a system in which the rate of evaporation is determined by a resistance r_V so that the heat budget can be expressed as

$$\mathbf{R}_{ni} = \mathbf{C} + \boldsymbol{\lambda}\mathbf{E} = \rho c_p (T_{e0}{}^* - T_e{}^*)/r_{HR} \qquad (13.4)$$

where $T_e{}^*$ is an apparent equivalent temperature (see p. 214). Equation (13.3) can now be re-written as

$$T_{e0}{}^* = T_{eR}{}^* \qquad (13.5)$$

where

$$T_{eR}{}^* = T_f + e/\gamma^* \qquad (13.6)$$

For simplicity, however, we shall return to equation (13.3) and postulate that, as a consequence of an increase either in $\mathbf{R}_{ni}$ or in T, T_f increases to a new value T_f'. If the system has a finite heat capacity, T_0 will not instantaneously increase to T_f' but will approach it at a rate depending on the physical properties of the system. If the heat capacity per unit *area* of the system is $\mathscr{C}$, heat will be stored at a rate $\mathscr{C}\partial T_0/\partial T$ so that the heat budget equation becomes

$$\mathbf{R}_{ni} = \mathbf{C} + \mathscr{C}\partial T_0/\partial t \qquad (13.7)$$

If $\mathbf{C}$ is obtained from equation (13.1) and T_f' from (13.2), then

$$\partial T_0/\partial t = (T_f' - T_0)/\tau \qquad (13.8)$$

where τ, known as the **'time constant'** of the system because it has dimensions of time, is given by

$$\tau = \mathscr{C} r_{HR}/(\rho c_p) \qquad (13.9)$$

In vegetation, the heat capacity per unit *volume* ranges from about 1 MJ m^{-3} K^{-1} in the heartwood of red pine to 2 to 3 MJ m^{-3} K^{-1} for organs such as leaves and fruits consisting mainly of water (which has a heat capacity of 4.2 MJ m^{-3} K^{-1}). Corresponding time constants are of the order of a few seconds for small leaves, a few minutes for large leaves and several hours for the trunks of trees (Monteith, 1981b). Values reported for animals range from about 9 min for a cockroach (Buatois and Croze, 1978), 0.5 h for a shrew, 2 h for a large Cardinal bird and 330 h for a sheep (Gates, 1980).

GENERAL CASES

Step change

When the effective temperature changes instantaneously from T_f to T_f', the boundary conditions for the solution of equation (13.8) are

$$T_0 = T_f \qquad t = 0$$

$$T_0 = T_f' \qquad t = \infty$$

and the solution is

$$T_0 = T_f' - (T_f' - T_f) \exp(-t/\tau) \tag{13.10}$$

Linacre (1972) measured the mean temperature of vine leaves which he covered with vaseline to stop transpiration and then suddenly shaded or unshaded to produce a step change of $\mathbf{R}_{ni}$ and therefore of $\mathbf{C}$ (Fig. 13.2). In one set of measurements the mean value of τ was 20 s and the value of $\mathscr{C}/\rho c_p$ was 720 (J m^{-2} K^{-1})/1200 (J m^{-3} K^{-1}) = 0.6 m. The corresponding resistance was

$$r_{HR} = \tau \rho c_p / \mathscr{C} = 33 \text{ s m}^{-1}$$

Linacre also measured time constants for leaves sprayed with water. Analysis of this case can proceed by replacing T by the equivalent tempera-

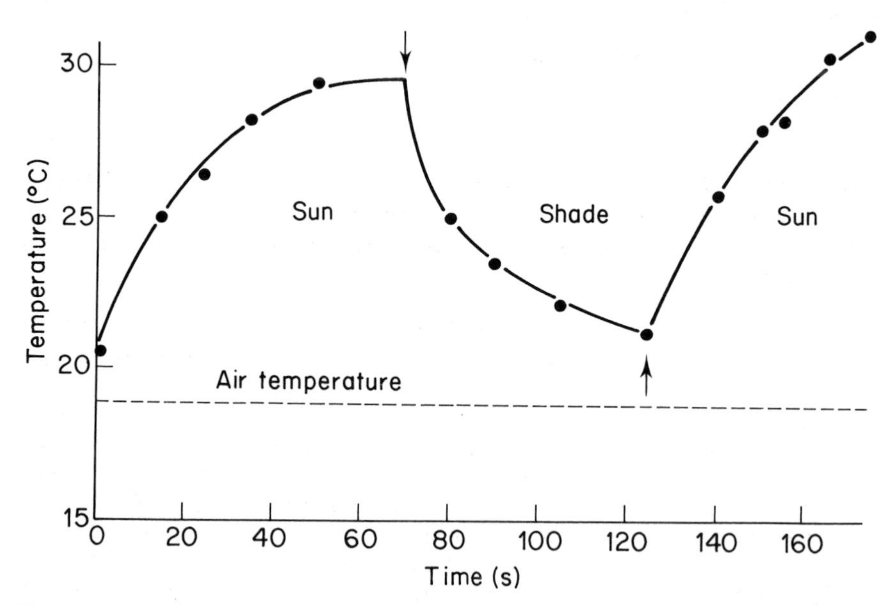

Fig. 13.2 Heating and cooling of a pepper leaf in sun and shade. (Redrawn by Gates (1980) from an example by Ansari and Loomis.)

ture $T + e/\gamma$, assuming that resistances for heat and vapour transfer are the same. For a transpiring leaf with $r_V > r_H$, it would be possible to determine by trial the value of γ^* needed to obtain the same time constant as for a non-transpiring leaf and the stomatal resistance could then be calculated as

$$r_V - r_H = r_H(\gamma^*/\gamma - 1)$$

Grigg *et al.* (1979) exposed specimens of the Eastern Water Dragon to step changes of temperature and measured time constants (for heating) which increased from about 4 to 8 min as body mass increased from 140 to 590 g. In conflict with elementary theory, the time constant for cooling was longer by about 30%, a difference accounted for by a more realistic model in which an isothermal body core is surrounded by an insulating layer.

The theory which describes the response of an organism to step changes of temperature is particularly relevant to the thermal regime of an animal which 'shuttles' between a cool and a hot microclimate in order to obtain food. Bakken and Gates (1975) and Porter *et al.* (1973) have explored the thermal implications of shuttling by lizards in the desert but few ecologists outside the USA seem to have appreciated the significance of this aspect of animal ecology.

Ramp change

When the rate of change of air temperature (or effective temperature) is α, so that the temperature at time t is

$$T(t) = T(0) + \alpha t \tag{13.11}$$

the solution of equation (13.8) is

$$T_0(t) = T(0) + \alpha t - \alpha\tau\{1 - \exp(-t/\tau)\} \tag{13.12}$$

where the heating rate is

$$\partial T_0/\partial t = \alpha\{1 - \exp(-t/\tau)\} \tag{13.13}$$

This equation shows that the rate of heating of the system is zero initially, increasing to α when t/τ is large. To be more specific, when t/τ exceeds 3, the exponential term is less than 0.05 and can therefore be neglected. Equation (13.12) then becomes

$$T_0(t) = T(0) + \alpha(t - \tau) \tag{13.14}$$

showing that the system heats at the same rate as its environment but with a time-lag of τ and therefore with a temperature lag of αT.

Equation (13.12) has been used to investigate the phenomenon of condensation on cocoa pods in Bahia occurring after sunrise as a consequence of thermal inertia (Monteith and Butler, 1979). In humid weather, the increase in pod temperature after dawn is initially much slower than the increase in air temperature (1.5 to 2.5 K h^{-1}) and pod surface temperature can therefore fall below the dew-point of the air which increases at about 1 to 2 K h^{-1}. Condensation then begins and continues until the temperature of the pod surface eventually rises above the dew-point after which the film of dew evaporates (Fig. 13.3). Values of τ determined experimentally ranged from

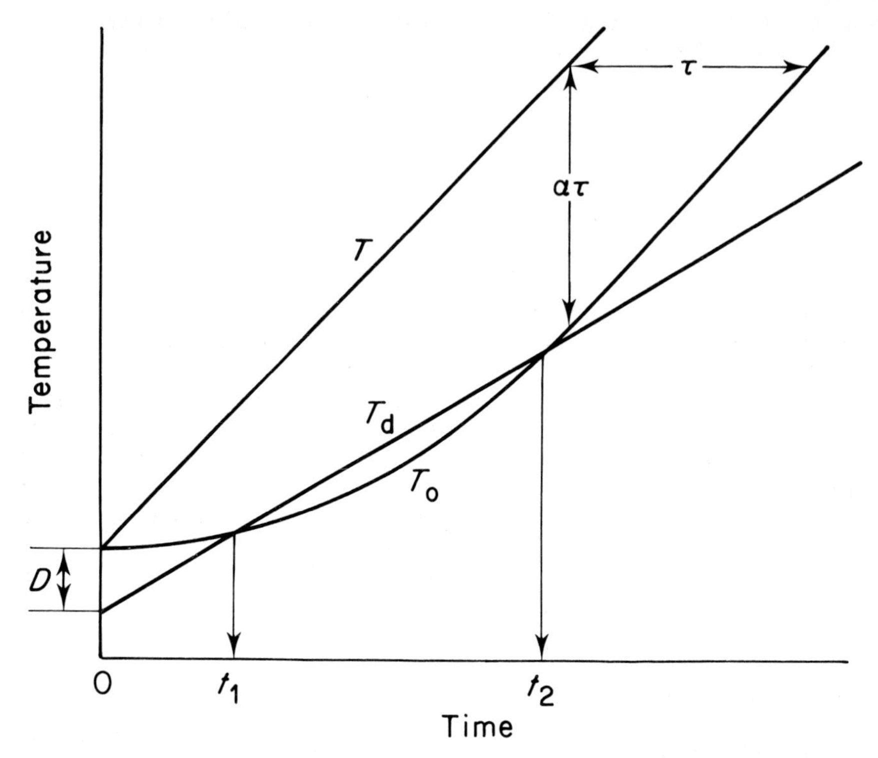

Fig. 13.3 Idealized representation of the increase of air temperature (T), dew-point temperature (T_d) and pod temperature (T_0) when the dew-point depression is D at sunrise ($t = 0$). Condensation begins at t_1 and stops at t_2. For other symbols, see text (from Monteith and Butler, 1979).

0.5 to 1.5 h depending on pod size and windspeed. By calculating the rate of condensation from the heat balance of a pod, it was possible to show that the mean thickness of the wet layer would often be between 10 and 20 μm, persisting for several hours. The existence of such a layer may play an important role in the germination of zoospores of the Black Pod fungus responsible for large losses of yield in some areas.

The same type of condensation must often occur on the trunks of trees but is rarely visible because of the rough nature of the surface.

Harmonic change

A harmonic change of air temperature has the form

$$T = \bar{T} + A \sin 2\pi t/P \qquad (13.15)$$

where $\bar{T}$ is a mean temperature and T oscillates between $\bar{T}+A$ and $\bar{T}-A$ (see Fig. 13.1) with a period P, expressed in the same units as time. For an isothermal system whose surface temperature is governed by equation (13.8), the solution of the equation is

$$T_0 = \bar{T} + A' \sin\{(2\pi t/P) - \phi\} \tag{13.16}$$

where the amplitude of the surface temperature is

$$A' = A \cos \phi \tag{13.17}$$

and ϕ, known as a 'phase lag', has a value of

$$\phi = \tan^{-1}(2\pi\tau/P) \tag{13.18}$$

In the natural environment, the diurnal change of temperature, though not a true sine wave, may be treated as one to demonstrate principles. For $P = 24$ h, ϕ is small and $A' \approx A$ unless τ is at least several hours, a figure appropriate for tree trunks or large fruit. In such cases, however, it is unrealistic to assume that the system is isothermal and although simple theory may give an approximate solution for conditions at the surface, it is not appropriate elsewhere. Harrington (1969) solved more complex equations for the radial flow of heat in a tree trunk and compared predictions with measurements in a pine tree. For the trunk surface temperature, A'/A was about 0.75 compared with 0.63 from equation (13.17) but, at the centre of the trunk, A'/A was much larger than the estimated value.

HEAT FLOW IN SOIL

The vertical flow of heat in soil provides a relevant and very important example of how a transient heat balance can be established in a system where temperature is a function of position as well as changing harmonically with time. As an introduction to the subject, the dependence of soil thermal properties on water content and mineral composition will be considered before deriving the appropriate differential equation for a soil in which thermal properties are assumed to be uniform with depth.

Soil thermal properties

By use of the symbol ρ for density and c for specific heat, the solid, liquid, and gaseous components of soil can be distinguished by subscripts s, l and g. If the volume fraction x of each component is expressed per unit volume of bulk soil

$$x_s + x_l + x_g = 1 \tag{13.19}$$

For a completely dry soil ($x_l = 0$), x_g is the space occupied by pores. In many sandy and clay soils x_g is between 0.3 and 0.4 and it increases with organic matter content, reaching 0.8 in peaty soils.

The bulk density of a soil ρ' is found by adding the mass of each component, i.e.

$$\rho' = \rho_s x_s + \rho_l x_l + \rho_g x_g = \Sigma(\rho x) \tag{13.20}$$

Because ρ_g for the soil atmosphere is much smaller than ρ_s or ρ_l, the term $\rho_g x_g$ can be neglected. When ρ_s and x_l are constant, soil density increases linearly

Table 13.1 Thermal properties of soils and their components (after van Wijk and de Vries, 1963)

		Density ρ 10^6g	Specific heat c $\mathrm{J\,g^{-1}\,K^{-1}}$	Thermal conductivity k' $\mathrm{W\,m^{-1}\,K^{-1}}$	Thermal diffusivity κ' $10^{-6}\,\mathrm{m^2\,s^{-1}}$
(a) *Soil components*					
Quartz		2.66	0.80	8.80	4.18
Clay minerals		2.65	0.90	2.92	1.22
Organic matter		1.30	1.92	0.25	0.10
Water		1.00	4.18	0.57	0.14
Air (20°C)		1.20×10^{-3}	1.01	0.025	20.50
(b) *Soils*	Water content x_1				
Sandy soil (40% pore space)	0.0	1.60	0.80	0.30	0.24
	0.2	1.80	1.18	1.80	0.85
	0.4	2.00	1.48	2.20	0.74
Clay soil (40% pore space)	0.0	1.60	0.89	0.25	0.18
	0.2	1.80	1.25	1.18	0.53
	0.4	2.00	1.55	1.58	0.51
Peat soil (80% pore space)	0.0	0.30	1.92	0.06	0.10
	0.4	0.70	3.30	0.29	0.13
	0.8	1.10	3.65	0.50	0.12

with the liquid fraction x_l, but, in a soil which swells when it is wetted, the relation is not strictly linear.

The volumetric specific heat ($\mathrm{J\,m^{-3}\,K^{-1}}$) is the product of bulk density ρ' and bulk specific heat c'. It can be found by adding the heat capacity of soil components to give

$$\rho' c' = \rho_s c_s x_s + \rho_l c_l x_l + \rho_g c_g x_g = \Sigma(\rho c x) \tag{13.21}$$

and this quantity increases linearly with water content in a non-swelling soil. The specific heat is therefore

$$c' = \Sigma \rho c x / \Sigma \rho x \tag{13.22}$$

Thermal properties of soil constituents and of three representative soils are listed in Table 13.1. Quartz and clay minerals, which are the main solid components of sandy and clay soils, have similar densities and specific heats. Organic matter has about half the density of quartz but about twice the specific heat. As a result, most soils have volumetric specific heats between 2.0 and 2.5 $\mathrm{MJ\,m^{-3}\,K^{-1}}$. As the specific heat of water is 4.18 $\mathrm{MJ\,m^{-3}\,K^{-1}}$, the heat capacity of a dry soil increases substantially when it is saturated with water.

The dependence of thermal conductivity (k') on water content is more complex. The thermal conductivity of a very dry soil may increase by an order of magnitude when a small amount of water is added because relatively large amounts of heat are transferred by the evaporation and condensation of water in the pores. For a clay soil, for example, k' may increase from 0.3 to 1.8 W m^{-1} K^{-1} when x_1 increases from zero to 0.2. With a further increase of x_1, from 0.2 to 0.4, the corresponding increase of k' is much smaller because the diffusion of vapour becomes increasingly restricted as more and more pores are filled with water. The conductivity of very wet soils is therefore almost independent of water content.

When water is added to a very dry soil, k' increases more rapidly than $\rho'c'$ initially so that the diffusivity $\kappa' = k'/\rho'c'$ also increases with water content. In a wet soil, however, the increase of k' with water content is much less rapid than the increase of $\rho'c'$ so that κ' decreases with water content. Between these two regimes, κ' reaches its maximum at a point where an increase of water content is responsible for equal fractional increases of k' and $\rho'c'$. Table 13.1 shows that sandy soils tend to have larger thermal diffusivities than other soil types because quartz has a much larger conductivity than clay minerals. Peat soils have the smallest diffusivities because the conductivity of organic matter is relatively small.

Formal analysis of heat flow

At a depth z below the soil surface, the downward flux of heat can be written

$$\mathbf{G}(z) = -k'(z)(\partial T/\partial z) \tag{13.23}$$

In any thin layer of thickness Δz, say, the difference between the flux entering the layer at level z and leaving at $z+\Delta z$ is $\mathbf{G}(z)-\mathbf{G}(z+\Delta z)$ or, in the notation of calculus, $\Delta z(\partial\mathbf{G}(z)/\partial z)$. The sign of this quantity determines whether there is net gain of flux or 'convergence' in the layer producing a local increase of soil temperature, or a net loss of flux or 'divergence' producing a fall in temperature. In general, the rate at which the heat content of the layer changes can be written $\partial(\rho'c'T\Delta z)/\partial t$ and this quantity must be equal to the change of flux with depth, i.e.

$$\frac{\partial G(z)}{\partial z}\Delta z = \frac{\partial}{\partial z}\left(-k'\frac{\partial T}{\partial z}\right) = -\frac{\partial(\rho'c'T)}{\partial t}\Delta z \tag{13.24}$$

For the special case in which the physical properties of the soil are constant with depth, the equation of heat conduction reduces to

$$\frac{\partial T}{\partial t} = \kappa'\frac{\partial^2 T}{\partial z^2} \tag{13.25}$$

Figure 13.4 is a graphical demonstration of this equation in terms of an imaginary temperature profile in a soil with constant diffusivity. Figure 13.5 shows the real change of temperature beneath a bare soil surface and beneath a crop. Observed changes of temperature at different depths can be compared

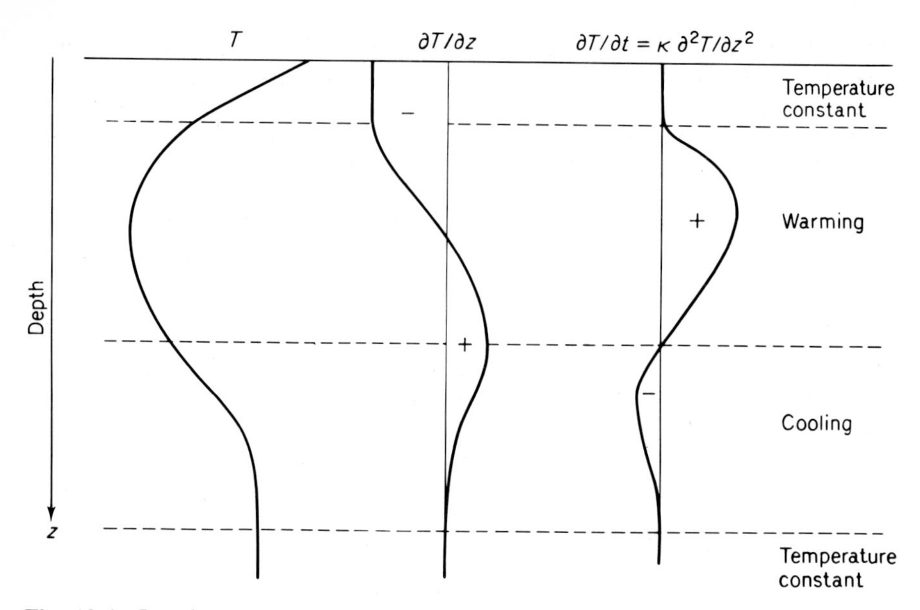

Fig. 13.4 Imaginary temperature gradient in soil (left-hand curve), and the corresponding first and second differentials of temperature with respect to depth, i.e. $\partial T/\partial z$ and $\partial^2 T/\partial z^2$. The second differential is proportional to the rate of temperature change $\partial T/\partial t$.

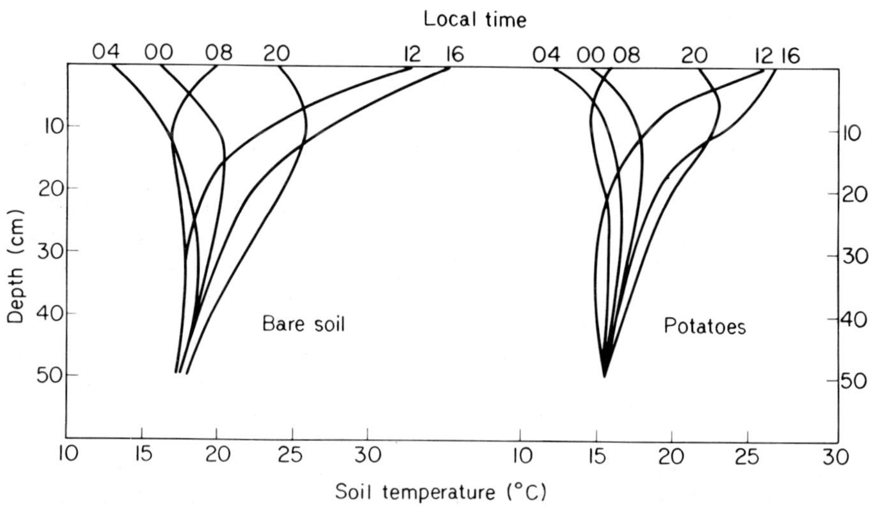

Fig. 13.5 Diurnal change of soil temperature measured below a bare soil surface and below potatoes (from van Eimern, 1964).

with the changes predicted from temperature gradients.

In most soils, composition, water content and compaction change with depth, and in cultivated soils substantial changes often occur near the surface. Precise measurements of the bulk density and thermal conductivity of soils are therefore difficult to acquire *in situ*. The theory of heat transfer in soils has nevertheless been used to determine average thermal properties from an observed temperature regime as well as for predicting daily and seasonal changes of soil temperature. In most analyses, κ' is assumed independent of depth but McCulloch and Penman (1956) derived a solution of equation (13.25) when κ' was a linear function of depth.

If temperature at depth z and time t is $T(z,t)$, the boundary condition which describes an harmonic oscillation of temperature at the surface can be written as

$$T(0,t) = \bar{T} + A(0) \sin \omega t \tag{13.26}$$

where $A(0)$ is the amplitude at the surface and $\omega = 2\pi/P$ is the angular frequency of the oscillation, i.e. for daily cycles $\omega = (2\pi/24)\ \mathrm{h}^{-1}$ and for annual cycles $\omega = (2\pi/365)\ \mathrm{d}^{-1}$.

The solution of equation (13.25) satisfying this boundary condition is

$$T(z,t) = \bar{T} + A(z) \sin (\omega t - z/D) \tag{13.27}$$

where the amplitude at depth z is

$$A(z) = A(0) \exp(-z/D) \tag{13.28}$$

and D is a depth defined by

$$D = (2\kappa'/\omega)^{0.5} \tag{13.29}$$

Several important features of conduction in soils can be related to the value of D as follows:

(i) at a depth $z = D$ (often called the **'damping depth'**) the amplitude of the temperature wave is $\exp(-1)$ or 0.37 times the amplitude at the surface;

(ii) the position of any fixed point on a temperature wave is specified by a fixed value of the phase angle $(\omega t - z/D)$, e.g. the angle is $\pi/2$ for maximum temperature and $-\pi/2$ for minimum temperatures. Differentiation of the simple equation $\omega t - z/D = \text{constant}$ gives $\partial z/\partial t = \omega D$ and this is the velocity with which temperature maxima and minima appear to move downwards into the soil;

(iii) at a depth $z = \pi D$, the phase angle is π less than the angle at the surface, i.e. the temperature wave is exactly out of phase with the wave at the surface. When the surface temperature reaches a maximum, the temperature at πD reaches a minimum and *vice versa*;

(iv) by differentiating equation (13.27) with respect to z and putting $z = 0$, it can be shown that the heat flux at the surface at time t is

$$\mathbf{G}(0,t) = \frac{\sqrt{2}A(0)k' \sin (\omega t + \pi/4)}{D} \tag{13.30}$$

The maximum heat flux is $\sqrt{2}A(0)k'/D$ which is the flow of heat that would be maintained through a slab of soil with thickness $\sqrt{2}D$ if one face were maintained at the maximum and the other at the minimum temperature of the surface. The quantity $\sqrt{2}D$ can

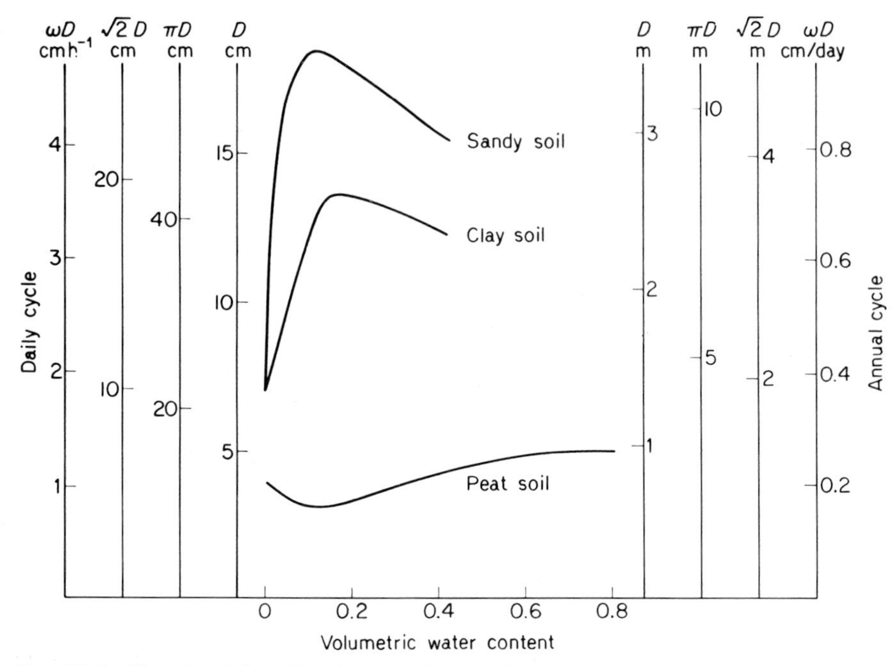

Fig. 13.6 Change of damping depth and related quantities for three soils over a wide range of water contents. Left-hand axes refer to a daily and right-hand axes to an annual cycle (data from van Wijk and de Vries, 1963).

therefore be regarded as an effective depth for heat flow. (Note that the flux reaches a maximum $\pi/4$ or one-eighth of a cycle before the temperature, i.e. 3 hours for the diurnal wave and 1.5 months for the annual wave);

(v) the amount of heat flowing into the soil during one half cycle is found by integrating $\mathbf{G}(0,t)$ from $\omega t = -\pi/4$ to $+3\pi/4$ and is $\sqrt{2}\rho' c' D A(0)$. This is the amount of heat needed to raise through $A(0)$ degrees kelvin a layer of soil equal to the effective depth $\sqrt{2}D$.

Values of D for three types of soil are plotted in Fig. 13.6 for daily and annual cycles and as a function of volumetric water content. In sandy and clay soils, D increases rapidly when x_1 increases from 0.0 to 0.1, reaching values between 12 and 18 cm for the daily cycle. For the peat soil, D lies between 3 and 5 cm over the whole range of water content, consistent with the slow heating or cooling of organic soil in response to changes of radiation or of air temperature. Corresponding values for the out-of-phase depth πD, the effective depth and the rate of penetration of the temperature wave can be read from the appropriate axes.

Figure 13.7 shows the type of record from which a damping depth and heat fluxes can be estimated. From the range of the temperature waves at 2.5 and 30 cm, it can be shown that the surface amplitude $A(0)$ was about 20 K. The value of D is 10 cm giving $\kappa' = 0.36 \times 10^{-6}\ \mathrm{m^2\ s^{-1}}$ from equation (13.29) and,

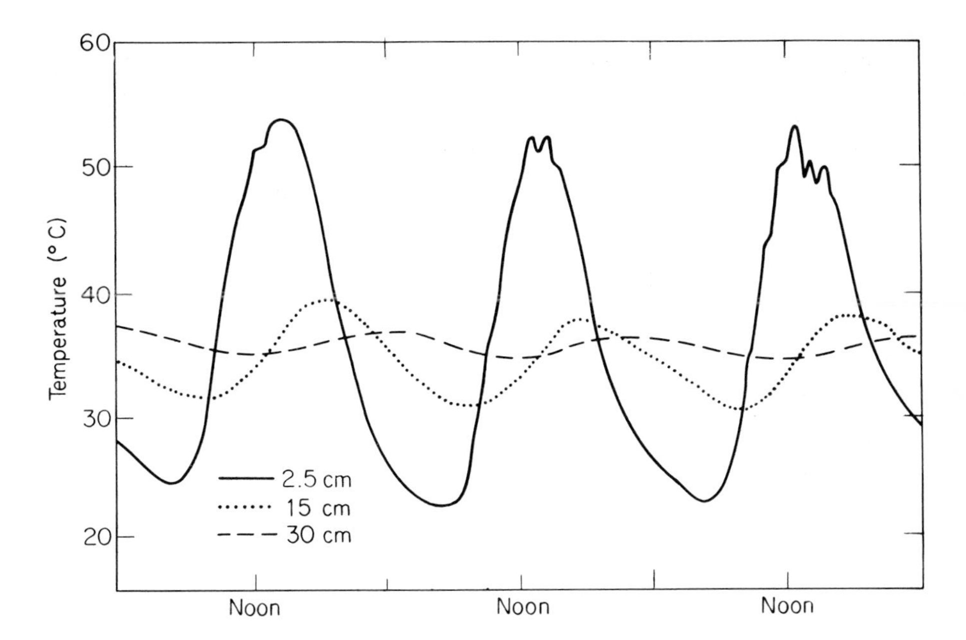

Fig. 13.7 Diurnal course of temperature at three depths in a sandy loam beneath a bare uncultivated surface; Griffith, New South Wales, 17–19 January 1939 (from Deacon (1969) after West).

as the soil was a sandy loam, these values imply that it was very dry (see Fig. 13.6 and Table 13.1). The volumetric heat capacity for a dry sandy soil is about 1.6 MJ m^{-3} K^{-1} and the thermal conductivity $k' = \kappa'\rho'c'$ would therefore be about 0.6 W m^{-1} K^{-1}. If the effective depth is taken as 14 cm ($\sqrt{2}D$), the maximum heat flux into the soil is $40k'/(\sqrt{2}D) = 170$ W m^{-2} and the amount of heat stored in the soil during a half cycle is $\sqrt{2}D\rho'c'A(0) = 4.6$ MJ m^{-2}.

In this analysis, the surface amplitude $A(0)$ has been treated as the independent variable in the system. In practice, $A(0)$ depends on the heat balance of the soil surface and on the relative thermal properties of soil and atmosphere. To compare the behaviour of different soils exposed to the same weather it is necessary to begin by calculating the amplitude of the heat flux. The surface amplitude and the soil temperature distribution can then be derived as a function of D. In practice, it is usually much easier to measure temperature and soil heat flux directly than to attempt calculations based on the idealized model of a uniform soil.

Modification of thermal regimes

Many biological processes depend on soil temperature: the metabolism and behaviour of microorganisms and many invertebrates; the germination of seeds and extension of root systems; shoot extension of seedlings. Because it is difficult to observe the behaviour of an undisturbed root system or of the fauna and flora in a natural soil, relatively little is known about physiological responses to changes of soil temperature or about the behavioural significance of soil temperature gradients. Ignorance of fundamental processes has not prevented agronomists, horticulturists and foresters from developing empirical methods of modifying the thermal regime of soils to help the establishment and growth of crops and trees. Their methods include mulching soils with layers of organic matter in the form of peat or straw to reduce heat losses in winter, covering peat soils with a layer of sand to inhibit evaporation and to reduce the risk of frost at the soil–air interface; covering the soil with polythene sheets to increase soil surface temperature in spring; and the use of black or white powders to raise or lower the temperature of the surface by changing its reflectivity.

Apart from direct intervention by man, the temperature regime of any soil is profoundly modified by the growth of vegetation because the surface becomes increasingly shaded as the canopy develops. The presence of shade reduces the maximum and increases the minimum temperature at any depth and there is usually a small decrease in average temperature. Although this effect is well documented, the implication for root and rhizosphere activity have not been explored.

14

CROP MICROMETEOROLOGY

(i) Profiles and fluxes

The physical principles that govern the transfer of radiant energy, momentum and mass in the atmosphere come together in attempts to apply micrometeorology to the study of how crop stands, including forests, respond to the weather. This branch of environmental physics has made rapid progress over the last thirty years. Four main problems can be distinguished:

(i) What is the rate at which a crop loses water in a given environment, how does it depend on air, soil and plant factors, and how can the rate be minimized?

(ii) What is the rate at which a crop absorbs carbon dioxide in a given environment, how does it depend on air, soil and plant factors and how can the rate be maximized?

(iii) What determines the temperature, humidity and wind regime of a crop and what part do these factors play in determining growth and development apart from their effects on rates of transpiration and of photosynthesis?

(iv) How does the microclimate of individual leaves (as distinct from the microclimate of the canopy as a whole) determine the activity of insects and fungi and the development of disorders and diseases which they cause?

Problem (i) has received much attention and is at a stage where the water use of many crops is fairly well understood. Although problem (ii) was first tackled more than thirty years ago there have have been few extensive studies of the relations between carbon dioxide exchange and crop growth. Micrometeorological research on problem (iii) 'stalled' for a number of years when it became clear that simple one-dimensional approaches to transfer in canopies were inadequate; new methods of describing and measuring transfer are now available, but practical applications are limited at present. Problem (iv) has also received attention in the past decade but there is scope for more research on this important topic.

This chapter primarily concerns the theory needed to attack problems (i) and (ii); in Chapter 15 the interpretation of exchange measurements above and within plant canopies is considered.

PROFILES

The change of potential with height above or within a crop canopy is called the *profile* of that entity. With sufficiently precise instrumentation, profiles of windspeed, temperature and gas concentrations can be measured. Figure 14.1 shows some idealized profiles representative of a cereal crop growing to a height of $h = 1$ m with most of its green foliage between $h/2$ and h. The shape of the profiles above the canopy is determined partly by the turbulent eddies which are produced by the drag of the crop elements on the wind blowing over it and partly by the fluxes to the crop. In Chapter 3 it was shown that in *laminar* boundary layers the transfer of momentum, mass and heat was determined by gradients of potential (profiles) and by diffusivities associated with molecular agitation. In the *turbulent* boundary layers above crops the same principles apply, but the diffusivities are associated with turbulent eddies. The size of turbulent eddies decreases as the surface is approached until finally they merge with molecular agitation. The change of eddy size with height, and the dependence of turbulent mixing on windspeed cause the shapes of profiles to be influenced both by windspeed and by the surface properties that generate turbulence.

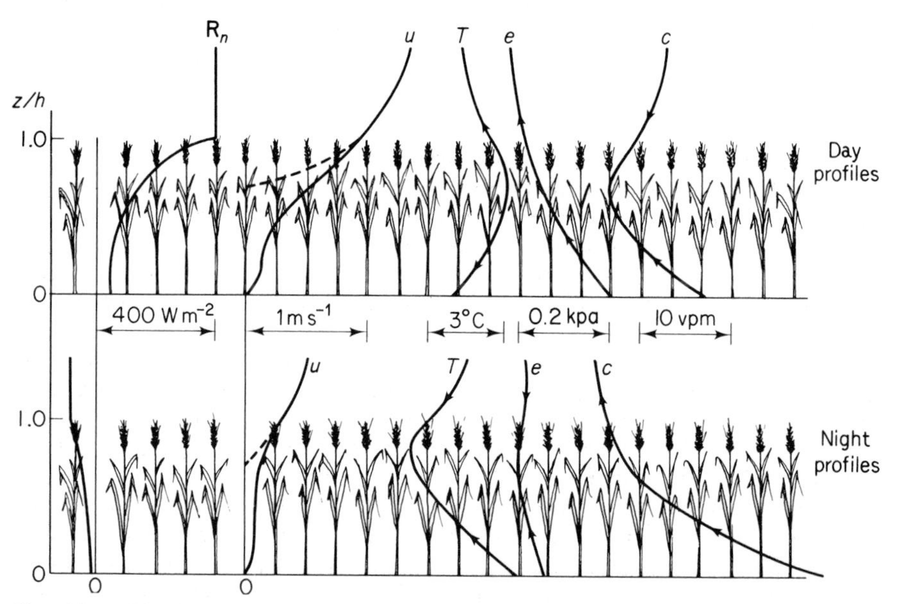

Fig. 14.1 Idealized profiles of net radiation ($\mathbf{R}_n$), windspeed (u), air temperature (T), vapour pressure (e) and CO_2 concentration (c) in a field crop growing to a height h plotted as a function of z/h. The pecked wind profiles represent an extrapolation of the logarithmic relation between u and $(z-d)$ above the canopy (see Chapter 7).

Boundary layer development and structure

Just as the depth of a laminar boundary layer on a flat plate depends on the distance from the leading edge (p. 103), the depth of a turbulent boundary

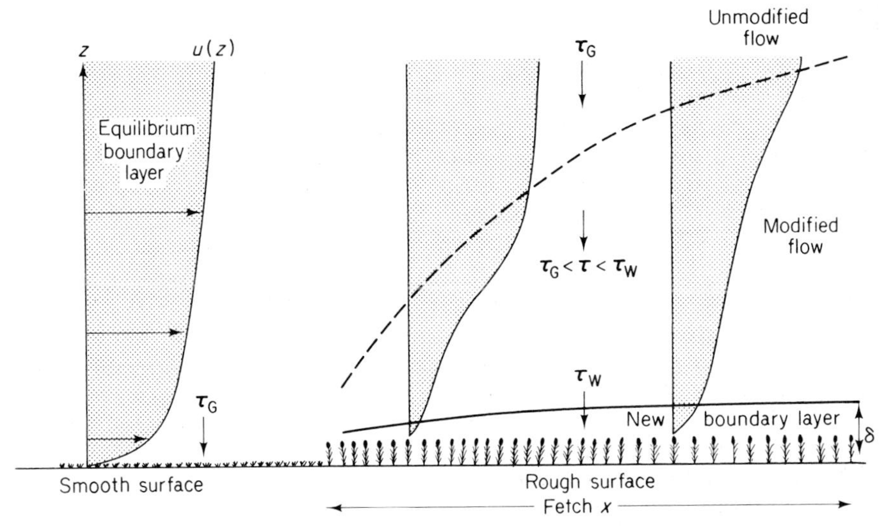

Fig. 14.2 Development of a new equilibrium boundary layer when air moves from a relatively smooth to a rougher surface. The ratio of the vertical to the horizontal scale is about 20:1. The broken line is the boundary between unmodified flow in which the vertical momentum flux is τ_G and modified flow in which the flux is between τ_G and τ_W. The flux is τ_W below the height δ.

layer can be related to the *fetch* or distance of traverse across a uniformly rough surface. Figure 14.2 illustrates the cross section of a flat field of wheat adjacent to an equally flat field of short grass. The wheat will exert a substantially larger drag on the air than the grass, so the air will be decelerated as it moves from the smoother to the rougher surface. The boundary layer which was in equilibrium with the short grass (i.e. the profile of windspeed was characteristic of the grass) is disturbed by the change in roughness so that a layer of modified flow develops in which the flow characteristics are intermediate between those corresponding to the two surfaces. Below the modified layer a new layer, characteristic of the new roughness, and in which fluxes are approximately constant, develops. Munro and Oke (1975) studied the development in depth δ of the new equilibrium wind profile after a change of roughness and concluded that the commonly applied simple ratios of fetch x to δ, e.g. 100:1 or 200:1, were too large. Gash (1986) measured turbulent fluxes of momentum in the extreme transitions from heathland about 0.25 m high to forest about 10 m high; for the heath–forest transition the flux measured 3.5 m over the forest and apparently reached equilibrium at about 120 m from the forest edge. Expressing δ as the height above the zero plane displacement which was about 7.5 m gives $x/\delta \simeq 20$ for this case; for the forest–heath transition the value of x/δ was about 70 when equilibrium was reached, larger than for the reverse case because the smoother heath generated less turbulence.

Very close to the roughness elements of the crop the turbulent structure is influenced by the wakes generated by the elements, establishing a *roughness sublayer* (Raupach and Legg, 1984). Within the roughness sublayer, the

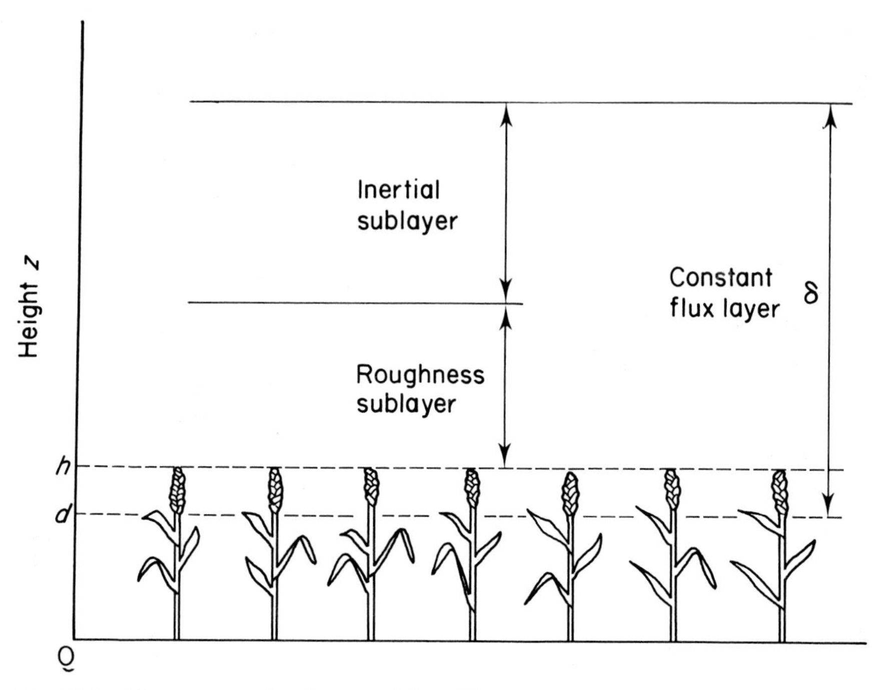

Fig. 14.3 The constant flux layer and its sublayers. The depth δ is about 15% of the surface boundary layer.

boundary layer structure is influenced by factors such as the distribution and structure of foliage elements and the spacing between plants (Fig. 14.3). Above the roughness sublayer is the *inertial sublayer*, in which fluxes are constant with height and the boundary layer structure depends only on scales such as friction velocity (see p. 112) and height. The constant flux layer is usually only about the lowest 15% of the overall boundary layer in which, by definition, flow is influenced by the surface.

It is in the inertial sublayer that micrometeorological measurements can be used most easily to deduce fluxes from profiles. The lower limit of the inertial sublayer is generally defined by $(z-d) \gg z_0$ (Garratt, 1980). If $z-d = 10z_0$, $z \gtrsim 3$ m above the ground for a maize crop 2.3 m high ($z_0 = 0.2$ m, $d = 0.95$ m) and $z \gtrsim 21$ m for a forest 16 m high ($z_0 = 0.9$ m, $d = 12.0$ m). Implications of taking measurements in the roughness sublayer are discussed on p. 239.

Profile equations and stability

For the special conditions listed on p. 113, mean windspeed increases logarithmically with height above the zero plane, giving the standard wind

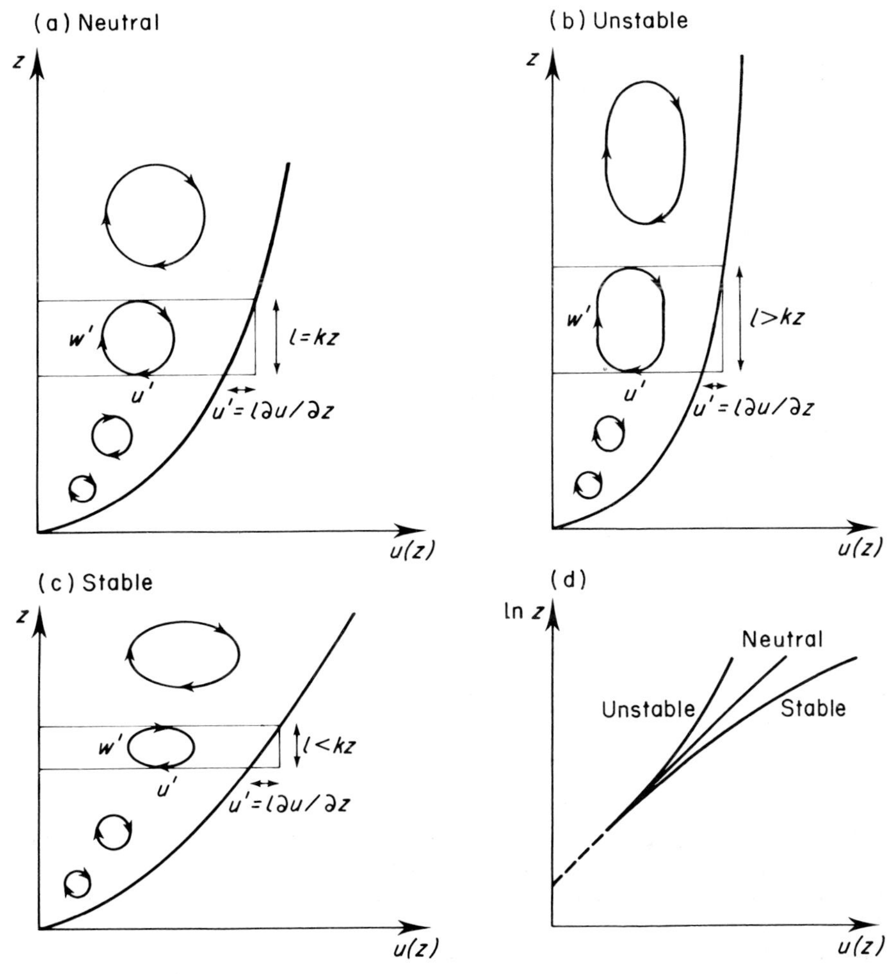

Fig. 14.4 Windspeed profiles and simplified eddy structures characteristic of the three basic stability states in air flow near the ground (from Thom, 1975).

profile equation, which by differentiation gives

$$\frac{\partial u}{\partial z} = \frac{u_*}{k(z-d)} \tag{14.1}$$

The friction velocity u_* quantifies the turbulent velocity fluctuations in the air.

The simplest interpretation of the relation between mean wind profile and turbulence is illustrated in Fig. 14.4. The eddy structure can be envisaged as a set of circular eddies with diameters increasing with height and given by the mixing length $l = kz$, rotating with tangential speed equal to the friction velocity u_*, i.e.

$$u' = w' = u_* = l\partial u/\partial z$$

where w' and u' represent the vertical and horizontal velocity fluctuations respectively (Thom, 1975).

In unstable (lapse) conditions which occur when the surface is strongly heated, vertical motion is enhanced by buoyancy. The amount of enhancement increases as the wind shear (depending on viscosity) decreases, and this is illustrated in Fig. 14.4b, where the eddies are progressively stretched vertically. Thus w' exceeds u', where u' is still given by $l\partial u/\partial z$ but l is greater than kz.

Conversely, in stable conditions (inversion), for example on a clear night with light winds, vertical eddy velocities are damped and so (Fig. 14.4c) $w' < u'$, with $u' = l\partial u/\partial z$ but $l < kz$.

The qualititative effect of stability on the shape of the wind profiles is apparent in Fig. 14.4a–c, and is summarized in Fig. 14.4d. In each of the examples in Fig. 14.4 the momentum flux transmitted to the surface is assumed the same, and so u_* is constant. Since, from equation (7.17),

$$u_* = k\partial u/\partial[\ln(z-d],$$

this requires the gradient of each profile at the lowest levels to be the same. As height increases, velocity gradients become smaller in unstable conditions and larger in stable conditions than those for the neutral case. The differential wind profile (equation (14.1)) can therefore be written in generalized form as

$$\frac{\partial u}{\partial z} = \frac{u_*}{k(z-d)}\phi_M \tag{14.2}$$

where ϕ_M is a dimensionless stability function with a value larger or smaller than unity in stable or unstable conditions respectively.

Because the relation between momentum flux and gradient can also be written

$$\tau = \rho u_*^2 = K_M \partial(\rho u)/\partial z \tag{14.3}$$

it can be readily shown that

$$K_M = ku_*(z-d)\phi_M^{-1} \tag{14.4}$$

Similar flux gradient relations for heat, water vapour, and gas concentration give

$$K_H = -\mathbf{C}/[\partial(\rho c_p T)/\partial z]$$

$$K_V = -\mathbf{E}/(\partial\chi/\partial z)$$

$$K_S = -\mathbf{F}/(\partial S/\partial z)$$

Stability functions can be defined for these entities by

$$K_H = ku_*(z-d)\phi_H^{-1} \tag{14.5}$$

and similarly for ϕ_V and ϕ_S.

The relations between K_M, K_H and K_V (or equivalent functions ϕ_M, ϕ_H and ϕ_V) have been a source of considerable argument in micrometeorology. In neutral stability, all entities are transported equally effectively, all profiles are logarithmic in the constant flux layer, $\phi_M = \phi_H = \phi_V = \phi_S$, and

$K_M = K_H = K_V = K_S$. In unstable conditions K_H exceeds K_M because there is preferential upward transport of heat.

Measurements (reviewed by Dyer, 1974) support the view that $K_H = K_V = K_S$ in unstable conditions; Dyer inferred that $K_M = K_H = K_V$ in stable conditions but there are few measurements to confirm this assumption.

The dependence of ϕ on stability is generally expressed as a function of parameters that depend on the ratio of the production of energy by buoyancy forces to the dissipation of energy by mechanical turbulence. The two best established parameters are the Richardson number Ri, calculated from gradients of temperature and windspeed, and the Monin-Obukhov length L which is a function of fluxes of heat and momentum. In symbols

$$L = \frac{-\rho c_p T u_*^3}{kg\mathbf{C}} \tag{14.6}$$

where T = absolute temperature (K) and g = gravitational acceleration; and

$$\mathrm{Ri} = (gT^{-1}\partial T/\partial z)/(\partial u/\partial z)^2 \tag{14.7}$$

When measurements are made over rough surfaces where gradients are small, or at heights of more than a few metres over any vegetated surface it is important to allow for the decrease in temperature with height that arises from adiabatic expansion when a parcel of air rises, i.e. the dry adiabatic lapse rate Γ which is approximately -0.01 K m^{-1} (see Chapter 2). Temperature T in the preceding equations (and in other heat flux equations elsewhere in this chapter) should be replaced by potential temperature $\theta = T - \Gamma z$, and temperature gradient $\partial T/\partial z$ by $\partial\theta/\partial z = (\partial T/\partial z) - \Gamma$. In a neutral atmosphere θ remains constant with height.

From the definitions of Ri and L it can be shown that

$$(z-d)/L = (\phi_M^2/\phi_H)\mathrm{Ri} \tag{14.8}$$

In unstable conditions, Dyer and Hicks (1970) concluded that

$$\phi_M^2 = \phi_H = \phi_V = [1 - 16(z-d)/L]^{-0.5} \tag{14.9}$$

i.e.

$$\phi_M^2 = \phi_H = \phi_V = (1 - 16\mathrm{Ri})^{-0.5} \qquad \mathrm{Ri} < -0.1 \tag{14.10}$$

In near-neutral conditions when $16(z-d)/L$ is only slightly less than zero, equation (14.9) can be written as

$$\phi_M \approx [1 + 4(z-d)/L] \tag{14.11}$$

From measurements in stable and slightly unstable conditions, Webb (1970) deduced a slightly different relation

$$\phi_M = \phi_H = \phi_V = [1 + 5(z-d)/L] \tag{14.12}$$

i.e.

$$\phi_M = \phi_H = \phi_V = (1 - 5\mathrm{Ri})^{-1} \qquad -0.1 \leqslant \mathrm{Ri} \leqslant 1 \tag{14.13}$$

As $\phi_V/\phi_H = K_H/K_V$ (equation (14.5)), equality of ϕ_V and ϕ_H in both stability states implies that turbulent exchanges of water and heat are always similar to each other, and presumably also to any other entity entrained in the atmosphere.

For correcting flux measurements it is useful to define the product $(\phi_V\phi_M)^{-1} \equiv (\phi_H\phi_M)^{-1} = F$ where F is called a **generalized stability factor** (Thom, 1975). From equations (14.10) and (14.13)

$$F = (1 - 5\mathrm{Ri})^2 \qquad -0.1 \leqslant \mathrm{Ri} \leqslant 1 \tag{14.14}$$

and

$$F = (1 - 16\mathrm{Ri})^{0.75} \qquad \mathrm{Ri} < -0.1 \tag{14.15}$$

Figure 14.5 shows these relationships plotted against Ri on a logarithmic scale.

When $-0.01 < \mathrm{Ri} < +0.01$, F is within 10% of unity, and this range is often taken to define 'fully forced' convection. As Ri approaches +0.2, F tends to zero and turbulent exchange is completely inhibited. When Ri is less than −1, it is generally assumed that 'free' convection dominates and the value of F then exceeds 8.

The relations shown in Fig. 14.5 were all derived from measurements over extensive, and relatively smooth, flat surfaces such as short grass, and at heights where $(z - d)/z_0$ was generally $10^2 - 10^3$, i.e. the measurements were well above the division between the roughness sublayer and the inertial

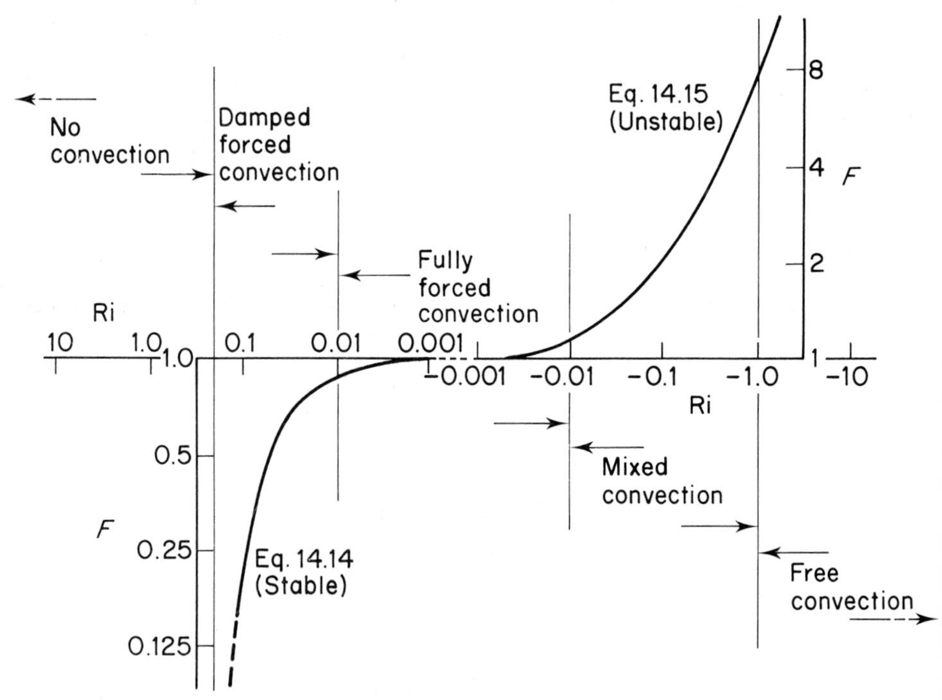

Fig. 14.5 The 'stability factor' F plotted logarithmically against the Richardson number Ri. Fluxes calculated in non-neutral conditions, using profile-gradient equations valid for neutral conditions must be multiplied by F (from Thom, 1975).

sublayer. Recently several studies over aerodynamically rough surfaces (forests and scrub) have suggested that, in unstable and neutral conditions, values of F measured with $(z-d)/z_0$ in the range 10–50 were up to twice the values in Fig. 14.5 (Garratt, 1978; Thom *et al.*, 1975). The causes of the discrepancy seem to be the wakes generated by the roughness elements, and thermals rising between the elements, e.g. between the trees in a forest. In stable conditions the discrepancy appears much less. Such problems arise when measurements are made in the roughness sublayer rather than the inertial sublayer, but it is often difficult to satisfy the strict fetch requirements over tall, rough crops.

MEASUREMENT OF FLUX ABOVE THE CANOPY

Over an extensive uniform stand of level vegetation there is a section of the turbulent boundary layer (about the lowest 15%) where fluxes of momentum, heat, water vapour and any other entrained gas are constant with height. Bulk rates of exchange between the canopy and the air flowing over it can be determined by measuring vertical fluxes in this part of the boundary layer.

There are three principal methods of determining fluxes above a uniform stand of vegetation from measurements in the boundary layer, two indirect and one direct. The indirect methods rely on the measurement of mean potentials and their gradients in the atmosphere, and they are usually referred to as the 'aerodynamic' and 'Bowen ratio' methods. The direct method, known as 'eddy correlation', requires simultaneous measurements of rapid fluctuations of vertical windspeed and the entity in question. The product of the magnitude of the fluctuations is a direct measure of the instantaneous flux at the measurement point. Because the indirect 'gradient' methods have been more widely applied, they are considered first.

Aerodynamic method

The method relies on the existence of relations between fluxes and gradients of the form described earlier (equations (14.3), etc.)

Momentum $\boldsymbol{\tau} = K_{\mathrm{M}}\partial(\rho u)/\partial z$

Heat $\mathbf{C} = -K_{\mathrm{H}}\partial(\rho c_{\mathrm{p}}T)/\partial z$

Water vapour $\mathbf{E} = -K_{\mathrm{V}}\partial\chi/\partial z$

Gas (e.g. CO_2) $\mathbf{F} = -K_{\mathrm{S}}\partial S/\partial z$

The similarity hypothesis states that, in neutral stability,

$$K_{\mathrm{M}} = K_{\mathrm{H}} = K_{\mathrm{V}} = K_{\mathrm{S}}$$

Consequently,

$$\frac{-\rho c_{\mathrm{p}}\partial T/\partial z}{\mathbf{C}} = \frac{\rho\partial u/\partial z}{\boldsymbol{\tau}} \qquad (14.16)$$

Similar equalities may be written between $\mathbf{E}$ and $\mathbf{C}$ or $\mathbf{F}$ and $\boldsymbol{\tau}$. By rearranging equation (14.16) and setting $\boldsymbol{\tau} = \rho u_*^2$ it can be shown that

$$\mathbf{C} = -c_p(\partial T/\partial u)\boldsymbol{\tau}$$
$$= -\rho c_p(\partial T/\partial u)u_*^2 \quad (14.17)$$

and it follows that

$$\mathbf{E} = -(\partial \chi/\partial u)u_*^2 \quad (14.18)$$

and

$$\mathbf{F} = -(\partial S/\partial u)u_*^2 \quad (14.19)$$

In neutral stability, u_* can be estimated from the wind profile alone, and so the aerodynamic method requires only two sets of profiles: temperatures or concentrations of water vapour or gas are measured at a series of heights above the crop, and windspeed is measured at identical heights. The friction velocity is found from the wind profile, and the gradient $\partial T/\partial u$ is found by plotting values of T against u (similarly for χ or S). Flux is then calculated from equation (14.17) (or equations 14.18–19).

An alternative way of applying the aerodynamic method eliminates u_* by differentiating the wind profile equation so that

$$\frac{\partial u}{\partial[\ln(z-d)]} = \frac{u_*}{k}$$

Substituting in equation (14.17) for u_* gives

$$\mathbf{C} = -\rho c_p k^2 \frac{\partial u}{\partial[\ln(z-d)]}\frac{\partial T}{\partial[\ln(z-d)]} \quad (14.20)$$

The minimum number of heights over which the gradients may be determined is two. If the heights are distinguished by subscripts 1 and 2, equation (14.20) becomes

$$\mathbf{C} = \rho c_p k^2 \frac{(u_1-u_2)(T_2-T_1)}{\{\ln[(z_2-d)/(z_1-d)]\}^2} \quad (14.21)$$

An equation of this form was first derived to calculate water vapour transfer by Thornthwaite and Holzman (1942) and has been used in many subsequent studies of transfer in the turbulent boundary layer. Its main defect is the dependence on wind and temperature (or vapour pressure) at two heights only, so that the estimate of flux is sensitive to the error in a single instrument or to local irregularities of the site. More accurate estimates of flux can therefore be obtained by using equation (14.17) with temperature and windspeed measured at four or more heights than from equation (14.21) with only two heights.

In non-neutral conditions it is necessary to know profiles of u and T to estimate u_*, and equality of K_M, K_H and K_V cannot be assumed. It can be shown that equation (14.21) takes the generalized form

$$\mathbf{C} = \rho c_p k^2 \frac{(u_1-u_2)(T_2-T_1)}{\{\ln[(z_2-d)/(z_1-d)]\}^2}(\phi_H\phi_M)^{-1} \quad (14.22)$$

and a similar equation may be written for **E**.

An alternative approach (Biscoe *et al.*, 1975), avoiding the errors inherent in 'two point' profiles, makes use of Webb's (1970) relation for near neutral and stable conditions (i.e. $-0.03<(z-d)/L<+1$) as given in equation (14.12).

This equation allows three steps to be taken. First, equality of ϕ_M and ϕ_H means that the profiles of windspeed and temperature have the same shape, so a plot of temperature versus windspeed will give a straight line with slope $\partial T/\partial u$. Second, ϕ_M is a linear function of $(z-d)/L$, and so equation (14.2) can be integrated to give the wind profile equation

$$u = (u_*/k)\{\ln[(z-d)/z_0] + 5(z-d-z_0)/L\} \tag{14.23}$$

Third, because $\mathbf{C} = -\rho c_p(\partial T/\partial u)u_*^2$, the Monin-Obukhov length (equation (14.6)) may be written $L = u_*T/kg(\partial T/\partial u)$ and so equation (14.23) may be written

$$u' = (u_*/k)\ln[(z-d)/z_0] \tag{14.24}$$

where

$$u' = u - 5(z-d-z_0)(\partial T/\partial u)g/T$$

Equation (14.24) can be used to find u_*/k as the slope of the straight line defined by plotting u' against $\ln(z-d)$. When u_* is known,

$$\mathbf{C} = -\rho c_p(\partial T/\partial u)u_*^2$$

Similar expressions can be used to find fluxes of water vapour, carbon dioxide and other gases in stable and slightly unstable conditions. Paulson (1970) derived profile equations for the more complicated case when the atmosphere is more unstable.

[It was stressed on p. 237 that temperatures and temperature gradients in flux equations should strictly be replaced by potential temperature and potential temperature gradients. Furthermore, humidity should be expressed as mixing ratio r (mass per unit mass of dry air) because **E** is strictly proportional to $\partial r/\partial z$ and r is conserved when pressure changes with height. The specific humidity q (mass per unit mass of moist air) is also conserved, and is often used in meteorological literature, but Webb *et al.* (1980) showed that r is strictly the correct ratio for flux calculations because it defines fluxes in a coordinate system fixed relative to the surface (i.e. there is no net vertical flux of dry air). However, for consistency with earlier chapters, and because the errors involved are usually small, temperature and vapour pressure are retained in the analysis which follows.]

Bowen ratio method

The Bowen ratio formula for flux measurement is derived from the energy balance of the underlying surface

$$\mathbf{R}_n - \mathbf{G} = \mathbf{C} + \lambda\mathbf{E}$$

which can be rewritten in the form

$$\boldsymbol{\lambda}\mathbf{E} = \frac{(\mathbf{R_n} - \mathbf{G})}{1 + \beta} \tag{14.25}$$

where β is the Bowen ratio $\mathbf{C}/\boldsymbol{\lambda}\mathbf{E}$. Measurements of the net radiation ($\mathbf{R_n}$) and soil heat flux ($\mathbf{G}$) are needed to establish ($\mathbf{R_n} - \mathbf{G}$) and β is found from measurements of temperature and vapour pressure at a series of heights within the constant flux layer. Assuming that the transfer coefficients of heat and water vapour are equal, it can be shown that

$$\beta = \mathbf{C}/\boldsymbol{\lambda}\mathbf{E} = \gamma \partial T/\partial e \tag{14.26}$$

and $\partial T/\partial e$ is found by plotting the temperature at each height against vapour pressure at the same height.

The Bowen ratio method can be generalized by writing the heat balance equation as

$$\begin{aligned} \mathbf{R_n} - \mathbf{G} &= -K\rho c_p(\partial T/\partial z) - K\rho c_p(\partial e/\gamma \partial z) \\ &= K\rho c_p(\partial T_e/\partial z) \end{aligned} \tag{14.27}$$

where K is a turbulent transfer coefficient assumed identical for heat and water vapour and T_e is the equivalent temperature $T + (e/\gamma)$. The Bowen ratio equation (14.26) is derived from equation (14.27) by writing

$$T_e = T + (e/\gamma)$$

By writing the sensible heat flux as $\mathbf{C} = -K\rho c_p(\partial T/\partial z)$ and forming similar expressions relating the latent heat flux $\boldsymbol{\lambda}\mathbf{E}$ to $\gamma^{-1}(\partial e/\partial z)$ and the flux $\mathbf{F}$ of any other gas to $\partial S/\partial z$, it can be shown that, combining each expression in turn with equation (14.27)

$$\begin{aligned} \mathbf{C} &= (\mathbf{R_n} - \mathbf{G})(\partial T/\partial T_e) \\ \boldsymbol{\lambda}\mathbf{E} &= (\mathbf{R_n} - \mathbf{G})(\partial e/\partial T_e)/\gamma \\ \mathbf{F} &= (\mathbf{R_n} - \mathbf{G})(\partial S/\partial T_e)/\rho c_p \end{aligned} \tag{14.28}$$

The aerodynamic and Bowen ratio methods of flux determinations are usually applied to potentials which have been averaged for periods of a half to one hour. Fluctuations in the potentials, especially on a day of intermittent cloud, often preclude the estimation of mean fluxes for shorter periods. On the other hand, diurnal changes make time-averaging undesirable for periods of more than two hours, particularly near sunrise and sunset.

Eddy correlation method

This method relies on measurements of the fluctuating components of wind in the constant flux region of the surface boundary layer, and of the associated fluctuations in temperature, humidity or gas concentration. The principles of transfer by eddies acting as 'carriers' were discussed in Chapter 3.

The mean wind velocity at any height in the constant flux layer is horizontal, but instantaneous values may be in any direction and generally have a vertical component that may be towards or away from the surface. The mean vertical flux of dry air is zero. The fluctuating vertical component carries with it other entities—we shall consider sensible heat. For there to be *net transport* of sensible heat by vertical eddies there must also be fluctuations of the temperature, and these fluctuations must be correlated to some extent with the fluctuations in vertical velocity. A flux of heat towards the surface arises when, on average, eddies moving towards the surface contain air at higher than average temperature, and those moving away from the surface are at lower temperatures.

Denoting the vertical velocity at time t by $w(t)$ m s^{-1} and temperature by $T(t)$, the instantaneous sensible heat flux $\mathbf{C}(t)$ is

$$\mathbf{C}(t) = \rho c_p w(t) T(t) \qquad \mathrm{W\ m^{-2}}$$

The average vertical flux density is the time average of $\mathbf{C}(t)$, i.e. $\mathbf{C} = \overline{\rho c_p w T}$.

Writing mean values of the quantities as $\overline{T}$ and $\overline{\rho c_p w}$, and their fluctuations as T' and $\rho c_p w'$

$$\begin{aligned} \mathbf{C} &= \overline{(\overline{\rho c_p w} + \rho c_p w')(\overline{T} + T')} \\ &= \overline{\rho c_p w}\,\overline{T} + \overline{\overline{\rho c_p w}\,T'} + \overline{\rho c_p w'\overline{T}} + \overline{\rho c_p w' T'} \end{aligned} \qquad (14.29)$$

The third term in equation (14.29) is zero because the fluctuations associated with $\overline{T}$ can make no net transport. The first and second terms are sometimes taken as zero, arguing that $\overline{\rho c_p w}$ is zero, but strictly it is the vertical flux of dry air $\overline{\rho_a c_p w}$ that is zero. Webb *et al.* (1980) showed that, allowing for this distinction, equation (14.29) reduces to

$$\mathbf{C} = \bar{\rho} c_p \overline{w' T'} \qquad (14.30)$$

which can be measured directly if detectors for windspeed and temperature with sufficiently fast responses are available, and if their signals can be sampled and averaged sufficiently rapidly.

The necessary response time of the sensors depends on the range of eddy sizes that carry the flux. Eddy sizes grow with height over the surface (see Fig. 14.3) and increase with increasing surface roughness and windspeed. Consequently sensors capable of operating between 0.1 and 10 Hz would often be adequate for use several metres above a rough forest canopy whereas a frequency response of 0.001 Hz might be needed for eddy flux measurements close to a smooth surface. To avoid sampling artefacts, readings must be taken at a time interval that is not more than half the fastest sensor response time. With recent developments in electronics and computing, signal processing for eddy correlation is seldom a limitation in using the method. Leuning *et al.* (1982), in a study of CO_2 fluxes by eddy correlation, describe some of the practical and theoretical difficulties of the method.

Relative merits of methods of flux measurement

In spite of the elegance and theoretical attractions of the eddy correlation technique, it has only recently been used over crops in studies that were sufficiently long to be useful in understanding relationships between vegetation and the atmosphere. The limitations have generally been in the availability of reliable, durable sensors for water vapour and CO_2.

The aerodynamic method is straightforward in neutral stability, when profiles of only windspeed and the entity in question are required. It is unreliable at low windspeeds when anemometers may stall for part of the measuring period. Empirical correlations for stability necessitate measurements of temperature profiles, and are not well defined in strongly stable conditions (e.g. still nights with low windspeeds): the accepted corrections appear valid over short crops, but the aerodynamic method appears to underestimate fluxes seriously over tall rough crops unless it is possible to make measurements well above the roughness sublayer but still within the inertial sublayer.

The Bowen ratio method, assuming that $K_H = K_V$, does not require stability corrections and so is often the preferred of the two gradient techniques, but it becomes indeterminate when $\mathbf{R}_n - \mathbf{G}$ tends to zero, and is generally difficult to apply at night or in other conditions when net radiation is small.

Density corrections to flux measurements

When fluxes of any atmospheric constituent are measured by gradient methods or eddy correlation it is sometimes necessary to take into account the simultaneous fluxes of any other entities, in particular heat and water vapour. Sources of other entities result in expansion of the air and so affect the density of the constituents (but not its mixing ratio r). For example, if gradients of CO_2 above a crop were measured by drawing air through a gas analyser from various heights sequentially, without altering its temperature or humidity, the apparent CO_2 gradient would be in error, because the density of CO_2 in the air at each height would be influenced by the content of water vapour and by the temperature at that height. Correction is not necessary if the flux is evaluated from measurements of the mixing ratio of the constituent (mass of constituent per unit mass of dry air).

Webb, Pearman and Leuning (1980) derived expressions relating the correction $\boldsymbol{\delta}\mathbf{F}$ required to allow for fluxes of sensible and latent heat when a mass flux is deduced from density measurements. Their analysis shows that when $\mathbf{C}$ and $\boldsymbol{\lambda}\mathbf{E}$ are in W m^{-2}, and $\boldsymbol{\delta}\mathbf{F}$ is in kg m^{-2} s^{-1}

$$\boldsymbol{\delta}\mathbf{F} = (\bar{\rho}_c/\bar{\rho}_a)(0.65 \times 10^{-6}\boldsymbol{\lambda}\mathbf{E} + 3.36 \times 10^{-6}\mathbf{C}) \qquad (14.31)$$

for a typical situation with $T_a = 20$°C and $e = 1.0$ kPa. The mean density of the constituent is $\bar{\rho}_c$, and $\bar{\rho}_a$ is the density of dry air. Equation (14.31) shows that the correction for sensible heat flux is about five times larger than for an equivalent flux of latent heat. For CO_2 with a mean content of 330 vpm in dry air, $\bar{\rho}_c/\bar{\rho}_a = 0.502 \times 10^{-3}$ so that, when $\boldsymbol{\lambda}\mathbf{E} = \mathbf{C} = 250$ W m^{-2}, $\boldsymbol{\delta}\mathbf{F} = 0.5$ mg m^{-2} s^{-1}, a value that is comparable with typical CO_2 fluxes over crops of 1–2 mg m^{-2} s^{-1}. The correction is therefore important if appropriate methods of measuring CO_2 have not been adopted. Similarly, for SO_2 at 0.010 vpm, $\bar{\rho}_c/\bar{\rho}_a = 0.22 \times 10^{-6}$ and so $\boldsymbol{\delta}\mathbf{F} = 0.2$ μg m^{-2} s^{-1}, comparable with reported fluxes (Fowler and Unsworth, 1979).

15

CROP MICROMETEOROLOGY

(ii) Interpretation of measurements

RESISTANCE ANALOGUES

Measurements of fluxes by micrometeorological methods are of relatively little value to the agricultural scientist or ecologist unless they can be associated with some factor or group of factors that describes how the surface controlled or responded to the flux. A useful way of extending the study of transfer from single leaves to complex canopies, and of pointing out the shortcomings of certain approaches, is to consider the canopy as an electrical analogue as in Fig. 15.1. The rate of exchange of an entity between a single leaf and its environment can be estimated (a) when the potential of the entity (e.g. the vapour pressure or CO_2 concentration) is known at the leaf and in the surrounding air *and* (b) the relevant resistances (e.g. stomatal and boundary layer) can be measured or estimated. In the same way, the bulk exchange of any entity between the canopy and the air above it can be estimated by measuring the potentials at two or more heights above the canopy (z_1, z_2, etc.) if the resistances across these potentials are also known. Within the canopy, resistances corresponding to the stomata and boundary layers of individual leaves have a clear physical significance, but the validity of describing transfer in the air within the canopy by Ohm's Law analogues is discussed later in this chapter.

It is possible to derive a parameter which plays the same part in equations for the water vapour exchange of a canopy as the stomatal resistance plays in similar equations for a single leaf. This parameter will be given the symbol r_c, where the subscript denotes canopy, crop or cover.

It was shown in Chapter 14 that the sensible heat loss from a surface can be written in the form

$$\mathbf{C} = -\rho c_p u_*^2 (\partial T/\partial u)$$

where T is a linear function of u. As a special case, the gradient $\partial T/\partial u$ can be written $[T(z) - T(0)]/[u(z) - 0]$ where $T(0)$, obtained by the extrapolation shown in Fig. 15.2, is the air temperature at the height where $u = 0$, i.e. $z = d + z_0$. The above equation can therefore be written as

$$\begin{aligned} \mathbf{C} &= -\rho c_p u_*^2 [T(z) - T(0)]/u(z) \\ &= -\rho c_p [T(z) - T(0)]/r_{aH} \end{aligned} \tag{15.1}$$

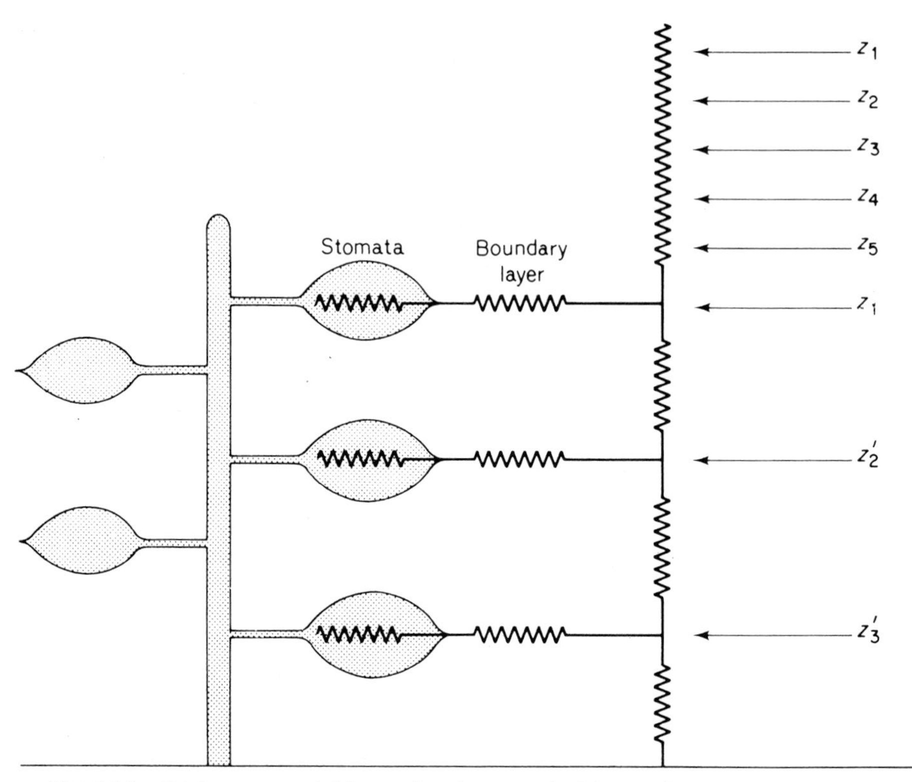

Fig. 15.1 Resistance model for a plant in a stand of vegetation.

where $r_{aH} = u(z)/u_*^2$ can be regarded as an aerodynamic resistance between a fictitious surface at the height $d + z_0$, and the height z. Similarly, it can be shown that

$$\lambda \mathbf{E} = \frac{-\rho c_p}{\gamma} \frac{\{e(z) - e(0)\}}{r_{aV}} \tag{15.2}$$

where $e(0)$ is the value of the vapour pressure extrapolated to $u = 0$ and $r_{aV} = r_{aH} = u(z)/u_*^2$. The diffusion of water vapour between the intercellular spaces of leaves and the atmosphere at height z can now be described formally by the equation

$$\lambda \mathbf{E} = \frac{-\rho c_p}{\gamma} \frac{\{e(z) - e_s(T(0))\}}{r_{aV} + r_c} \tag{15.3}$$

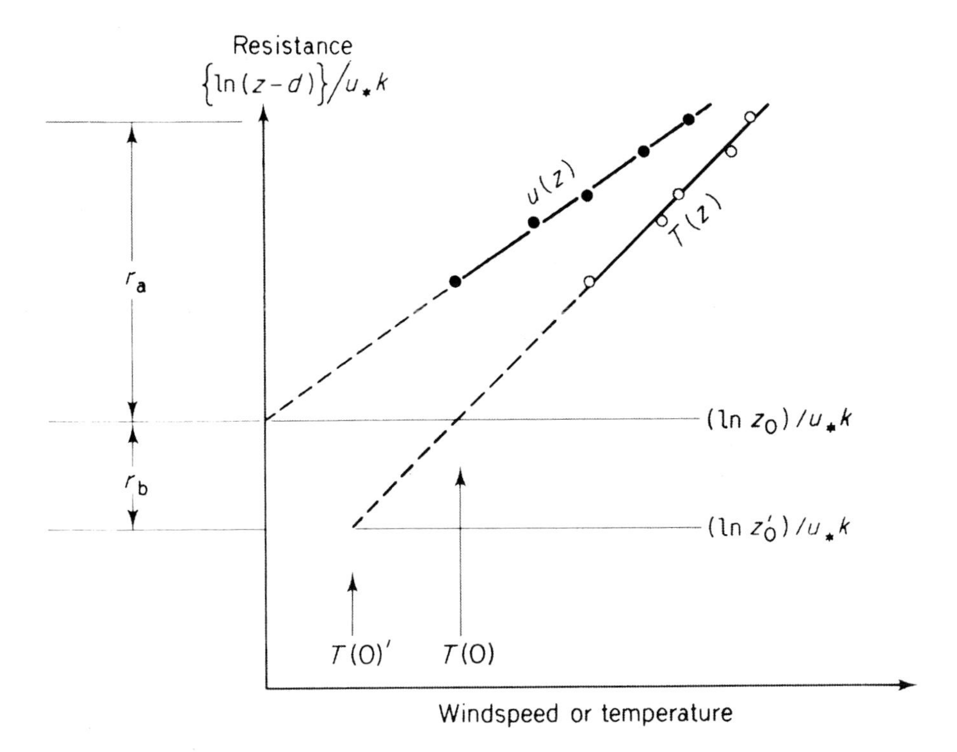

Fig. 15.2 Diagram showing how $T(0)$ and $T(0')$ are determined by plotting temperature against $[\ln(z-d)]/u_*k$. $T(0)$ is the value of temperature extrapolated to $z-d=z_0$ and $T(0')$ is the value extrapolated to $z-d=z_0'$. The significance of r_a and r_b is shown on the left-hand axis and is discussed on p. 248.

This relation defines the canopy resistance r_c and is identical to the corresponding equation for an amphistomatous leaf with r_c replacing the stomatal resistance r_s (p. 188) and r_{aV} replacing r_V. Introducing the energy balance equation, writing $r_{aH}=r_{aV}=r_a$ and eliminating $T(0)$ (see pp. 185–187) gives

$$\boldsymbol{\lambda}\mathbf{E}=\frac{\Delta(\mathbf{R}_n-\mathbf{G})+\rho c_p\{e_s(T(z))-e(z)\}/r_a}{\Delta+\gamma^*} \qquad (15.4)$$

where $\gamma^*=\gamma(r_a+r_c)/r_a$. Values of r_c for a given stand can therefore be derived either directly from profiles of temperature, humidity and windspeed using equations (15.1) and (15.2) or indirectly from the Penman-Monteith equation (15.4) when the relevant climatological parameters are known and $\boldsymbol{\lambda}\mathbf{E}$ is measured or estimated independently.

Two objections can be raised to this apparently straightforward method of separating the aerodynamic and physiological resistances of a crop canopy. In the first place, the values of r_c derived from measurements are not unique unless the sources (or sinks) of sensible and latent heat have the same spatial distribution. In a closed canopy, fluxes of both heat and water vapour are dictated by the absorption of radiation by the foliage and, provided the stomatal resistance of leaves does not change violently with depth in the part of the canopy where most of the radiation is absorbed, the distributions of heat and vapour sources will usually be similar but will seldom be identical. Conversely, anomalous values of r_c are likely to be obtained in a crop with little foliage if evaporation from bare soil beneath the leaves makes a substantial contribution to the total flux of water vapour.

In the second place, the analysis cannot yield values of r_c which are strictly independent of r_a unless the apparent sources of heat and water vapour, as determined from the relevant profiles, are at the same level $d+z_0$ as the apparent sink for momentum. This is a more serious restriction. Form drag, rather than skin friction (p. 103), is often the dominant mechanism for the absorption of momentum by vegetation so that the resistance to the exchange of momentum between a leaf and the surrounding air is smaller than the corresponding resistances to the exchange of heat and vapour which depend on molecular diffusion alone. It follows that the *apparent* sources of heat and water vapour will, in general, be found at a lower level in the canopy than the *apparent* sink of momentum, say at $z = d + z_0'$ rather than at $z = d + z_0$ where z_0' is smaller than z_0. Atmospheric resistances to transfer may therefore be described in terms of r_a, the resistance to momentum transfer and r_b, an additional resistance, assumed to be the same for heat and water vapour.

From the definition of r_a and the wind profile equation in neutral stability

$$\begin{aligned} r_a = \rho u(z)/\tau &= u(z)/u_*^2 \\ &= \{\ln[(z-d)/z_0]\}/ku_* \\ &= \{\ln[(z-d)/z_0]\}^2/k^2u(z) \end{aligned} \qquad (15.5)$$

In the derivation of this expression the effective height for the sink of momentum is $z = d + z_0$ for which $(z-d)/z_0 = 1$ and so $r_a = 0$. Similarly, the resistance between a height z above the ground and the apparent source (or sink) of heat and vapour at a height $d + z_0'$ can be written as

$$\begin{aligned} r_a' = \frac{\ln[(z-d)/z_0']}{ku_*} &= \frac{\ln[(z-d)/z_0]}{ku_*} + \frac{\ln[(z_0/z_0']}{ku_*} \\ &= \quad r_a \quad + \quad r_b \end{aligned} \qquad (15.6)$$

where r_b is the additional boundary layer resistance, assumed to be the same for heat and water vapour. The implications of equation (15.6) are shown in Fig. 15.2.

If z_0 were independent of windspeed, equation (15.5) shows that $1/r_a$ would be proportional to $u(z)$. Figure 15.3 illustrates this for values of roughness length appropriate for short grass, cereal crops, and forests.

In practice the roughness length of many crops decreases as wind increases (p. 117) and so $1/r_a$ is approximately constant over a range of low windspeeds.

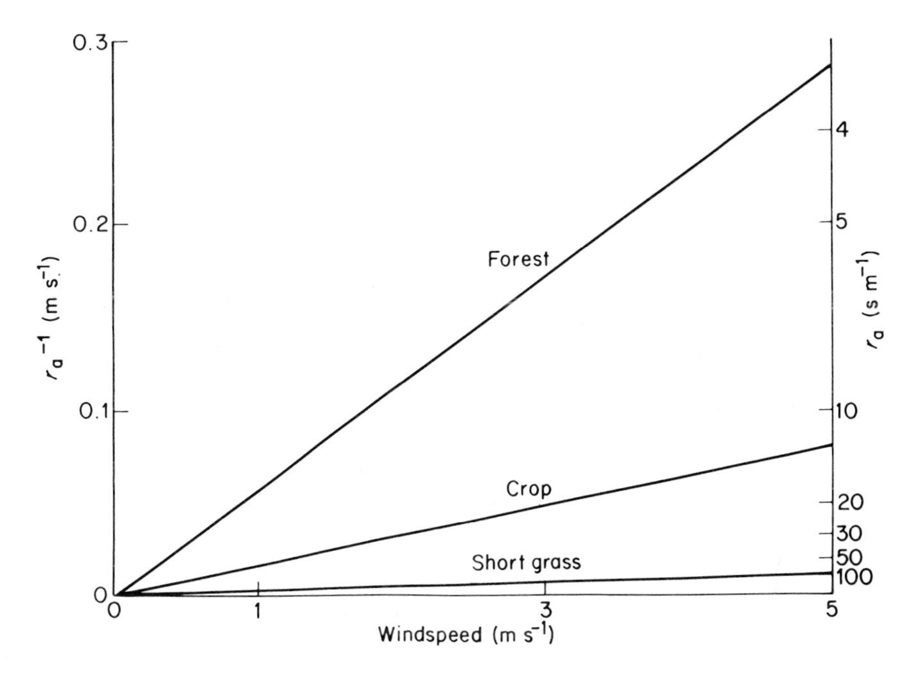

Fig. 15.3 Calculated values (from equation (15.5)) of resistance r_a in relation to windspeed over surfaces with roughnesses characteristic of: short grass ($z_0 = 1$ mm, $d = 7$ mm); cereal crop ($z_0 = 0.2$ m, $d = 0.95$ m); forest ($z_0 = 0.9$ m, $d = 11.8$ m). Windspeeds are referred to a standard height $z - d = 5$ m for each surface.

In near neutral and stable conditions the resistance derived from equation (14.23) is

$$[ku_*]^{-1}\left[\ln\left(\frac{z-d}{z_0}\right) + \frac{5(z-d)}{L}\right] \qquad (15.7)$$

provided that $z - d$ is at least an order of magnitude greater than z_0. When $[\ln(z-d)/z_0]/ku_*$ is identified as r_a, the additional component $+[5(z-d)/L]/ku_*$ can be regarded as a stability resistance. Similarly, equation (14.11) can be used to show that the stability resistance in more unstable conditions (L negative) is approximately $[4(z-d)/L]/ku_*$. The tendency for $1/r_a$ to become independent of windspeed as the speed decreases is more pronounced in unstable than in neutral or stable conditions because the decrease in turbulent energy associated with decreased friction is compensated by an increase in the supply of energy from buoyancy.

The resistance r_b is $[\ln(z_0/z_0')]/ku_*$ (equation 15.6) and $u_* r_b$ is identical to the parameter B^{-1} which a number of workers have used to analyse processes of exchange at rough surfaces.

The size of r_b for real and for model vegetation has been estimated by Chamberlain (1966) and Thom (1972) and the results of their work can be summarized as follows:

(i) For a given value of u_*, z_0/z_0' and r_b are almost constant over a wide range of surface roughnesses. For example, a set of measurements of evaporation from an artificial grass surface gave $z_0 = 1$ cm, $u_* = 25$ cm s^{-1}, $B^{-1} = 4.5$. At the same value of u_*, achieved at a higher windspeed over towelling with $z_0 = 0.045$ cm, B^{-1} was 4.7. The corresponding resistances r_b are 0.18 and 0.19 s cm^{-1}.

(ii) For a given value of z_0, z_0/z_0' increases with windspeed and therefore with u_*. For a fourfold increase of u_* from 25 to 100 cm s^{-1}, $u_* r_b$ increased by a factor of 1.3 for the grass and 1.7 for the towelling. For evaporation from a bean crop, Thom found that $u_* r_b = A u_*^{0.33}$ where the constant A had the value 6.2 when u_* was in m s^{-1}. This implies that $r_b \propto u_*^{-0.67}$.

(iii) The value of z_0' and hence of r_b is expected to depend on the molecular diffusivity of the property being transferred. On the assumption that $r_b \propto (\text{diffusivity})^n$, values of n determined experimentally range from about -0.8 to about -0.3. For a stand of beans, n appeared to be about -0.66 implying that r_b for heat may be 10% greater than r_b for water vapour.

Values of r_b are seldom determined directly in crop micrometeorology, but are usually estimated from the results such as those above. Thom's empirical equation

$$r_b = 6.2 u_*^{-0.67} \quad (15.8)$$

is an adequate approximation in estimating r_b for heat and water vapour transfer to crops, at least over the usual range of u_*, 0.1–0.5 m s^{-1}. Values of r_b for a gas with diffusivity much different from the value for water vapour may be estimated from equation (15.8) using the approximations $r_b \propto D^{-0.67}$ and $D_g = D_v[M_v/M_g]^{0.5}$ where M is molecular weight, D diffusivity and the subscripts g and v refer to the gas and water vapour respectively.

When vegetated surfaces are unusually rough or fibrous (e.g. pine needles), equation (15.8) may not be a good approximation to estimating r_b. Thom (1972) discussed more detailed treatments. For vapour transfer to rigid rough surfaces Fig. 15.4 shows that the expression

$$B^{-1} = 7.3 \mathrm{Re}_*^{0.25} \mathrm{Sc}^{0.5} - 5.0$$

proposed by Chamberlain *et al.* (1984) on the basis of analysis by Brutsaert (1982) fits observations well, and may be useful for estimating r_b for ploughed fields or urban areas; Re_* is the roughness Reynolds number $u_* z_0/\nu$ and Sc is the Schmidt number ν/D. The expression is also applicable to the diffusive transfer of particles in the size range where impaction and sedimentation are unimportant, i.e. $r \lesssim 0.5$ μm (see Chapter 10).

The surface resistance of vegetation has sometimes been determined from relations such as

$$\lambda \mathbf{E} = \rho c_p \frac{[e_s(T_0) - e]}{\gamma(r_a + r_c)} \quad (15.9)$$

where $r_a = u/u_*^2$, r_c is the 'apparent' surface resistance and T_0 the 'apparent'

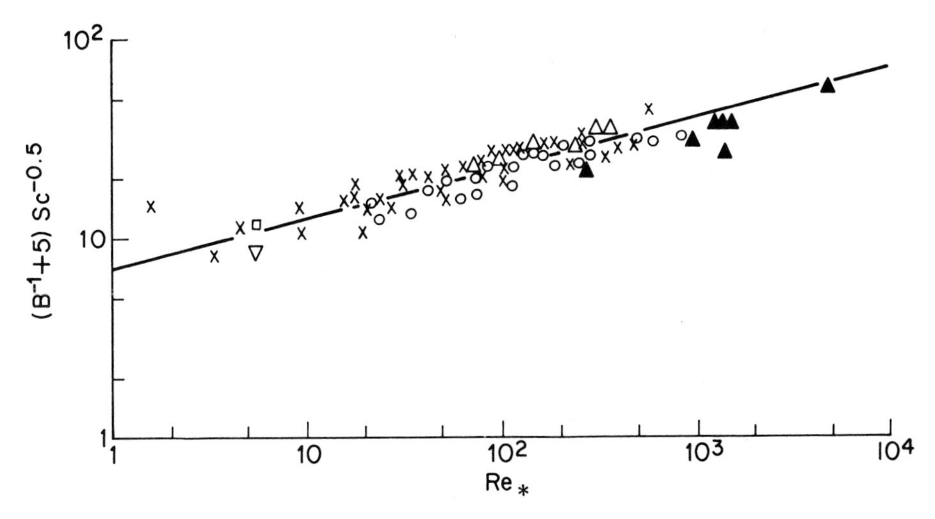

Fig. 15.4 Diffusive transport to rough surfaces (from Chamberlain *et al.*, 1984). Symbols refer to different surface structures, vapours and particles: ^{212}Pb vapour ▽, △, x; ^{123}I vapour ▲; water vapour ○; Aitken nuclei ($r = 0.08$ μm) □. The straight line has slope 0.25 and intercept 7.3.

surface temperature. More rigorous analysis gives

$$\lambda \mathbf{E} = \rho c_p \frac{[e_s(T_0') - e]}{\gamma(r_a + r_c')} \tag{15.10}$$

where r_c' is the 'true' resistance allowing for the existence of the additional boundary layer resistance r_b, and T_0' is the 'true' surface temperature. By manipulating these equations and using the relation $\mathbf{C} = \rho c_p(T_0' - T_0)/r_b$, it can be shown that the error in calculating r_c without allowing for r_b is

$$\delta r_c = r_c' - r_c = r_b \left(\frac{\Delta}{\gamma} \frac{\mathbf{C}}{\lambda \mathbf{E}} - 1 \right) \tag{15.11}$$

This error is zero when the Bowen ratio $\mathbf{C}/\lambda\mathbf{E}$ is equal to γ/Δ. For a well watered crop growing in a temperate climate, the average value of $\mathbf{C}/\lambda\mathbf{E}$ is usually about 0.1 and with $(\Delta/\gamma) = 2.0$, $\delta r_c = -0.8 r_b$. The absolute magnitude of this error is less important than the fact that it changes in size and sign during the day as the Bowen ratio changes. For example if $\mathbf{C}/\lambda\mathbf{E}$ decreases from $+0.3$ in the early morning to -0.3 in the late afternoon and r_b is 0.2 s cm^{-1}, the value of r_c will change during the day from $(r_c' + 0.08)$ to $(r_c' + 0.32)$ s cm^{-1}.

Surface resistances to the uptake of pollutant gases have also been determined from equations analogous to equation (15.9), i.e.

$$\mathbf{F} = \frac{S - 0}{r_a + r_b + r_c} \tag{15.12}$$

assuming that there are sinks within the crop canopy where the gas is absorbed and hence where S may be assumed zero. An example of this analysis is given later.

CASE STUDIES

Water vapour and transpiration

One of the earliest practical applications of micrometeorology was for measuring the water use of agricultural crops. The relatively simple instrumentation required, and the importance of efficient planning of irrigation in many parts of the world ensure that this remains an active area of research. Water use of forests has also been studied; gradients are much smaller over these tall rough canopies than over agricultural crops, and the small aerodynamic resistance of forest canopies has important consequences for the evaporation of intercepted rainfall, i.e. water trapped on the canopy after rain.

Stewart and his colleagues (Stewart and Thom, 1973; Thom *et al.*, 1975) used the aerodynamic and Bowen ratio techniques to study evaporation from a forest of Scots and Corsican pine at Thetford in south-east England. In determining the available energy for the Bowen ratio method it was necessary to allow for the storage of heat in the trunks and branches and in the air within the canopy; this term **J** (W m^{-2}) was about $18\delta T$ where δT is the rate of temperature change of air in the canopy (K h^{-1}). The maximum value of **J** was ± 55 W m^{-2}.

Comparison of flux measurements at Thetford by the aerodynamic and Bowen ratio methods identified the discrepancy discussed on p. 239, requiring a large empirical correction to aerodynamic estimates. Bowen ratios on fine days ranged from near 1 to 4 or more.

Stewart and Thom (1973) analysed flux measurements at Thetford using resistance analogues. The aerodynamic resistance r_a, derived from wind profiles, was about 5–10 s m^{-1}, and the excess resistance r_b (equation (15.8)) was estimated as 3–4 s m^{-1}. They then used the Penman-Monteith equation (15.4) to deduce the canopy resistance, making the small correction (equation (15.11)) for excess boundary layer resistance. Figure 15.5 shows the variation of canopy resistance on a fine day, and compares it with values found at other forest sites. Near dawn the foliage was wet with dew and so r_c was small. Once foliage had dried, r_c was about 100–150 s m^{-1} in the middle part of the day, implying that the average stomatal resistance of needles in the canopy (which had leaf area index about 10) was about 1000–1500 s m^{-1}. Later in the day r_c increased, probably as a result of stomatal closure in response to water stress.

Figure 15.5 shows that minimum canopy resistances of other mature forests are also typically about 100 s m^{-1}. The ratio r_c/r_a for forests is often about 50, and equation (15.4) can be used to show that evaporation from forest canopies wet with rain would proceed much faster than from dry canopies exposed to the same weather. This contrasts with the situation for many agricultural crops for which minimum values of r_c are typically 100 s m^{-1} but r_c/r_a is often close to unity. Consequently, forests in regions where rain is frequent tend to use more water by evaporation from foliage and transpiration than shorter crops growing nearby.

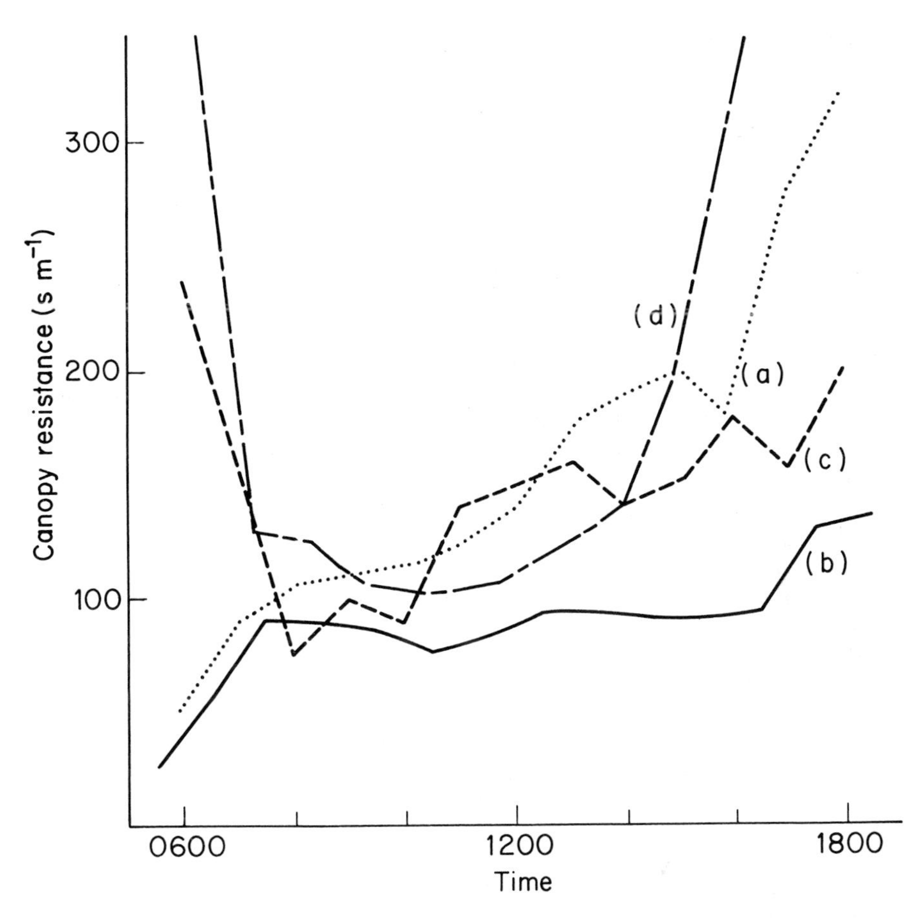

Fig. 15.5 Diurnal variation of the canopy resistance of forests (after Jarvis, James and Landsberg, 1976): (a) *Pinus sylvestris*, Thetford, UK (Stewart and Thom, 1973), (b) *Picea sitchensis*, Fetteresso, UK (Jarvis, James and Landsberg, 1976), (c) *Pseudotsuga menziesii*, British Columbia (McNaughton and Black, 1973), and (d) Amazonian forest, Brazil (Shuttleworth *et al.*, 1984).

Carbon dioxide and growth

The flux of CO_2 in the air above a crop canopy is a measure of the net exchange of CO_2 between the soil-plant system and the atmosphere. Figure 15.6 shows how the components of this exchange are likely to alter over a period of 24 hours (Monteith, 1962). The line zasbz′ represents the flux in the air above a crop, directed upwards during the night when there is a net loss of CO_2 from the system and downwards during the day when there is a net gain. The axis OO′ represents zero flux. Respiration from three sources contributes to the flux of CO_2 at night: plant tops, plant roots and soil organisms. Total respiration between midnight and sunrise is represented by za. At sunrise (a) the photosynthetic system begins to assimilate part of the respired CO_2 and

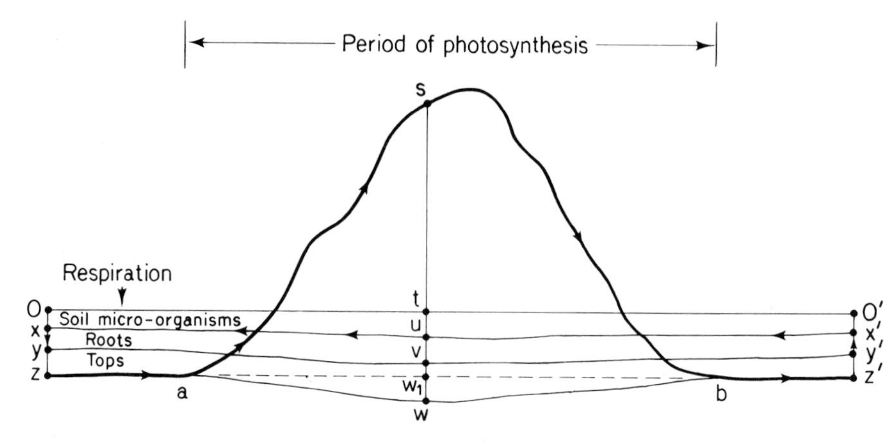

Fig. 15.6 The diurnal change of CO_2 flux above a stand of vegetation, shown by the bold line zasbz′. For the significance of other components, see text.

the upward flux decreases to zero when solar irradiance reaches the light compensation point for the stand, usually about 1 to 2 hours after sunrise over actively growing vegetation. After the irradiance exceeds the compensation point, there is a downward flux of CO_2 representing the atmospheric contribution to photosynthesis. Shortly before sunset (b) the compensation point is reached again and after sunset the rate of respiration is shown by bz′.

Components of the CO_2 balance during the day are given by segments of the line sw:

st = uptake of CO_2 from atmosphere
tw = uptake of CO_2 from plant and soil respiration
sw = gross uptake of CO_2
uw = plant respiration
su = net photosynthesis

Only one of these quantities, st, can be readily measured. The respiration during the day is not known but can be estimated from the average flux at night obtained by drawing a straight line through za and bz′ intersecting sw at w_1. The segment ww_1 represents the increase in total respiration as a result of the higher soil and air temperature during the day. The proportion of total respiration attributable to soil organisms is very difficult to establish experimentally because the presence of plant roots stimulates microbial activity in the rhizosphere. If β is the ratio of plant respiration to the total respiration of the system, the instantaneous rate of net photosynthesis is $(sw_1) - \beta(tw_1)$. During the life of a crop, the value of β will increase from zero at germination to a maximum which will usually lie between 0.5 and 0.9 when the crop is mature.

The integrated rates of photosynthesis for the 24 hour period are

gross photosynthesis: area asbwa $\approx$ asbw_1a
plant respiration: area xux′z′wzx

net photosynthesis: area zsz′x′uxz (which would be measured by a planimeter following the arrows on Fig. 15.6)

In practice, the net photosynthesis would need to be found from the area zsz′O′Oz plus the nocturnal respiration from soil organisms which is $(1-\beta)$ times the total nocturnal respiration.

Measurement of the flux $\mathbf{P}_a$ of CO_2 from the atmosphere to crops has been possible for about thirty years, using gradient methods of micrometeorology and sensitive infra-red gas analysers. More recently fast response CO_2 analysers have been developed to allow eddy correlation measurements over crops (Ohtaki, 1984; Anderson and Verma, 1986).

Biscoe, Scott and Monteith (1975) used aerodynamic and Bowen ratio methods to measure $\mathbf{P}_a$ to barley throughout a complete growing season, and supplemented their study with measurements of soil and root respiration to enable net photosynthesis to be calculated on an hourly basis. Figure 15.7 shows the relationship between photosynthesis rate and irradiance for five consecutive weeks from the stage of maximum green leaf area up to a time approaching harvest. At the beginning of the period, photosynthesis rate increased with irradiance, even in strong sunlight. Later, as the foliage

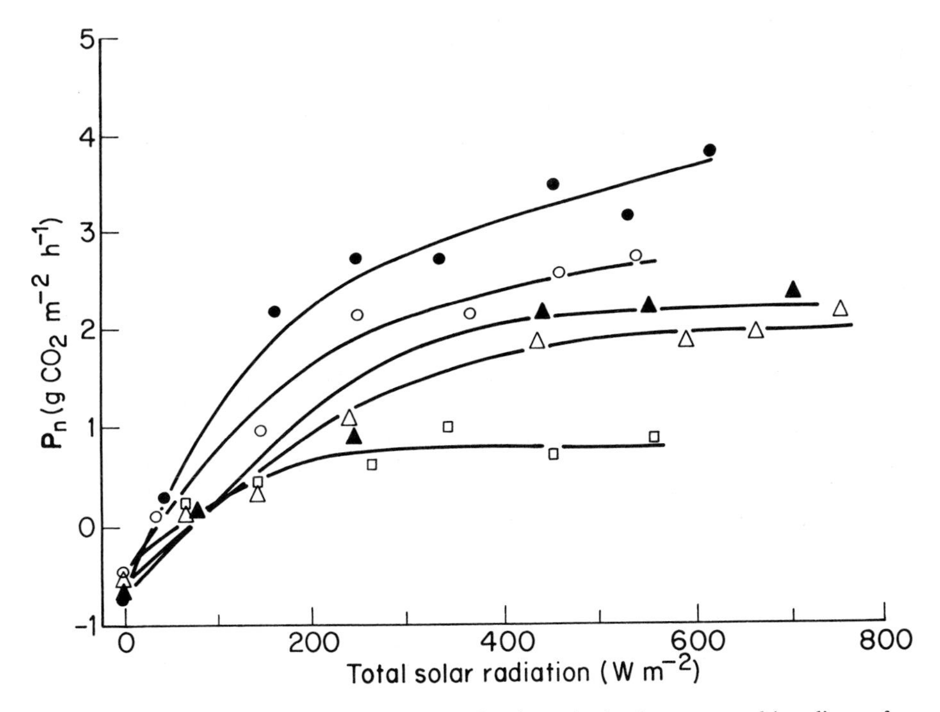

Fig. 15.7 The relation between net CO_2 fixation of a barley crop and irradiance for the five weeks after anthesis in 1972. Dates, total green leaf area indices and symbols are as follows: 28 June, 5.95 (●), 5 July, 5.69 (○), 12 July, 5.59 (▲), 19 July, 4.02 (△), 26 July, 2.68 (□) (from Biscoe, Scott and Monteith, 1975).

senesced, maximum rates of photosynthesis declined and were achieved at steadily lower irradiances. The decrease was partly a consequence of decreasing green leaf area, but changes in the photosynthetic activity of individual organs also affected the crop photosynthesis.

Figure 15.8 shows the result of summing the hourly fluxes of carbon dioxide over a period of eight days when the canopy was fully established but before sensescence. The accumulation of carbon progresses in cycles corresponding to the succession of light and dark periods when the crop gains and loses carbon. Large differences in photosynthesis from day to day are correlated with the daily insolation shown in the histogram. There have been surprisingly few studies of CO_2 fluxes over long enough periods to demonstrate, as in Fig. 15.8, the importance of short-term changes in photosynthetic activity in determining crop growth over periods of a week, the shortest time for which growth can be determined by conventional destructive methods. Micrometeorological methods allow studies on a time scale that enables physiological responses to the weather to be studied in the field.

Sulphur dioxide and pollutant fluxes to crops

In the same way that carbon dioxide is transported from the atmosphere to

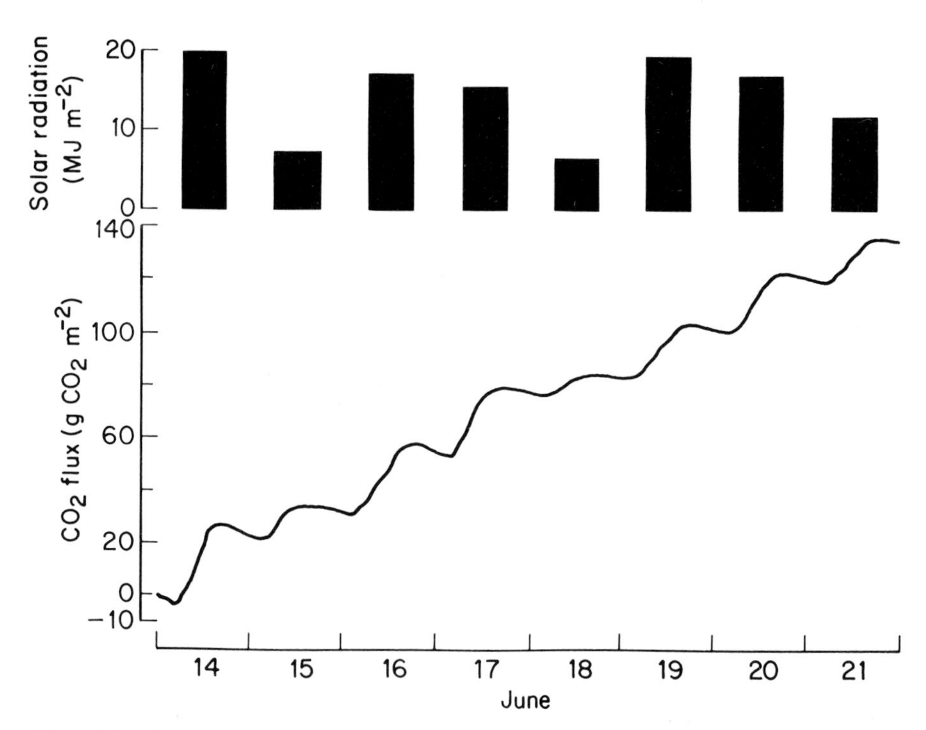

Fig. 15.8 The hourly rates of net CO_2 fixation by a barley crop summed over the period 14–21 June 1972. The histogram shows the total solar radiation for each day (from Biscoe, Scott and Monteith, 1975).

crops, to be absorbed at sites of photosynthesis, pollutant gases are transferred and absorbed within a crop canopy. This process is known as *dry deposition* to distinguish it from the *wet deposition* of pollutants in rain and snow.

In general, pollutant gases can be absorbed (or adsorbed) at various sites in the canopy, depending on their solubility and affinity for materials on the surface of, and within, leaves, and in soils. Resistance analogues can be used to establish the importance of these several pathways, applying the analysis to micrometeorological measurements of fluxes by gradient or eddy correlation techniques.

Figure 15.9 shows a resistance analogue of dry deposition of sulphur

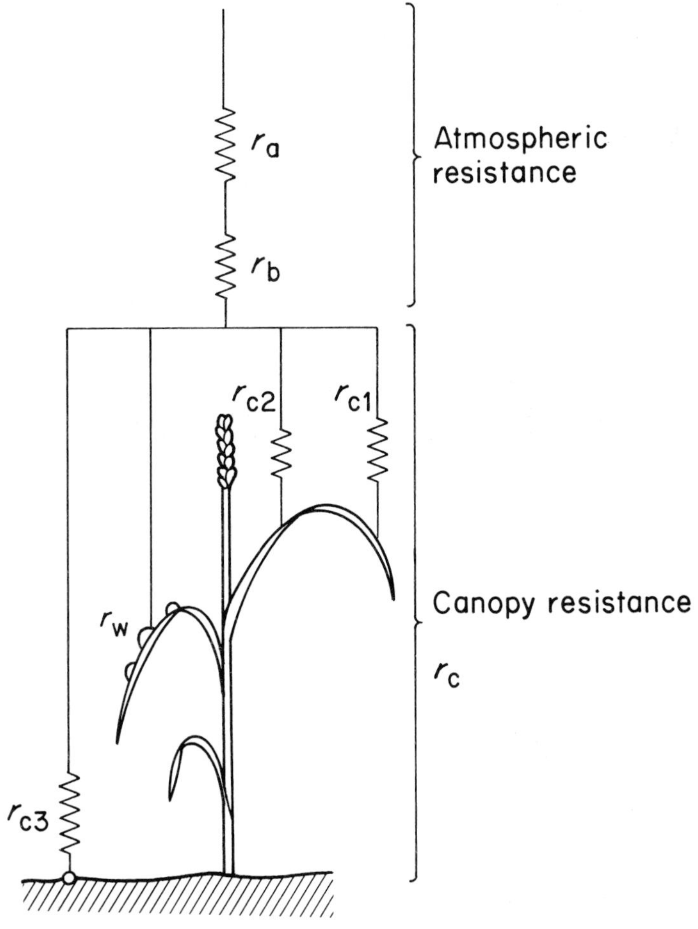

Fig. 15.9 Resistance analogue of dry deposition of SO_2 to a wheat canopy, showing aerodynamic resistance, r_a, additional boundary layer resistance, r_b, and resistances in canopy, r_{c1}, to stomatal uptake, r_{c2}, to surface deposition, r_{c3}, to uptake by soil, and, r_w, to uptake by surface moisture (from Fowler and Unsworth, 1979).

dioxide to a wheat canopy (Fowler and Unsworth, 1979). Resistances to turbulent transfer within the canopy are much smaller than resistances associated with the sinks and are ignored. There are four possible sinks in the canopy: (i) SO_2 may diffuse in through stomata, dissolve in the substomatal cavity and ultimately be used as sulphate in plant metabolism. The canopy resistance component r_{c1} for the stomatal pathway is therefore similar to the canopy resistance for water vapour loss, but a correction is required for the smaller diffusivity of SO_2; (ii) SO_2 may be absorbed or adsorbed onto the surface of leaves; the controlling resistance r_{c2} probably depends on the surface structure, and on any deposited particles, dust, etc.; (iii) drops of water on leaf surfaces absorb SO_2; the resistance r_w is influenced by other soluble substances, and increases as the liquid increases in acidity, eventually halting any further uptake of SO_2; (iv) SO_2 transported through the canopy can be absorbed by the soil; the resistance r_{c3} is smaller for chalky soils than for clays.

The flux $\mathbf{F}_s$ of SO_2 to a crop can be described by equation (15.12),

$$\mathbf{F}_s = \frac{S}{r_a + r_b + r_c}$$

where S is the SO_2 concentration at a reference height above the crop, and r_c, the canopy resistance, is the resultant of the resistances in Fig. 15.9 acting in parallel.

Fowler and Unsworth (1979) made measurements of $\mathbf{F}_s$ to a wheat crop in central England throughout a growing season, using the aerodynamic method and a chemical method for measuring SO_2 concentration at five heights. Values of r_a ranged from 10–200 s m^{-1} and r_b was 20–100 s m^{-1} (the larger values in both cases being for light winds at night). Using equation (15.12), values of r_c were derived, and the component resistances were estimated by interpreting diurnal changes as in Fig. 15.9. During the day, provided that the crop was dry and not senescent, r_c was dominated by the stomatal pathway; consequently the minimum values of r_c during the day, 50–100 s m^{-1}, give an approximate value to r_{c1}. At night, when stomata closed, r_c increased to about 250–300 s m^{-1} and this is an estimate of r_{c2}. This value is about an order of magnitude lower than cuticular resistances for water vapour loss (allowing for a leaf area index of 4.5), indicating that there was an effective sink for SO_2 on leaf surfaces. Figure 15.10 shows an occasion when dew formed from about 0300 to 0600, and r_c decreased rapidly to about 100 s m^{-1}, indicating that r_w was the controlling resistance in this case. Analysis of flux measurements when the crop was senescent suggested that absorption of SO_2 by soil below the canopy was not significant. From a knowledge of the components of r_c and of the seasonal mean SO_2 concentration (50 μg m^{-3}), it was estimated that the wheat crop absorbed about 11 kg sulphur $(ha)^{-1}$ (1.1 g sulphur m^{-2}) in the period May–July; 5 kg ha^{-1} entering through stomata and 6 kg ha^{-1} deposited on leaf surfaces.

These methods for measuring and analysing SO_2 fluxes are applicable to other gaseous pollutants; Wesely *et al.* (1978) used an eddy correlation technique to study ozone deposition on maize, demonstrating that stomatal

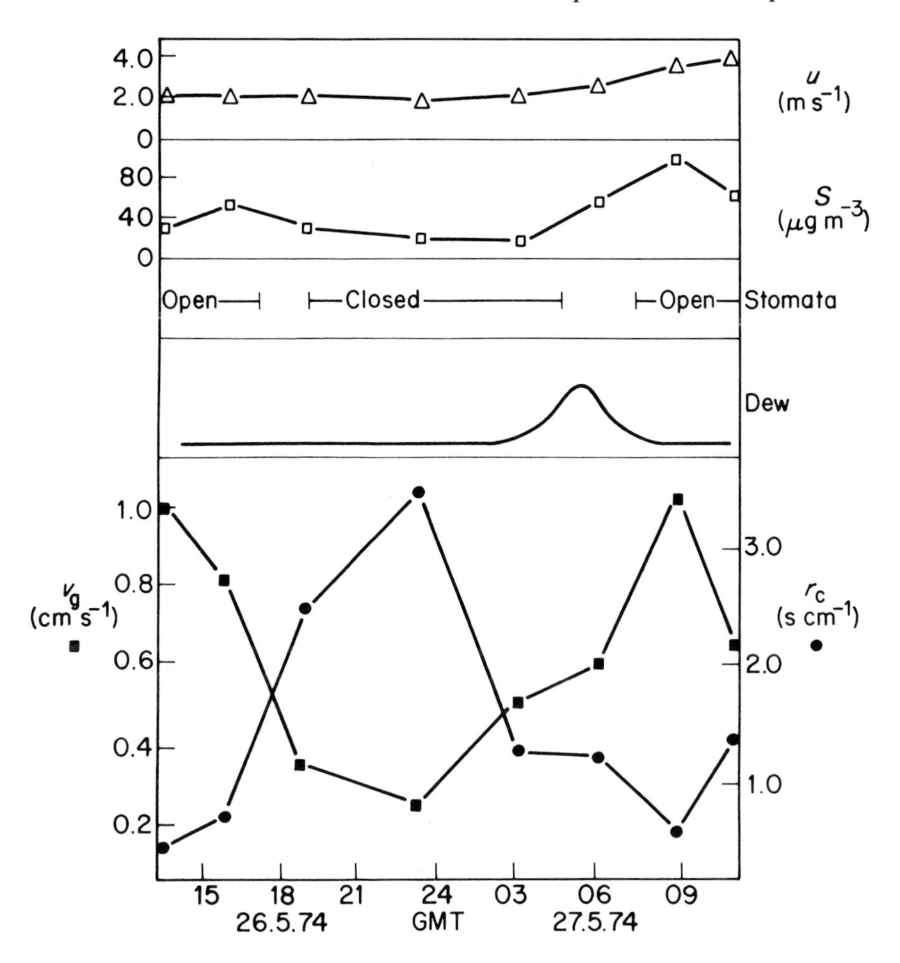

Fig. 15.10 Diurnal variations of SO_2 deposition velocity v_g, canopy resistance, r_c, windspeed, u, and SO_2 concentration, S, over a wheat crop. Durations of dew deposition and of estimated stomatal opening are also shown. All height-dependent parameters are referred to 1 m above the zero plane (from Fowler and Unsworth, 1979).

control formed the main component of r_c by day, but that absorption of ozone by soil below the canopy was also an important pathway.

TRANSPORT WITHIN CANOPIES

Much of the previous part of this chapter has been concerned with the measurement and interpretation of profiles of heat, mass and momentum *above* crop canopies to deduce appropriate fluxes and to identify controlling resistances. A logical extension is to deduce fluxes within canopies by analysing profiles and hence to investigate how different layers in the canopy contribute to the total flux. Most attempts to derive fluxes in canopies have

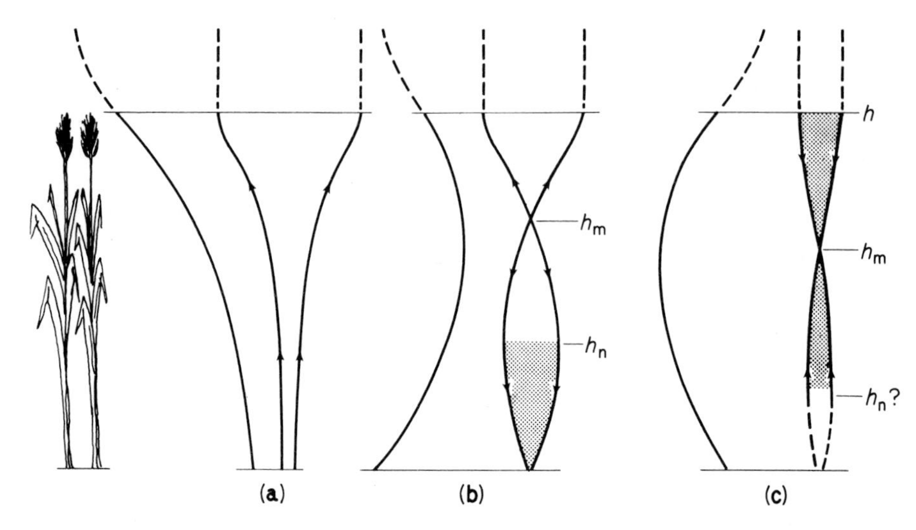

Fig. 15.11 Hypothetical distribution of a potential and flux in a cereal crop. In each of the three sections (a), (b) and (c), the left-hand curve represents the potential and the separation of the right-hand pair of curves is proportional to the flux; h is crop height, h_m is height at which flux is zero and h_n is height where flux divergence is zero. The stippled areas represent heights at which the flux of an entity is being absorbed by the vegetation.

assumed that transport is one-dimensional and that fluxes **F** and potentials S are related by K-theory so that

$$\mathbf{F} = -K\partial S/\partial z$$

The source strength of **F** is the flux per unit depth of canopy

$$\frac{\partial \mathbf{F}}{\partial z} = \frac{-\partial}{\partial z}\left(K\frac{\partial S}{\partial z}\right) = -\left\{\frac{\partial K}{\partial z}\frac{\partial S}{\partial z} + K\frac{\partial^2 S}{\partial z^2}\right\} \tag{15.13}$$

The position of sources can often be determined without evaluating the terms in this equation if K is assumed to increase between the soil surface and the top of the canopy, i.e. $\partial K/\partial z$ is positive throughout the canopy. A layer in which the two terms in curly brackets are both negative must be a source because $\partial \mathbf{F}/\partial z$ is positive, and conversely a layer in which both terms are positive must be a sink. When the terms have opposite signs, they must be evaluated to determine the sign of $(\partial \mathbf{F}/\partial z)$.

The interpretation of the profile of any potential in terms of the location of sources and sinks is shown graphically in Fig. 15.11. Beside the three profiles (a), (b) and (c) there are three pairs of lines which are mirror images. The arrows on these lines show the direction of the fluxes depending on the way in which the potential changes with height, and the separation of the lines is proportional to the size of the flux.

In (a), the flux increases steadily with height to the top of the canopy (cf. the profile for water vapour during the day or for CO_2 at night, p. 232). In (b), the flux is zero at a height h_m and a layer of foliage at this level acts as a source from which heat or mass diffuses upwards and downwards (cf. the profile for temperature during the day). The profile below the level h_n acts as a sink. In profile (c) a layer of foliage at a height h_m acts as sink (cf. the profile for CO_2 during the day). A source *may* exist below h_n but equation (15.13) is indeterminate in this case.

It has long been realized that one-dimensional diffusion is unlikely to be able to relate the profiles of potentials to fluxes within canopies to any great degree of accuracy. There are two principal reasons for the breakdown of K theory: first, within canopies, the presence of sources of heat and mass results in large changes in gradients of potential over distances that are smaller than the eddy sizes that determine K, whereas above canopies K can be treated as a constant over the height range where $\partial S/\partial z$ is determined. Consequently, within canopies the fluxes of heat and mass at a specific level cannot be uniquely related to a K value and a potential gradient; the local K value is itself likely to depend on the potential gradient. The second reason is that the existence of sources and sinks in the canopy affects the correlations between vertical velocities and instantaneous values of potentials so that there are likely to be substantial differences in K values for different entities.

Experimental evidence suggesting that flux-gradient methods are not generally applicable within canopies can be seen in the wind profiles of Fig. 15.12. The vertical flux of momentum must be downwards throughout a canopy because there are no sources of mean momentum in the canopy. Consequently $\partial u/\partial z$ should decrease monotonically throughout the canopy. Whilst this is the case in Fig. 15.12 for maize, and hence momentum fluxes could be estimated, the wind profile in the pine forest shows a secondary maximum in the trunk space. It therefore appears that the assumed downward flux of momentum is in the opposite direction to the gradient of windspeed in part of the forest canopy. Whilst there are other possible explanations of maxima in wind profiles in forests, such as air blowing

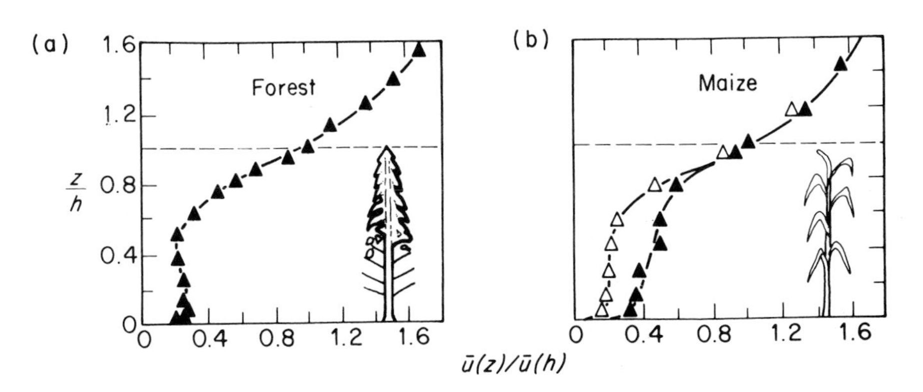

Fig. 15.12 (a) Profile of mean windspeed in a pine forest canopy (h = 16 m); data averaged from 18 very-near-neutral one-hour runs. (b) Profiles of mean windspeed in a maize canopy (h = 2.1 m) during periods of light wind [$\bar{u}(h)$ = 0.88 m s^{-1}, ▲] and strong wind [$\bar{u}(h)$ = 2.66 m s^{-1}, △] (from Raupach and Thom, 1981).

through the trunk space from the upwind edge of the forest, it is difficult to account for all observations of this type. Measurements of heat and mass transport by eddy correlation in forests also indicate that fluxes may be counter-gradient in the lower regions of the canopy (Denmead and Bradley, 1985). Until such methods can be more widely applied, and instrumentation can be further miniaturized for use in dense crop canopies, it seems unlikely that it will be clear under what circumstances fluxes in canopies can be estimated with any reliability from measurements of gradients.

The synthesis of profiles in canopies

Although the nature of transport in crop canopies is not fully understood, it is still possible to develop mathematical models which generate realistic canopy microclimates. This apparent inconsistency arises for two reasons: (i) the source and sink strengths deduced by differentiating potential gradients in equation (15.13) are extremely sensitive to the shape of the measured profiles. Although this limits the scope of equation (15.13) for deducing the distribution of flux from measurements of profiles in a situation where K theory applies, it means that plausible profiles can be generated by integration from an approximate distribution of sources; (ii) the main restrictions on the transfer of water vapour, CO_2 and some pollutant gases between the atmosphere and leaves are associated with stomata, so a realistic distribution of stomatal resistance in various layers of a canopy can be combined with a relatively uncertain distribution of the atmospheric resistances to transfer between layers to produce canopy profiles.

Several stand models in the literature follow the same procedure:

(i) the distribution of radiant energy in the canopy is expressed as a function of cumulative leaf area index.
(ii) the net radiation absorbed by each leaf is partitioned between sensible and latent heat, making assumptions about the distribution of stomatal and leaf boundary layer resistances.
(iii) the resistance between each layer of leaves is estimated as $\int dz/K$, assuming that K theory applies.
(iv) the increase of the relevant potentials from the lowest layer to the second lowest layer is calculated from the product of flux and resistance. This procedure is repeated layer by layer until the top of the canopy is reached.

Models of this type, readily solved by computer, were developed by Waggoner and Reifsnyder (1968), Goudriaan (1977), Jarvis, Miranda and Muetzelfeldt (1985) and Chen (1984). As a generalization, the models demonstrate that the exchange of heat and water vapour between a canopy and the air above it depends much more critically on the behaviour of stomata than on the structure of the microclimate within the canopy.

Whilst models provide useful insight into the distribution of sources and sinks in canopies they are weak in simulating exchange between the atmosphere and the soil below the canopy, because the resistances associated with the soil boundary layer are difficult to define in a convenient form. This is seldom of major importance below dense canopies, but merits further work for widely spaced systems and for early stages of crop development. An attempt to solve the heat balance equations for a canopy and soil divided into four discrete layers has been described by Choudhury and Monteith (1988).

Models of transfer in canopies based on K theory are particularly unlikely to be successful in simulating profiles when dealing with phenomena which are not controlled by surface resistances. Evaporation from wet foliage and transport of the spores of plant pathogens are examples. Legg (1983) reviewed alternative methods with particular reference to spore transport. A more complex treatment of the conservation of matter and momentum can be used to obtain sets of equations including second order terms such as pressure-concentration correlations, and with suitable approximations these second-order closure models can be solved (Wilson and Shaw, 1977; Raupach and Thom, 1981). The equations contain so many unknown constants that it seems unlikely that they will find general application in the near future for the interpretation of measurements.

A currently more promising approach is Markov chain simulation (Wilson *et al.*, 1981; Legg, 1983) in which the trajectories of individual 'carriers' (see Chapter 3) within the fluid are simulated as a random walk. The method appears to be suitable for the highly turbulent conditions that occur in crop canopies, and to be capable of explaining the types of profiles shown in Fig. 15.12, but there have been few sets of measurements of the turbulent structure within canopies from which rigorous tests of the method can be made.

APPENDIX

Table A1 Système International (SI) units with c.g.s. equivalents

Quantity	Dimensions	SI	c.g.s.
Length	L	$1\ m$	$= 10^2\ cm$
Area	L^2	$1\ m^2$	$= 10^4\ cm^2$
Volume	L^3	$1\ m^3$	$= 10^6\ cm^3$
Mass	M	$1\ kg$	$= 10^3\ g$
Density	$M\ L^{-3}$	$1\ kg\ m^{-3}$	$= 10^{-3}\ g\ cm^{-3}$
Time	T	1 s (or min, h, etc.)	= 1 s
Velocity	$L\ T^{-1}$	$1\ m\ s^{-1}$	$= 10^2\ cm\ s^{-1}$
Acceleration	$L\ T^{-2}$	$1\ m\ s^{-2}$	$= 10^2\ cm\ s^{-2}$
Force	$M\ L\ T^{-2}$	$1\ kg\ m\ s^{-2} = 1\ N$ (Newton)	$= 10^5\ g\ cm\ s^{-2} = 10^5$ dynes
Pressure	$M\ L^{-1}\ T^{-2}$	$1\ kg\ m^{-1}\ s^{-2} = 1\ N\ m^{-2}$ (Pascal)	$= 10\ g\ cm^{-1}\ s^{-2} = 10^{-2}$ mbar
Work, energy	$M\ L^2\ T^{-2}$	$1\ kg\ m^2\ s^{-2} = 1\ J$ (Joule)	$= 10^7\ g\ cm^2\ s^{-2} = 10^7$ ergs
Power	$M\ L^2\ T^{-3}$	$1\ kg\ m^2\ s^{-3} = 1\ W$ (Watt)	$= 10^7\ g\ cm^2\ s^{-3} = 10^7\ ergs\ s^{-1}$
Dynamic viscosity	$M\ L^{-1}\ T^{-1}$	$1\ N\ s\ m^{-2}$	$= 10\ dynes\ s\ cm^{-2} = 10$ Poise
Kinematic viscosity	$L^2\ T^{-1}$	$1\ m^2\ s^{-1}$	$= 10^4\ cm^2\ s^{-1} = 10^4$ Stokes
Temperature		1 °C (or 1 K)	= 1°C (or 1 K)
Heat energy	H (or $M\ L^2\ T^{-2}$)	1 J	= 0.2388 cal
Heat or radiation flux	$H\ T^{-1}$	1 W	$= 0.2388\ cal\ s^{-1}$
Heat flux density	$H\ L^{-2}\ T^{-1}$	$1\ W\ m^{-2}$	$= 2.388 \times 10^{-5}\ cal\ cm^{-2}\ s^{-1}$
Latent heat	$H\ M^{-1}$	$1\ J\ kg^{-1}$	$= 2.388 \times 10^{-4}\ cal\ g^{-1}$
Specific heat	$H\ M^{-1}\ \theta^{-1}$	$1\ J\ kg^{-1}\ K^{-1}$	$= 2.388 \times 10^{-4}\ cal\ g^{-1}\ K^{-1}$
Thermal conductivity	$H\ L^{-1}\ \theta^{-1}\ T^{-1}$	$1\ W\ m^{-1}\ K^{-1}$	$= 2.388 \times 10^{-3}\ cal\ cm^{-1}\ s^{-1}\ K^{-1}$
Thermal diffusivity (and other diffusion coefficients)	$L^2\ T^{-1}$	$1\ m^2\ s^{-1}$	$= 10^4\ cm^2\ s^{-1}$

Table A.2 Properties of air, water vapour and CO_2 (treated as constant between −5 and 45°C)

		Air	Water vapour	Carbon dioxide
Specific heat	($J\ g^{-1}\ K^{-1}$)	1.01	1.88	0.85
Prandtl number	$Pr = (\nu/\kappa)$	0.70_5	—	—
	$Pr^{0.67}$	0.79	—	—
	$Pr^{0.33}$	0.89	—	—
	$Pr^{0.25}$	0.92	—	—
Schmidt number	$Sc = (\nu/D)$	—	0.63	1.04
	$Sc^{0.67}$	—	0.74	1.02
	$Sc^{0.33}$	—	0.86	1.01
	$Sc^{0.25}$	—	0.89	1.01
Lewis number	$Le = (\kappa/D)$	—	0.89	1.48
	$Le^{0.67}$	—	0.93	1.32
	$Le^{0.33}$	—	0.96	1.14
	$Le^{0.25}$	—	0.97	1.11

Table A.3 Properties of air, water vapour and CO_2 (changing by less than 1% per K)

	Temperature	Densities of air		Virtual temperature of air	Latent heat of vaporization of water		Thermal conductivity of air	Molecular diffusion coefficients of air			
Symbol	T	ρ_a	$\rho_{as}(T)$	T_v	λ	γ	k	κ	ν	D_v	D_c
Unit °C	K	kg m^{-3}		°C	J g^{-1}	Pa K^{-1}	mW m^{-1} K^{-1}	10^{-6} m^2 s^{-1}			
−5	268.2	1.316	1.314	−4.57	2513	64.3	24.0	18.3	12.9	20.5	12.4
0	273.2	1.292	1.286	0.64	2501	64.6	24.3	18.9	13.3	21.2	12.9
5	278.2	1.269	1.265	5.92	2489	64.9	24.6	19.5	13.7	22.0	13.3
10	283.2	1.246	1.240	11.32	2477	65.2	25.0	20.2	14.2	22.7	13.8
15	288.2	1.225	1.217	16.87	2465	65.5	25.3	20.8	14.6	23.4	14.2
20	293.2	1.204	1.194	22.62	2442	66.2	26.0	22.2	15.5	24.9	15.1
30	303.2	1.164	1.145	34.97	2430	66.5	26.4	22.8	16.0	25.7	15.6
35	308.2	1.146	1.121	41.73	2418	66.8	26.7	23.5	16.4	26.4	16.0
40	313.2	1.128	1.096	49.03	2406	67.1	27.0	24.2	16.9	27.2	16.5
45	318.2	1.110	1.068	57.02	2394	67.5	27.4	24.9	17.4	28.0	17.0

ρ_a density of dry air
$\rho_{as}(T)$ density of air saturated with water vapour at temperature
T_v virtual temperature of saturated air
λ latent heat of vaporization of water
γ $c_p\rho/\lambda\varepsilon$—'psychrometer constant'
k thermal conductivity of dry air
κ thermal diffusivity of dry air
ν kinematic viscosity of dry air
D_v diffusion coefficient of water vapour in air
D_c diffusion coefficient of CO_2 in air

Table A.4 Quantities changing by more than 1% per K. $e_s(T)$ saturation vapour pressure at temperature T (°C); Δ change of saturation vapour pressure per K, i.e. $\partial e_s/\partial T$; σT^4 full radiation at temperature T (K); $4\sigma T^3$ change of full radiation per K. Note that the quantities Δ and $4\sigma T^3$ can be used as mean differences to interpolate between the tabulated values of e_s and σT^4 respectively

T (°C)	T (K)	e_s(T) (kPa)	Δ(T) Pa K^{-1}	σT^4 W m^{-2}	$4\sigma T^3$ W m^{-2} K^{-1}
−5	268.2	0.421	32	293.4	4.4
−4	269.2	0.455	34	297.8	4.4
−3	270.2	0.490	37	302.2	4.5
−2	271.2	0.528	39	306.7	4.5
−1	272.2	0.568	42	311.3	4.6
0	273.2	0.611	45	315.9	4.6
1	274.2	0.657	48	320.5	4.7
2	275.2	0.705	51	325.2	4.7
3	276.2	0.758	54	330.0	4.8
4	277.2	0.813	57	334.8	4.8
5	278.2	0.872	61	339.6	4.9
6	279.2	0.935	65	344.5	5.0
7	280.2	1.001	69	349.5	5.0
8	281.2	1.072	73	354.5	5.1
9	282.2	1.147	78	359.6	5.1
10	283.2	1.227	83	364.7	5.2
11	284.2	1.312	88	369.9	5.2
12	285.2	1.402	93	375.1	5.3
13	286.2	1.497	98	380.4	5.3
14	287.2	1.598	104	385.8	5.4
15	288.2	1.704	110	391.2	5.4
16	289.2	1.817	117	396.6	5.5
17	290.2	1.937	123	402.1	5.6
18	291.2	2.063	130	407.7	5.6
19	292.2	2.196	137	413.3	5.7
20	293.2	2.337	145	419.0	5.7
21	294.2	2.486	153	424.8	5.8
22	295.2	2.643	162	430.6	5.8
23	296.2	2.809	170	436.4	5.9
24	297.2	2.983	179	442.4	6.0
25	298.2	3.167	189	448.3	6.0
26	299.2	3.361	199	454.4	6.1
27	300.2	3.565	210	460.5	6.2
28	301.2	3.780	221	466.7	6.2
29	302.2	4.006	232	472.9	6.3
30	303.2	4.243	244	479.2	6.3
31	304.2	4.493	257	485.5	6.4
32	305.2	4.755	269	492.0	6.5
33	306.2	5.031	283	498.4	6.5
34	307.2	5.320	297	505.0	6.6
35	308.2	5.624	312	511.6	6.7
36	309.2	5.942	327	518.3	6.7
37	310.2	6.276	343	525.0	6.8
38	311.2	6.262	357	531.8	6.9
39	312.2	6.993	376	538.7	6.9
40	313.2	7.378	394	545.6	7.0
41	314.2	7.780	413	552.6	7.1
42	315.2	8.202	432	559.7	7.1
43	316.2	8.642	452	566.8	7.2
44	317.2	9.103	473	574.0	7.3
45	318.2	9.586	494	581.3	7.3

Table A.5 Nusselt numbers for air

(a) Forced convection

Shape	Case	Range of Re	Nu
(1) Flat plates			
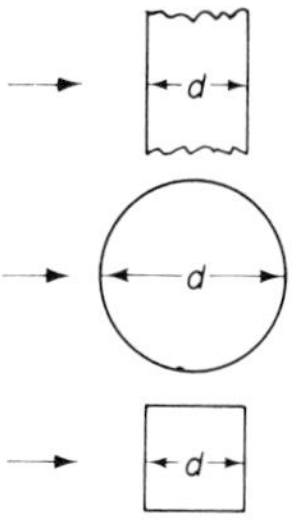	Streamline flow	$<2\times10^4$	$0.60\ Re^{0.5}$
	Turbulent flow	$>2\times10^4$	$0.032\ Re^{0.8}$
(2) Cylinders			
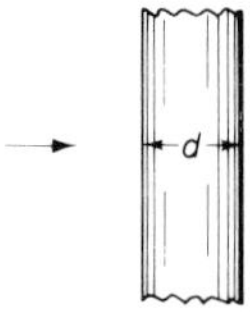	Narrow range of Reynolds numbers	$1-4$	$0.89\ Re^{0.33}$
		$4-40$	$0.82\ Re^{0.39}$
		$40-4\times10^3$	$0.62\ Re^{0.47}$
		$4\times10^3-4\times10^4$	$0.17\ Re^{0.62}$
		$4\times10^4-4\times10^5$	$0.024\ Re^{0.81}$
		or	
	Wide range of Reynolds numbers	$10^{-1}-10^3$	$0.32+0.51\ Re^{0.52}$
		$10^3-5>10^4$	$0.24\ Re^{0.60}$
(3) Spheres			
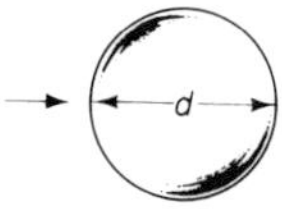		$0-300$	$2+0.54\ Re^{0.5}$
		$50-1.5\times10^5$	$0.34\ Re^{0.6}$

Notes

(i) Arrows show direction of airflow

(ii) d is characteristic dimension; take width of a long crosswind strut as shown or mean side for a rectangle whose width and length are comparable

(iii) To find corresponding Sherwood numbers multiply Nu by $Le^{0.33}$ (see values in Table A.1)

(iv) Sources—Ede (1967), Fishenden and Saunders (1950), Bird, Stewart and Lightfoot (1960)

Table A.5—(*continued*)
(b) Free convection

Shape and relative temperature	Range		Nu
	Laminar flow	Turbulent flow	
(1) Horizontal flat plates or cylinders			
(i) 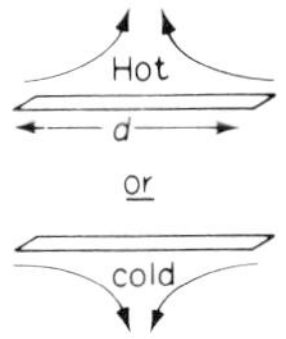 Hot d or cold	$Gr < 10^5$		$0.50\ Gr^{0.25}$
		$Gr > 10^5$	$0.13\ Gr^{0.33}$
(ii) 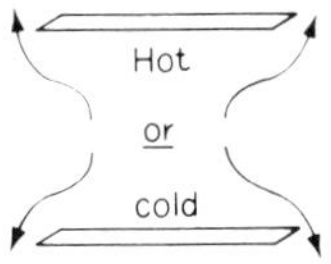 Hot or cold		Arrangement not conducive to turbulence	$0.23\ Gr^{0.25}$
(iii) 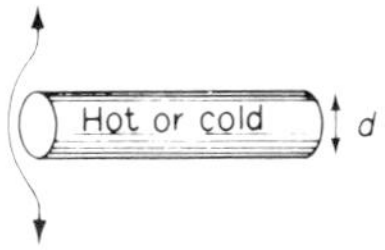 Hot or cold d	$10^4 < Gr < 10^9$		$0.48\ Gr^{0.25}$
		$Gr > 10^9$	$0.09\ Gr^{0.33}$
(2) Vertical flat plates or cylinders 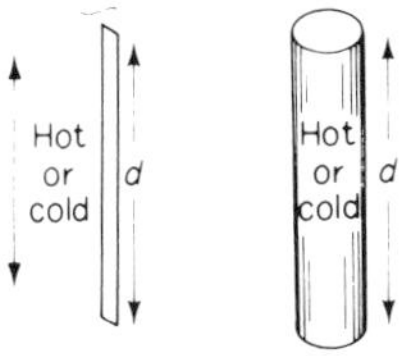 Hot or cold d Hot or cold d	$10^4 < Gr < 10^9$		$0.58\ Gr^{0.25}$
		$10^9 < Gr < 10^{12}$	$0.11\ Gr^{0.33}$
(3) Spheres 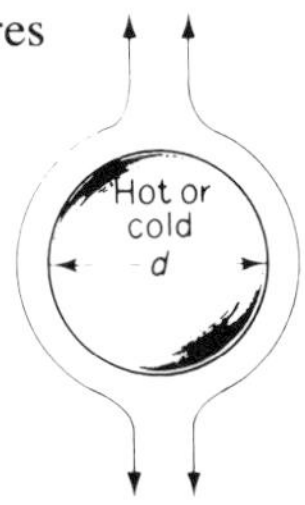 Hot or cold d	$Gr^{0.25} < 220$		$2 + 0.54\ Gr^{0.25}$

Notes

(i) Arrows indicate direction of air circulation

(ii) d is characteristic dimensions for calculation of Gr: take height for vertical plate and average chord for horizontal plate

(iii) To find corresponding Sherwood numbers, multiply Nu by $Le^{0.25}$ for laminar flow or tubulent flow (see values in Table A.1)

(iv) Sources—Ede (1967), Fishenden and Saunders (1950), Bird, Stewart and Lightfoot (1960)

Table A.6 Characteristic quantities for particle transfer in air*: D diffusion coefficient ($m^2\ s^{-1}$); τ Relaxation time (s); $\overline{\Delta x_B}$ Mean displacement in 1 s in a given direction $2(Dt/\pi)^{0.5}$ (μm); $\overline{\Delta x_s}$ Distance fallen in 1 s under gravity (μm)

Radius r (μm)	D ($10^{-9}\ m^2\ s^{-1}$)	τ (10^{-6} s)	$\overline{\Delta x_B}$ (μm)	$\overline{\Delta x_s}$ (μm)
1.0×10^{-3}	1.28×10^{3}	1.33×10^{-3}	1.28×10^{3}	1.31×10^{-2}
5.0×10^{-3}	5.24×10^{1}	6.76×10^{-3}	2.58×10^{2}	6.63×10^{-2}
1.0×10^{-2}	1.35×10^{1}	1.40×10^{-2}	1.31×10^{2}	1.37×10^{-1}
5.0×10^{-2}	6.82×10^{-1}	8.81×10^{-2}	2.95×10^{1}	8.64×10^{-1}
0.1	2.21×10^{-1}	0.23	1.68×10^{1}	2.24
0.5	2.7×10^{-2}	3.54	5.90	3.47×10^{1}
1.0	1.3×10^{-2}	1.31×10^{1}	4.02	1.28×10^{2}
5.0	2.4×10^{-3}	3.08×10^{2}	1.74	3.0×10^{3}
10.0	1.4×10^{-3}	1.23×10^{3}	1.23	1.2×10^{4}

* From Fuchs (1964), calculated at 23°C, standard atmospheric pressure; particle density 1 g cm^{-3}.

BIBLIOGRAPHY

GENERAL INTRODUCTORY TEXT BOOKS

The Climate near the Ground, R. Geiger. Harvard University Press, Cambridge, Mass., 1965.
A classic review and discussion of the early literature of microclimatology.

Vegetation and the Atmosphere. Volume I. Principles, ed. J.L. Monteith. Academic Press, London, 1975.
Radiation, transfer processes and general concepts.

An Introduction to Environmental Biophysics, G.S. Campbell. Springer-Verlag, New York, 1977.
Physics of plant and animal microclimates. Numerical examples and problems.

Boundary Layer Climates, T.R. Oke. Methuen, London, Second edition, 1987.
General review of microclimatology with geographical emphasis.

Biophysical Ecology, D.M. Gates. Springer-Verlag, New York, 1980.
Comprehensive treatment of microclimate in relation to leaves and small non-domesticated animals.

Plants and their Atmospheric Environment, eds. J. Grace, E.D. Ford and P.G. Jarvis. Blackwell Scientific, Oxford, 1981.
Conference proceedings with broad ecological coverage.

Microclimate: the Biological Environment, N.J. Rosenberg, B.L. Blad and S.B. Verma. Wiley, New York, 2nd edition, 1983.
Emphasis on agricultural crops and soils.

Principles and Measurements in Environmental Biology, F.I. Woodward and J.E. Sheehy. Butterworths, London, 1983.
Detailed discussion of physical theory and of instrumentation.

Plants and Microclimate, H.G. Jones. Cambridge University Press, Cambridge, 1983.
Microclimate and environmental physiology.

MORE SPECIALIZED BOOKS

Radiation

Solar and Terrestrial Radiation, K.L. Coulson. Academic Press, New York, 1975.
Reviews short- and long-wave radiation with emphasis on instrumentation.

Radiative Processes in Meteorology and Climatology, G.W. Paltridge and C.M.R. Platt. Elsevier, Amsterdam, 1976.
Principles and meteorological implications.
Plants and the Daylight Spectrum, ed. H. Smith. Academic Press, London, 1981.
Conference proceedings reviewing many aspects of plant-light relations.

Transfer processes

Plant Response to Wind, J. Grace. Academic Press, London, 1977.
Basic physics and physiological processes.
Fluid Behaviour in Biological Systems, L. Leyton. Oxford University Press, 1975.
Comprehensive review of flow and transfer in air and liquids. Deals with plants and animals.
Evaporation into the Atmosphere, W.H. Brutsaert. D. Reidel, Dordrecht, Holland, 1982.
Theoretical concepts and their applications.
Atmospheric Diffusion, F. Pasquill and F.B. Smith. Ellis Horwood, Chichester, 1982.
Physics and mathematics of diffusion, and examples of dispersion of particles and gases.
Windborne Pests and Diseases, D. Pedgeley. Ellis Horwood, Chichester, 1982.
Reviews role of weather in dispersion and transport.
Biophysical Plant Physiology and Ecology, P.S. Nobel. W.H. Freeman and Co., San Francisco, 1983.

Particles

Relevant chapters will be found in the general texts: *Plants and the Atmospheric Environment: Vegetation and the Atmosphere Volume I*. Particles are also considered in *Atmospheric Diffusion*, and *Windborne Pests and Diseases* (Transfer processes).
The Mechanics of Aerosols, N.A. Fuchs. Pergamon Press, Oxford, 1964.
Classic book translated from Russian covering all aspects of aerosol physics.
Aerosol Technology in Hazard Evaluation, T.T. Mercer. Academic Press, New York, 1973.
Principles and methods with emphasis on hazards of inhalation.

Plant environment

The Forest-Atmosphere Interaction, eds. B.A. Hutchison and B.B. Hicks. D. Reidel, Dordrecht, Holland, 1985.
Conference proceedings with main emphasis on measurements.
Physiological Ecology of Forest Production, J.J. Landsberg. Academic Press, London, 1986.
Short but comprehensive review of forest environment and tree physiology.
Vegetation and the Atmosphere, Volume II, Case Studies, ed. J.L. Monteith. Academic Press, London, 1976.
Summarizes field experiments on specific crops, types of forests and natural ecosystems.

Soil physics

Application of Soil Physics, D. Hillel. Academic Press, New York, 1980.
Practical aspects. Complements companion volume.

Fundamentals of Soil Physics, D. Hillel. Academic Press, New York, 1980.
Review of theoretical principles of heat and mass transport in soils.
Soil Physics with Basic, G.S. Campbell. Elsevier, Amsterdam, 1986.
Transport processes in soil-plant systems with BASIC computer programs.

Animal environment

Environmental Aspects of Housing for Animal Production, ed. J.A. Clark. Butterworths, London, 1981.
Conference proceedings with reviews of principles of heat balance and practical details of housing.
Adaptation to Thermal Environment, L.E. Mount. Edward Arnold, London, 1979.
Heat exchanges between animals (including man) and their environments.
Man and his Thermal Environment, R.P. Clark and O.G. Edholm. Edward Arnold, London, 1985.
Heat exchange and thermal physiology of man.

Instrumentation

Instrumentation for Environmental Physiology, eds. B. Marshall and F.I. Woodward. Cambridge University Press, 1985.
Conference proceedings with reviews of principles of measurements and techniques.

Remote sensing

Principles of Remote Sensing, P.J. Curran. Longmans, London, 1985.
Good basic text.
Manual of Remote Sensing (2nd edition) Volume I, Theory, Instruments and Techniques.
Volume II Interpretation and Applications, ed. R.N. Colwell. American Society of Photogrammetry, Falls Church, Virginia, 1983.
Comprehensive volumes useful for basic principles and practice.

REFERENCES

Achenbach, E. 1977. The effect of surface roughness on the heat transfer from a circular cylinder. *International Journal of Heat and Mass Transfer*, **20**, 359–369.

Anderson, D.E. and Verma, S.B. 1986. Carbon dioxide, water vapor and sensible heat exchanges of a grain sorghum canopy. *Boundary-Layer Meteorology*, **34**, 317–331.

Anderson, M.C. 1966. Stand structure and light penetration. II. A theoretical analysis. *Journal of Applied Ecology*, **3**, 41–54.

Ar, A., Pagenelli, C.V., Reeves, R.B., Greene, D.G. and Rahn, H. 1974. The avian egg; water vapour conductance, shell thickness and functional pore area. *Condor*, **76**, 153–158.

Asrar, G., Fuchs, M., Kanemasu, E.T. and Hatfield, J.L. 1984. Estimating absorbed photosynthetic radiation and leaf area index from spectral reflectance in wheat. *Agronomy Journal*, **76**, 300–306.

Aylor, D.E. 1975. Force required to detach conidia of *Helminthosporium maydis*. *Plant Physiology, Lancaster*, **55**, 99–101.

Aylor, D.E. and Ferrandino, F.J. 1985. Rebound of pollen and spores during deposition on cylinders by inertial impact. *Atmospheric Environment*, **19**, 803–806.

Bagnold, R.A. 1941. *The Physics of Blown Sand and Desert Dunes*. Chapman and Hall, London.

Bakken G.S. and Gates, D.M. 1975. Heat transfer analysis of animals. In: *Perspectives of Biophysical Ecology*, eds Gates, D.M. and Schmerl, R.G. Springer-Verlag, New York.

Baumgartner, A. 1953. Das Eindringen des Lichtes in den Boden. *Forstwissenschaftlisches Zentralblatt*, **72**, 172–184.

Beament, J.W.L. 1958. The effect of temperature on the water-proofing mechanism of an insect. *Journal of Experimental Biology*, **35**, 494–519.

Becker, F. 1981. Angular reflectivity and emissivity of natural media in the thermal infrared bands. *Proceedings of Conference on Signatures Spectrales D'objets en Télédetection, Avignon, 8–11 Sept*. 1981, 57–72.

Becker, F., Ngai, W. and Stoll, M.P. 1981. An active method for measuring thermal infrared effective emissities: implications and perspectives for remote sensing. *Advanced Space Research*, **1**, 193–210.

Bird, R.B., Stewart, W.D. and Lightfoot, E.N. 1960. *Transport Phenomena*. John Wiley, New York.

Biscoe, P.V. 1969. *Stomata and the Plant Environment*. Ph.D. thesis, University of Nottingham.

Biscoe, P.V., Clark, J.A., Gregson, K., McGowan, M., Monteith, J.L. and Scott,

R.K. 1975. Barley and its environment. I. Theory and practice. *Journal of Applied Ecology*, **12**, 227–257.

Biscoe, P.V., Scott, R.K. and Monteith, J.L. 1975. Barley and its environment. III. Carbon budget of the stand. *Journal of Applied Ecology*, **12**, 269–293.

Blaxter, K.L. 1967. *The Energy Metabolism of Ruminants*. Second edition. Hutchinson, London.

Bolin, B. (ed.) 1981. *Carbon Cycle Modelling*. John Wiley, New York.

Bowers, S.A. and Hanks, R.D. 1965. Reflection of radiant energy from soils. *Soil Scientist*, **100**, 130–138.

Brutsaert, W.H. 1982. *Evaporation into the Atmosphere*. D. Reidel Publishing Company, Dordrecht, Holland.

Buatois, A. and Crose, J.P. 1978. Thermal responses of an insect subjected to temperature variations. *Journal of Thermal Biology*, **3**, 51–56.

Burton, A.C. and Edholm, D.G. 1955. *Man in a Cold Environment*. Edward Arnold, London.

Buss, I.O. and Estes, J.A. 1971. The functional significance of movements and positions of the pinnae of the African elephant, *Loxodonta africana*. *Journal of Mammology*, **52**, 21–27.

Calder, I.R. 1977. A model of transpiration and interception loss from a spruce forest in Plynlimon, central Wales. *Journal of Hydrology*, **33**, 247–265.

Calder, I.R. and Neal, C. 1984. Evaporation from saline lakes: a combination equation approach. *Journal of Hydrological Sciences*, **29**, 89–97.

Campbell, G.S. 1986. Extinction coefficients for radiation in plant canopies calculated using an ellipsoidal inclination angle distribution. *Agricultural and Forest Meteorology*, **36**, 317–321.

Campbell, G.S., McArthur, A.J. and Monteith, J.L. 1980. Windspeed dependence of heat and mass transfer through coats and clothing. *Boundary-Layer Meteorology*, **18**, 485–493.

Cena, K. and Monteith, J.L. 1975a. Transfer processes in animal coats. I. Radiative transfer. *Proceedings of the Royal Society of London, B*, **188**, 377–393.

Cena, K. and Monteith, J.L. 1975b. Transfer processes in animal coats. II. Conduction and convection. *Proceedings of the Royal Society of London, B*, **188**, 395–411.

Cena, K. and Monteith, J.L. 1975c. Transfer processes in animal coats. III. Water vapour diffusion. *Proceedings of the Royal Society of London, B*, **188**, 413–423.

Chamberlain, A.C. 1966. Transport of gases to and from grass and grass-like surfaces. *Proceedings of the Royal Society of London, A*, **290**, 236–265.

Chamberlain, A.C. 1974. Mass transfer to bean leaves. *Boundary-Layer Meteorology*, **6**, 477–486.

Chamberlain, A.C. 1975. The movement of particles in plant communities. In: *Vegetation and the Atmosphere. Volume 1. Principles*, ed. Monteith, J.L., pp. 155–203. Academic Press, London.

Chamberlain, A.C. 1983. Roughness length of sea, sand and snow. *Boundary-Layer Meteorology*, **25**, 405–409.

Chamberlain, A.C., Garland, J.A. and Wells, A.C. 1984. Transport of gases and particles to surfaces with widely spaced roughness elements. *Boundary-Layer Meteorology*, **29**, 343–360.

Chamberlain, A.C. and Little, P. 1981. Transport and capture of particles by vegetation. In: *Plants and their Atmospheric Environment*, eds. Grace, J., Ford, E.D. and Jarvis, P.G., pp. 147–173. Blackwell Scientific, Oxford.

Chandrasekhar, S. 1960. *Radiative Transfer*. Dover, New York.

Chen, Jialin. 1984. *Mathematical Analysis and Simulation of Crop Micrometeorology*. Ph.D. thesis, Department of Theoretical Production Ecology, Agricultural University, Wageningen, Netherlands.

Choudhury, B.J. and Monteith, J.L. 1988. A four-layer model for the heat budget of homogeneous land surfaces. *Quarterly Journal of the Royal Meteorological Society*.

Chrenko, F.A. and Pugh, L.G.C.E. 1961. The contribution of solar radiation to the thermal environment of man in Antarctica. *Proceedings of the Royal Society of London, B*, **155**, 243–265.

Church, N.S. 1960. Heat loss and the body temperature of flying insects. *Journal of Experimental Biology*, **37**, 171–185.

Clapperton, J.L., Joyce, J.P. and Blaxter, K.L. 1965. Estimates of the contribution of solar radiation to thermal exchanges of sheep. *Journal of Agricultural Science*, Cambridge, **64**, 37–49.

Clark, J.A. 1976. Energy transfer and surface temperature over plants and animals. In: *Light as an Ecological Factor, II*, eds Evans, G.C., Bainbridge, R. and Rackham, O., pp. 451–463. Blackwell Scientific Publications, Oxford.

Clark, R.P. and Toy, N. 1975. Natural convection around the human head. *Journal of Physiology*, **244**, 283–293.

Deacon, E.L. 1969. Physical processes near the surface of the earth. In: *World Survey of Climatology, Vol. 2, General Climatology*, Landsberg, H.E. Elsevier, Amsterdam.

Denmead, O.T. and Bradley, E.F. 1985. Flux-gradient relationships in a forest canopy. In: *The Forest-atmosphere Interaction*, eds Hutchinson, B.A. and Hicks, B.B., pp. 421–442. D. Reidel, New York.

Digby, P.S.B. 1955. Factors affecting the temperature excess of insects in sunshine. *Journal of Experimental Biology*, **32**, 279–298.

Dixon, M. and Grace, J. 1983. Natural convection from leaves at realistic Grashof numbers. *Plant, Cell and Environment*, **6**, 665–670.

Dyer, A.J. 1974. A review of flux-profile relationships. *Boundary-Layer Meteorology*, **7**, 363–372.

Dyer, A.J. and Hicks, B.B. 1970. Flux-gradient relationships in the constant flux layer. *Quarterly Journal of the Royal Meteorological Society*, **96**, 715–721.

Ede, A.J. 1967. *An Introduction to Heat Transfer Principles and Calculations*. Pergamon Press, Oxford.

Finch, V.A., Bennett, I.L. and Holmes, C.R. 1984. Coat colour in cattle. *Journal of Agricultural Science*, Cambridge, **102**, 141–147.

Fishenden, M. and Saunders, O.A. 1950. *An Introduction to Heat Transfer*. Clarendon Press, Oxford.

Fleischer, R. von 1955. Der Jahresgang der Strahlungsbilanz sowie ihrer lang-und kurzwelligen Komponenten. Sonderdruck aus: *Bericht des deutschen Wetterdienstes Nr. 22*, pp. 32–40. Die meteorologische Tagung in Frankfurt a.M.

Fowler, D. and Unsworth, M.H. 1979. Turbulent transfer of sulphur dioxide to a wheat crop. *Quarterly Journal of the Royal Meteorological Society*, **105**, 767–783.

Frankland, B. 1981. Germination in shade. In: *Plants and the Daylight Spectrum*, ed. Smith, H. Academic Press, London.

Fraser, A.I. 1962. *Wind Tunnel Studies of the Forces Acting on the Crowns of Small Trees*. Report of Forest Research, London.

Fuchs, N.A. 1964. *The Mechanics of Aerosols*. Pergamon Press, Oxford.

Funk, J.P. 1964. Direct measurement of radiative heat exchange of the human body. *Nature*, London, **201**, 904–905.

Gale, J. 1972. Elevation and transpiration: some theoretical considerations with special reference to Mediterranean-type climate. *Journal of Applied Ecology*, **9**, 691–701.

Garland, J.A. 1977. The dry deposition of sulphur dioxide to land and water surfaces. *Proceedings of the Royal Society of London, A*, **354**, 245–268.

Garnier, B.J. and Ohmura, A. 1968. A method of calculating the direct shortwave radiation income of slopes. *Journal of Applied Meteorology*, **7**, 796–800.

Garratt, J.R. 1978. Flux profile relations above tall vegetation. *Quarterly Journal of the Royal Meteorological Society*, **104**, 199–211.

Garratt, J.R. 1980. Surface influence upon vertical profiles in the atmospheric near-surface layer. *Quarterly Journal of the Royal Meteorological Society*, **106**, 803–819.

Gash, J.H.C. 1986. Observations of turbulence downwind of a forest-heath interface. *Boundary-Layer Meteorology*, **36**, 227–237.

Gatenby, R.M. 1977. Conduction of heat from sheep to ground. *Agricultural Meteorology*, **18**, 387–400.

Gatenby, R.M., Monteith, J.L. and Clark, J.A. 1983. Temperature and humidity gradients in the steady state. *Agricultural Meteorology*, **29**, 1–10.

Gates, D.M. 1980. *Biophysical Ecology*. Springer-Verlag, New York.

Gilby, A.R. 1980. Transpiration, temperature and lipids in insect cuticle. In: *Advances in Insect Physiology*, **15**, eds Berridge, M.J., Treherne, J.E. and Wigglesworth, V.B., pp. 1–33. Academic Press, New York.

Gloyne, R.W. 1972. The diurnal variation of global radiation on a horizontal surface—with special reference to Aberdeen. *Meteorological Magazine*, **101**, 44–51.

Goudriaan, J. 1977. *Crop Micrometeorology: a Simulation Study*. Centre for Agricultural Publishing and Documentation, Wageningen.

Grace, J. 1978. The turbulent boundary layer over a flapping *Populus* leaf. *Plant, Cell and Environment*, **1**, 35–38.

Grace, J. and Collins, M.A. 1976. Spore liberation from leaves by wind. In: *Microbiology of Aerial Plant Surfaces*, eds Dickinson, C.H. and Preece, T.F., pp. 185–198. Academic Press, London.

Grace, J. and Wilson, J. 1976. The boundary layer over a *Populus* leaf. *Journal of Experimental Botany*, **27**, 231–241.

Graser, E.A. and Bavel, C.H.M. van 1982. The effect of soil moisture upon soil albedo. *Agricultural Meteorology*, **27**, 17–26.

Green, C.F. 1987. Nitrogen nutrition and wheat growth in relation to absorbed radiation. *Agricultural and Forest Meteorology*, **41**, 207–248.

Grigg, G.C., Drane, C.R. and Courtice, G.P. 1979. Time constants of heating and cooling in the Eastern Water Dragon, *Physignathus lesueruii* and some generalizations about heating and cooling in reptiles. *Journal of Thermal Biology*, **4**, 95–103.

Hammel, H.T. 1955. Thermal properties of fur. *American Journal of Physiology*, **182**, 369–376.

Harrington, L.P. *On Temperature and Heat Flow in Tree Stems*. School of Forestry Bulletin No. 73, Yale University.

Haseba, T. 1973. Water vapour transfer from leaf-like surfaces within canopy models. *The Journal of Agricultural Meteorology*, Tokyo, **29**, 25–33.

Hatfield, J.L. 1983. Comparison of long-wave radiation calculation methods over the United States. *Water Resources Research*, **19**, 285–288.

Heagle, A.S., Body, D.E. and Heck, W.W. 1973. An open-top field chamber to assess the impact of air pollution on plants. *Journal of Environmental Quality*, **2**, 365–368.

Hemmingsen, A.M.'1960. Energy metabolism as related to body size and respiratory surfaces. *Report of the Steno Memorial Hospital*, Copenhagen, 9, Part 2.

Henderson, S.T. 1977. *Daylight and its Spectrum*. Adam Hilger, Bristol.

Hickey, J.R., Alton, B.M., Griffen, F.J., Jacobowitz, H., Pelegrino, P., Maschhoff, R.H., Smith, E.A. and von der Haar, T.H. 1982. Extraterrestrial solar irradiance

variabiiity. Two and one-half years of measurements from Nimbus 7. *Solar Energy*, **29**, No. 2, 127.
Howell, T.A., Meek, D.W. and Hatfield, J.L. 1983. Relationship of photosynthetically active radiation to shortwave radiation in the San Joaquin Valley. *Agricultural Meteorology*, **28**, 157–175.
Hutchinson, J.C.D., Allen, T.E. and Spence, F.B. 1975. Measurements of the reflectances for solar radiation of the coats of live animals. *Comparative Biochemistry and Physiology*, **52A**, 343–349.
Idso, S.B., Jackson, R.D., Reginato, R.J., Kimball, B.A. and Nakayama, F.S. 1975. The dependence of bare soil albedo on soil water content. *Journal of Applied Meteorology*, **14**, 109–113.
Impens, I. 1965. *Experimentele Studie van de Physische en Biologische Aspektera van de Transpiratie*. Ryklandbouwhogeschool, Ghent.
Jarvis, P.G., James, G.B. and Landsberg, J.J. 1976. Coniferous forest. In: *Vegetation and the Atmosphere. Volume 2, Cast studies*, ed. Montieth, J.L., pp. 171–240. Academic Press, London.
Jarvis, P.G. and McNaughton, K.G. 1986. Stomatal control of transpiration. In: *Advances in Ecological Research*, **15**, pp. 1–49. Academic Press, New York.
Jarvis, P.G., Miranda, H.S. and Muetzelfeldt, R.I. 1985. Modelling canopy exchanges of water vapor and carbon dioxide in coniferous forest plantations. In: *The Forest-Atmosphere Interaction*, eds Hutchison, B.A. and Hicks, B.B., pp. 521–542. D. Reidel, New York.
Johnson, G.T. and Watson, I.D. 1985. Modelling longwave radiation exchange between complex shapes. *Boundary-Layer Meteorology*, **33**, 363–378.
Kleiber, M. 1965. Metabolic body size. In: *Energy Metabolism*, ed. Blaxter, K.L., pp. 427–435. Academic Press, London.
Kondratyev, K.J. and Manolova, M.P. 1960. The radiation balance of slopes. *Solar Energy*, **4**, 14–19.
Lang, A.R.G., McNaughton, K.G., Fazu, C., Bradley, E.F. and Ohtaki, E. 1983. Inequality of eddy transfer coefficients for vertical transport of sensible and latent heats during advective inversions. *Boundary-Layer Meteorology*, **25**, 25–41.
Legg, B.J. 1983. Movement of plant pathogens in the crop canopy. *Philosophical Transactions of the Royal Society of London, B*, **302**, 559–574.
Legg, B.J., Long, I.F. and Zemroch, P.J. 1981. Aerodynamic properties of field bean and potato crops. *Agricultural Meteorology*, **23**, 21–43.
Leuning, R. 1983. Transport of gases into leaves. *Plant, Cell and Environment*, **6**, 181–194.
Leuning, R., Denmead, O.T. and Lang, A.R.G. 1982. Effects of heat and water vapor transport on eddy covariance measurement of CO_2 fluxes. *Boundary-Layer Meteorology*, **23**, 209–222.
Lewis, H.E., Forster, A.R., Mullan, B.J., Cox, R.N. and Clark, R.P. 1969. Aerodynamics of the human microenvironment. *Lancet*, **1**, 1273–1277.
Linacre, E.T. 1972. Leaf temperature, diffusion resistances, and transpiration. *Agricultural Meteorology*, **10**, 365–382.
List, R.J. (ed.) 1966. *Smithsonian Meteorological Tables*, 6th edition. Smithsonian Institution, Washington DC.
Little, P. and Whiffen, R.D. 1977. Emission and deposition of petrol engine exhaust lead. I. Deposition of exhaust lead to plant and soil surfaces. *Atmospheric Environment*, **11**, 437–447.
Lumb, F.E. 1964. The influence of cloud on hourly amounts of total solar radiation at the sea surface. *Quarterly Journal of the Royal Meteorological Society*, **90**, 43–56.
McAdams, W.H. 1954. *Heat transmission*, 3rd edition. McGraw Hill, New York.

McArthur, A.J. 1987. Thermal interaction between animal and microclimate—a comprehensive model. *Journal of Theoretical Biology*, **126**, 203–238.

McArthur, A.J. and Clark, J.A. 1988. Body temperature of homeotherms and the conservation of energy and water. *Journal of Thermal Biology*, **13**, 9–13.

McArthur, A.J. and Monteith, J.L. 1980a. Air movement and heat loss from sheep. I. Boundary layer insulation of a model sheep with and without fleece. *Proceedings of the Royal Society of London, B*, **209**, 187–208.

McArthur, A.J. and Monteith, J.L. 1980b. Air movement and heat loss from sheep. II. Thermal insulation of fleece in wind. *Proceedings of the Royal Society of London, B*, **209**, 209–217.

McCartney, H.A. 1978. Spectral distribution of solar radiation. II. Global and diffuse. *Quarterly Journal of the Royal Meteorological Society*, **104**, 911–926.

McCartney, H.A. and Unsworth, M.H. 1978. Spectral distribution of solar radiation. I. Direct radiation. *Quarterly Journal of the Royal Meteorological Society*, **104**, 699–718.

McCree, K.J. 1972. The action spectrum, absorptance and quantum yield of photosynthesis in crop plants. *Agricultural Meteorology*, **9**, 191–216.

McCulloch, J.S.G. and Penman, H.L. 1956. Heat flow in the soil. *Report of the 6th International Soil Science Congress, B*, 275–280.

McNaughton, K.G. and Jarvis, P.G. 1983. Predicting effects of vegetation changes on transpiration and evaporation. In: *Water Deficits and Plant Growth*, Vol. VII, ed. Kozlowski, T.T., pp. 1–47. Academic Press, New York.

McNaughton, K.G. and Spriggs, T.W. 1986. A mixed-layer model for regional evaporation. *Boundary-Layer Meteorology*, **34**, 243–262.

Mahoney, S.A. and King, J.R. 1977. The use of the equivalent black-body temperature in the thermal energetics of small birds. *Journal of Thermal Biology*, **2**, 115–120.

Meidner, H. and Mansfield, T.A. 1968. *Physiology of Stomata*. McGraw Hill, London.

Mellor, R.S., Salisbury, F.B. and Raschke, K. 1964. Leaf temperatures in controlled environments. *Planta*, **61**, 56–72.

Mercer, T.T. 1973. *Aerosol Technology in Hazard Evaluation*. Academic Press, New York.

Milthorpe, F.L. and Penman, H.L. 1967. The diffusive conductivity of the stomata of wheat leaves. *Journal of Experimental Botany*, **18**, 422–457.

Monteith, J.L. 1962. Measurement and interpretation of carbon dioxide fluxes in the field. *Netherlands Journal of Agricultural Science*, **10**, 334–346.

Monteith, J.L. 1969. Light interception and radiative exchange in crop stands. In: *Physiological Aspects of Crop Yield*, ed. Eastin, J.D. American Society of Agronomy, Madison, Wisconsin.

Monteith, J.L. 1972. Latent heat of vaporization in thermal physiology. *Nature*, London, **236**, 96.

Monteith, J.L. 1975. *Principles of Environmental Physics*. First edition. Edward Arnold, London.

Monteith, J.L. 1981a. Evaporation and surface temperature. *Quarterly Journal of the Royal Meteorological Society*, **107**, 1–27.

Monteith, J.L. 1981b. Coupling of plants to the atmosphere. In: *Plants and Their Atmospheric Environment*, eds Grace, J., Ford, E.D. and Jarvis, P.G., pp. 1–29. Blackwell Scientific Publications, Oxford.

Monteith, J.L. and Butler, D. 1979. Dew and thermal lag: a model for cocoa pods. *Quarterly Journal of the Royal Meteorological Society*, **105**, 207–215.

Monteith, J.L. and Campbell, G.S. 1980. Diffusion of water vapour through integuments—potential confusion. *Journal of Thermal Biology*, **5**, 7–9.

Monteith, J.L. and Elston, J. 1983. Performance and productivity of foliage in the field. In: *The Growth and Functioning of Leaves*, eds Dale, J.E. and Milthorpe, F.L., pp. 499–518. Cambridge University Press.
Moon, P. 1940. Proposed standard solar radiation curves for engineering use. *Journal of the Franklin Institute*, **230**, 583–617.
Mount, L.E. 1967. Heat loss from new-born pigs to the floor. *Research in Veterinary Science*, **8**, 175–186.
Mount, L.E. 1968. *The Climatic Physiology of the Pig*. Edward Arnold, London.
Mount, L.E. 1979. *Adaptation of Thermal Environment of Man and his Productive Animals*. Edward Arnold, London.
Munro, D.S. and Oke, T.R. 1975. Aerodynamic boundary-layer adjustment over a crop in neutral stability. *Boundary-Layer Meteorology*, **9**, 53–61.
Murray, F.W. 1967. On the computation of saturation vapour pressure. *Journal of Applied Meteorology*, **6**, 203–204.
Nobel, P.S. 1974. Boundary layers of air adjacent to cylinders. *Plant Physiology*, Lancaster, **54**, 177–181.
Nobel, P.S. 1975. Effective thickness and resistance of the air boundary layer adjacent to spherical plant parts. *Journal of Experimental Botany*, **26**, 120–130.
Ohtaki, E. 1984. Application of an infrared carbon dioxide and humidity instrument to studies of turbulent transport. *Boundary-Layer Meteorology*, **29**, 85–107.
Oosthuizen, P.H. and Madan, S. 1970. Combined convective heat transfer from horizontal cylinders in air. Transactions of the American Society of Mechanical Engineers, Series C: *Journal of Heat Transfer*, **92**, 194–196.
Parkhurst, D.F., Duncan, P.R., Gates, D.M. and Kreith, F. 1968. Wind-tunnel modelling of convection of heat between air and broad leaves of plants. *Agricultural Meteorology*, **5**, 33–47.
Parlange, J.-Y., Waggoner, P.E. and Heichel, G.H. 1971. Boundary layer resistance and temperature distribution on still and flapping leaves. *Plant Physiology*, Lancaster, **48**, 437–442.
Paulson, C.A. 1970. The mathematical representation of wind speed and temperature profiles in the unstable atmospheric surface layer. *Journal of Applied Meteorology*, **9**, 857–861.
Penman, H.L. 1948. Natural evaporation from open water, bare soil and grass. *Proceedings of the Royal Society of London, A*, **194**, 120–145.
Penman, H.L. and Schofield, R.K. 1951. Some physical aspects of assimilation and transpiration. *Symposium of the Society of Experimental Biology*, **5**, 115–129.
Pohlhausen, E. 1921. Der wärmestausch zwischen festen Körpen und Flüssigkeiten mit Reibung und kleiner Wärmeleitung. *Zeitschrift für angewandte Mathematik und Mechanik*, **1**, 115–121.
Porter, W.P., Mitchell, J.W., Beckman, W.A. and DeWitt, C.B. 1973. Behavioural implications of mechanistic ecology. Thermal and behavioural modelling of desert ectotherms and their microenvironment. *Oecologia*, **13**, 1–54.
Powell, R.W. 1940. Further experiments on the evaporation of water from saturated surfaces. *Transactions of the Institution of Chemical Engineers*, **18**, 36–55.
Priestley, C.H.B. 1957. The heat balance of sheep standing in the sun. *Australian Journal of Agricultural Research*, **8**, 271–280.
Priestley, C.H.B. and Taylor, R.J. 1972. On the assessment of surface heat flux and evaporation using large scale parameters. *Monthly Weather Review*, **100**, 81–92.
Prothero, J. 1984. Scaling of standard energy metabolism in mammals: I. Neglect of circadian rhythms. *Journal of Theoretical Biology*, **106**, 1–8.
Rainey, R.C., Waloff, Z. and Burnett, G.F. 1957. *The Behaviour of the Red Locust*. Anti-locust Research Centre, London.
Rapp, G.M. 1970. Convective mass transfer and the coefficient of evaporative heat

loss from human skin. In: *Physiological and Behavioural Temperature Regulation*, eds Hardy, J.D., Gagge, A.P. and Stolwiyk, J.A.J. C.C. Thomas, Illinois.

Raupach, M.R. and Legg, B.J. 1984. The uses and limitations of flux-gradient relationships in micrometeorology. *Agricultural Water Management*, **8**, 119–131.

Raupach, M.R. and Thom, A.S. 1981. Turbulence in and above plant canopies. *Annual Review of Fluid Mechanics*, **13**, 97–129.

Reeve, J.E. 1960. Appendix to 'Inclined Point Quadrats' by J. Warren Wilson. *The New Phytologist*, **59**, 1–8.

Ross, J. 1975. Radiative transfer in plant communities. In: *Vegetation and the Atmosphere. Volume 1. Principles*, ed. Monteith, J.L., pp. 13–55. Academic Press, London.

Russell, G., Jarvis, P.G. and Monteith, J.L. 1989. Absorption of radiation by canopies and stand growth. In: *Plant Canopies: their growth, form and function*, eds. Russell, G., Marshall, B. and Jarvin, P.G., pp. 21–39. Cambridge University Press, Cambridge.

Schmidt-Nielsen, K. 1965. *Desert Animals*. Oxford University Press, Oxford.

Scholander, P.F., Walters, V., Hock, R. and Irving, L. 1950. Body insulation of some arctic and tropical mammals and birds. *Biological Bulletin*, **99**, 225–234.

Shaw, R.H. and Pereira, A.R. 1982. Aerodynamic roughness of a plant canopy: a numerical experiment. *Agricultural Meteorology*, **26**, 51–65.

Shuttleworth, W.J. 1988. Evaporation from Amazonian rainforest. *Proceedings of the Royal Society of London, B*, **233**, 321–340.

Shuttleworth, W.J. *et al.* 1984. Eddy correlation measurements of energy partition for Amazonian forest. *Quarterly Journal of the Royal Meteorological Society*, **110**, 1143–1162.

Slatyer, R.O. and McIlroy, I.C. 1961. *Practical Microclimatology*. CSIRO Australia and UNESCO.

Smith, H. and Morgan, D.C. 1981. The spectral characteristics of the visible radiation incident upon the surface of the earth. In: *Plants and the Daylight Spectrum*, ed. Smith, H. Academic Press, London.

Spence, D.H.N. 1976. Light and plant response in fresh water. In: *Light as an Ecological Factor*, *II*, eds Evans, G.C., Bainbridge, R. and Rackham, O., pp. 93–133. Blackwell Scientific Publications, Oxford.

Stanhill, G. 1970. Some results of helicopter measurements of albedo. *Solar Energy*, **13**, 59.

Steven, M.D. 1977. Standard distributions of clear sky radiance. *Quarterly Journal of the Royal Meteorological Society*, **103**, 457–465.

Steven, M.D. 1983. *The Physical and Physiological Interpretation of Infrared/Red Spectral Ratios Over Crops*. 12 pp. Royal Chemical Society, London.

Steven, M.D., Biscoe, P.V. and Jaggard, R.W. 1983. Estimation of sugar beet productivity from reflection in the red and infrared spectral bands. *International Journal of Remote Sensing*, **4**, 325–334.

Steven, M.D., Moncrieff, J.B. and Mather, P.M. 1984. Atmospheric attenuation and scattering determined from multiheight multispectral scanner imagery. *International Journal of Remote Sensing*, **5**, 733–747.

Steven, M.D. and Unsworth, M.H. 1979. The diffuse solar irradiance of slopes under cloudless skies. *Quarterly Journal of the Royal Meteorological Society*, **105**, 593–602.

Stewart, J.B. and Thom, A.S. 1973. Energy budgets in pine forest. *Quarterly Journal of the Royal Meteorological Society*, **99**, 154–170.

Stigter, C.J. and Musabilha, V.M.M. 1982. The conservative ratio of photosynthetically active to total radiation in the tropics. *Journal of Applied Ecology*, **19**, 853–858.

Sunderland, R.A. 1968. *Experiments on Momentum and Heat Transfer with Artificial Leaves*. B.Sc. dissertation, University of Nottingham.

Swinbank, W.C. 1963. Long-wave radiation from clear skies. *Quarterly Journal of the Royal Meteorological Society*, **89**, 339–348.

Szeicz, G. 1974. Solar radiation for plant growth. *Journal of Applied Ecology*, **11**, 617–636.

Tageeva, S.V. and Brandt, A.B. 1961. Optical properties of leaves. In: *Progress in Photobiology*, ed. Cristenson, B.C. Elsevier, Amsterdam.

Tani, N. 1963. The wind over the cultivated field. *Bulletin of the National Institute of Agricultural Science*, Tokyo. A.10, 99 pp.

Tetens, O. 1930. Uber einige meteorologische Begriffe. *Zeitschrift Geophysic*, **6**, 297–309.

Thom, A.S. 1968. The exchange of momentum, mass and heat between an artificial leaf and the airflow in a wind-tunnel. *Quarterly Journal of the Royal Meteorological Society*, **94**, 44–55.

Thom, A.S. 1971. Momentum absorption by vegetation. *Quarterly Journal of the Royal Meteorological Society*, **97**, 414–428.

Thom, A.S. 1972. Momentum, mass, and heat exchange of vegetation. *Quarterly Journal of the Royal Meteorological Society*, **98**, 124–134.

Thom, A.S. 1975. Momentum, mass and heat exchange of plant communities. In: *Vegetation and the Atmosphere. Volume 1. Principles*, ed. Monteith, J.L., pp. 57–109. Academic Press, London.

Thom, A.S., Stewart, J.B., Oliver, H.R. and Gash, J.H.C. 1975. Comparison of aerodynamic and energy budget estimates of fluxes over a pine forest. *Quarterly Journal of the Royal Meteorological Society*, **101**, 93–105.

Thornthwaite, C.W. and Holzman, B. 1942. *Measurement of Evaporation from Land and Water Surfaces*. US Department of Agriculture Technical Bulletin No. 817, 75 pp.

Thorpe, M.R. 1978. Net radiation and transpiration of apple trees in rows. *Agricultural Meteorology*, **19**, 41–57.

Thorpe, M.R. and Butler, D.R. 1977. Heat transfer coefficients for leaves on orchard apple trees. *Boundary-Layer Meteorology*, **12**, 61–73.

Tibbals, E.C., Carr, E.K., Gates, D.M. and Kreith, F. 1964. Radiation and convection in conifers. *American Journal of Botany*, **51**, 529–538.

Tooming, H.G. and Gulyaev, B.E. 1967. *Methods of Measuring Photosynthetically Active Radiation* (in Russian). Nauka, Moscow.

Tucker, C.J., Townshend, J.R.G. and Goff, T.E. 1985. African land-cover classification using satellite data. *Science*, **227**, 369–375.

Tucker, V.A. 1969. The energetics of bird flight. *Scientific American*, **220**, 70–78.

Tullett, S.G. 1984. The porosity of avian eggshells. *Comparative Biochemistry and Physiology*, **78A**, 5–13.

Underwood, C.R. and Ward, E.J. 1966. The solar radiation area of man. *Ergonomics*, **9**, 155–168.

Unsworth, M.H. 1975. Geometry of long-wave radiation at the ground. II. Interception by slopes and solids. *Quarterly Journal of the Royal Meteorological Society*, **101**, 25–34.

Unsworth, M.H., Heagle, A.S. and Heck, W.W. 1984a. Gas exchange in open-top field chambers. I. Measurement and analysis of atmospheric resistances to gas exchange. *Atmospheric Environment*, **18**, 373–380.

Unsworth, M.H., Heagle, A.S. and Heck, W.W. 1984b. Gas exchange in open-top field chambers. II. Resistance to ozone uptake by soybeans. *Atmospheric Environment*, **18**, 381–385.

Unsworth, M.H. and Monteith, J.L. 1972. Aerosol and solar radiation in Britain.

Quarterly Journal of the Royal Meteorological Society, **99**, 778–797.
Unsworth, M.H. and Monteith, J.L. 1975. Geometry of long-wave radiation at the ground. I. Angular distribution of incoming radiation. *Quarterly Journal of the Royal Meteorological Society*, **101**, 13–24.
Van Eimern, J. 1964. *Untersuchungen über das Klima in Pflanzengestanden*. Bericht des Deutschen Wetterdienstes, Offenbach, No. 96.
Van Wijk, W.R. and De Vries, D.A. 1963. Periodic temperature variations. In: *Physics of Plant Environment*, ed. Van Wijk, W.R. North-Holland Publishing Company, Amsterdam.
Vogel, S. 1970. Convective cooling at low airspeeds and the shapes of broad leaves. *Journal of Experimental Botany*, **21**, 91–101.
Waggoner, P.E. and Reifsnyder, W.E. 1968. Simulation of the temperature, humidity and evaporation profiles in a leaf canopy. *Journal of Applied Meteorology*, **7**, 400–409.
Walsberg, G.E., Campbell, G.S. and King, J.R. 1978. Animal coat colour and radiative heat gain. *Journal of Comparative Physiology*, **126**, 211–212.
Wathes, C. and Clark, J.A. 1981a. Sensible heat transfer from the fowl: I. Boundary layer resistance of model fowl. *British Poultry Science*, **22**, 161–173.
Wathes, C. and Clark, J.A. 1981b. Sensible heat transfer from the fowl: II. Thermal resistance of the pelt. *British Poultry Science*, **22**, 175–183.
Webb, E.K. 1970. Profile relationships: the log-linear range, and extension to strong stability. *Quarterly Journal of the Royal Meteorological Society*, **96**, 67–90.
Webb, E.K., Pearman, G.I. and Leuning, R. 1980. Correction of flux measurements for density effects due to heat and water vapour transfer. *Quarterly Journal of the Royal Meteorological Society*, **106**, 85–100.
Webster, M.D., Campbell, G.S. and King, J.R. 1985. Cutaneous resistance to water vapour diffusion in pigeons, and the role of plumage. *Physiological Zoology*, **58**, 58–70.
Weiss, A. and Norman, J.M. 1985. Partitioning solar radiation into direct and diffuse, visible and near-infrared components. *Agricultural and Forest Meteorology*, **34**, 205–213.
Wesely, M.L., Eastman, J.A., Cook, D.R. and Hicks, B.B. 1978. Daytime variations of ozone eddy fluxes to maize. *Boundary-Layer Meteorology*, **15**, 361–373.
Wheldon, A.E. and Rutter, N. 1982. The heat balance of small babies nursed in incubators and under radiant warmers. *Early Human Development*, **6**, 131–143.
Wiersma, F. and Nelson, G.L. 1967. Nonevaporative convective heat transfer from the surface of a bovine. *Transactions of the American Society of Agricultural Engineers*, **10**, 733–737.
Wigley, G. and Clark, J.A. 1974. Heat transfer coefficients for constant energy flux models of broad leaves. *Boundary-Layer Meteorology*, **7**, 139–150.
Wilson, J.D., Thurtell, G.W. and Kidd, G.E. 1981. Numerical simulation of particle trajectories in inhomogeneous turbulence. II. Systems with variable turbulent velocity scale. *Boundary-Layer Meteorology*, **21**, 423–441.
Wilson, N.R. and Shaw, R.H. 1977. A higher-order closure model for canopy flow. *Journal of Applied Meteorology*, **16**, 1198–1205.
Wylie, R.G. 1979. Psychrometric wet-elements as a basis for precise physico-chemical measurements. *Journal of Research of the National Bureau of Standards* (US), **84**, 161–177.
Yuge, T. 1960. Experiments on heat transfer from spheres including combined natural and forced convection. Transactions of the American Society of Mechanical Engineers, Series C: *Journal of Heat Transfer*, **82**, 214–220.

INDEX

Bold symbols indicate main entries.